FROM QUASICRYSTALS
TO MORE COMPLEX SYSTEMS

Les Houches School, February 23 – March 6, 1998

Editors

F. AXEL
F. DÉNOYER
J.P. GAZEAU

Springer-Verlag France S.A.R.L

Centre de Physique des Houches

Books already published in this series

Book series coordinated by Michèle LEDUC

Editors of "From Quasicrystals to More Complex Systems " (No. 13)
F. Axel (LPS, Bât. 510, Université Paris-Sud, Orsay, France)
F. Dénoyer (LPS, Bât. 510, Université Paris-Sud, Orsay, France)
J.P. Gazeau (LPTMC, Université Paris 7 – Denis Diderot, Paris, France)

ISSN 1436-6452
ISBN 978-3-540-67464-1   ISBN 978-3-662-04253-3 (eBook)
DOI 10.1007/978-3-662-04253-3

© Springer-Verlag France 2000
Originally published by Springer-Verlag France in 2000

This school was supported by:

Université Joseph Fourier de Grenoble

Centre National de la Recherche Scientifique (CNRS – Formation permanente)

Ministère de l'Éducation Nationale de la Recherche et de la Technologie
(DRC4 – DRA7)

Université Paris 7 Denis Diderot

Laboratoire de Physique Théorique de la Matière Condensée (LPTMC)
de l'Université Paris 7 Denis Diderot

Direction des Recherches et Études Techniques (DGA – DRET)

Ministère des Affaires Étrangères

Ambassade de France à Ottawa

Ambassade de France à Washington

Office of Naval Research European Office (ONR – EUR)

Obituary

Shlomo Alexander

A few months after this school, on August 6, Shlomo Alexander was tragically killed in an automobile accident near Caesarea, Israel.

Shlomo was born in Freiburg, Germany, on September 4, 1930. His father, an eminent crystallographer, moved the family to the Holy Land shortly thereafter. Shlomo earned his BSc at the Hebrew University, Jerusalem, and then completed his doctorate under the guidance of Saul Meiboom at the Weizmann Institute on experimental investigations of NMR of organic molecules. Of course, Alexander is best known as a theorist; this career reorientation started to take place during his post-doctoral period at Bell Laboratories where he worked with P.W. Anderson on the coupling between localized moments (Anderson impurities) in metals.

After his post-doc, Shlomo went back to the Weizmann where he resumed resonance experiments, namely NQR, in ferroelectrics. In 1969, he moved to the Racah Institute of the Hebrew University where he set up a theory group in solid state physics. Beginning at this period, he never carried out experiments. This also marked the time when his horizons broadened explosively to many areas of condensed matter including phase transitions, percolation, transport in random systems, fractals, superconductivity, polymer solutions, microemulsions, colloidal crystals, quasi-crystals, liquid crystals, hydrodynamics of electrolytic solutions, etc. In most of these areas Shlomo has made seminal contributions: Alexander McTague theory of freezing, superconductivity on fractal nets, lattice model of microemulsions, Alexander de Gennes theory of polymer brushes, structure of the blue liquid crystalline phases, charge renormalization of charged colloids, fracton excitations on fractal structures to name a few. These accomplishments were recognized in 1993 by his award of the Israel Prize.

Except for sabbatical leaves in Paris with de Gennes and multiple visits to UCLA (where he also held a part-time faculty position), Shlomo remained at the Weizmann Institute from 1989. He completely retired from the Weizmann in 1995 and subsequently maintained an affiliation with Bar-Ilan University near Tel Aviv.

For the last decade of his life, Shlomo became deeply involved with the fundamental aspects of elasticity theory of disordered systems. He was especially concerned with swollen gels and foams where there is an interplay between isotropic, internal osmotic pressure and tensorial elasticity. This culminated in a seminal paper in Physics Reports.

Shlomo was a most unusual personality. He had extraordinary intuition on spatial organization and crystallography which most likely was, at least partially, derived from his heritage; he had physical feeling for and could visualize crystallographic groups and their transformations. He was a true physicist in the classical sense; he was quite willing and happy to discuss any field, always returning to the fundamental laws. Shlomo also had remarkable patience. Discussions with him could be never-ending (one usually had to prepare some external stop in advance) until the problem under consideration was solved. His many collaborators found this tough but rewarding. He was scientifically honest and did not suffer gladly those who, in T. Holstein's terminology, "publish in haste and repent at leisure".

In his spare time, Shlomo collaborated with his deeply beloved wife, Esther, on various aspects of economic theory. This is a subject on which he would happily lecture (to all who would listen) for many hours on his views of the "strange" world of economic models and actual cases.

Shlomo Alexander was a truly extraordinary human being, a brilliant scientist, and a wonderful friend and colleague. We do and will continue to miss him.

D. Levine (Technion, Haifa).
P. Pincus (UCSB, Santa Barbara).

FOREWORD

This book is a collection of part of the written versions of the Physics Courses given at the Winter School "Order, Chance and Risk: Aperiodic Phenomena from Solid State to Finance" held at the Les Houches Center for Physics, between February 23 and March 6, 1998. The School gathered lecturers and participants from all over the world.

On a thematic level, the content of the school can be viewed both as a continuation (aperiodic phenomena in solid state physics) and an extension (mathematical aspects of finance and economy) of the previous "Beyond Quasicrystals", also held at Les Houches, March 7-18 1994 and published in the same series. One of its important goals was to promote in-depth concrete scientific exchanges between theoretical physicists, experimental physicists and mathematicians on the one hand, and on the other hand practitioners of the economico-financial sphere and specialists of financial mathematics.

Therefore, besides the mathematical tools and concepts at work in theoretical descriptions, relevant experimental data were also presented together with methods allowing their interpretation. As a result of this choice, the School was stimulated by experimentalists and financial market operators who joined the theoretical physicists and mathematicians at the conference.

The present volume deals with the theoretical and experimental studies on aperiodic solids with long range order, incommensurate phases, quasicrystals, glasses, and more complex systems (fractal, chaotic), while a second volume to appear in the same series is devoted to the finance and economy facet.

The first part of this book is more specifically devoted to methods – experimental and theoretical – together with results on quasicrystals: Claire Berger develops the electronic properties of quasicrystals using a comparison with approximant phases and disordered systems; Francoise Dénoyer describes diffraction experiments on quasicrystals and related phases, for which the theoretical courses by Ted Janssen (Dynamics and transport properties of aperiodic systems) and Peter Kramer (Exact electron states in 1D (quasi-) periodic arrays of delta-potentials) present elements for interpretation and synthesis.

The courses which follow deal with certain models and concepts currently at work in the field, and, in particular, present the structures and mathematical objects which underlie the most generic study of aperiodic media and phenomena: Eric Cockayne describes random tiling models for quasicrystals, Jiri Patera the search for acceptance windows compatible with a quasicrystal fragment, Jean-Pierre Gazeau presents counting systems for quasicrystals and Robert Moody gives a survey on "model sets".

The transition to "more complex systems" begins with two courses on glasses: Eric Courtens gives an overview on acoustic-like excitations in strongly disordered media, and Jean-Philippe Bouchaud on intermittent dynamics and ageing in glassy systems. The application of the theory of dynamical systems to the study of condensed matter aperiodic systems is then presented by Michel Dekking in a short introduction to ergodic theory and its applications, and George Zaslavsky gives examples on fractality and the kinetics of chaos. Fractal and multifractal methods follow in a natural fashion: Michael Shlesinger surveys long-tailed distributions in physics and Roger Balian the scaling *vs.* fractality aspects of the distribution of galaxies.

The authors have gracefully made efforts at the editors' request, to render the contents of their course accessible. The deliberate pedagogical aim of this school stands out in the glossary written by the editors, with suggestions by the authors. It will help the readers – particularly the experimentalists – to find their way through the book.

We are particularly grateful to the colleagues who generously and kindly provided help in shaping this book, particularly T. Janssen and J. Peyrière.

Neither the School, nor this book could have existed without the constant encouragement of Michèle Leduc, Director of the Les Houches Center for Physics, her very competent staff at Les Houches, Mesdames B. Rousset, I. Lelièvre and G. d'Henry, as well as the efficiency and patience shown us by Mesdames Agnès Henri and I. Houlbert (EDP Sciences/Springer-Verlag). We wish to express to them our sincerest gratitude.

Particular acknowledgements and thanks are due to Madame Françoise Kakou. She took care of the entire administrative organisation of the school, as well as of the preparation of this book and its annexes, all with her well-known skill, competence, efficiency and kindness, without which we would not have succeeded.

Françoise Axel
Françoise Dénoyer
Jean-Pierre Gazeau

List of Participants

Alexander, Esther Dept. of Chemical Physics, Weizmann Institute of Science, 76100 Rehovot, Israël

Alexander, Shlomo Dept. of Chemical Physics, Weizmann Institute of Science, 76100 Rehovot, Israël

Aneva, Boyka Bulgarian Academy of Sciences, Institut for Nuclear Research, 1784 Sofia, Bulgarie, blan@inrne.acad.bg

Antoni, Mickaël Max Planck-Institut für Physik komplexer Systeme, Nöthnitzer Strasse, 38, D-01187 Dresden, Allemagne, antoni@mpipks-dresden.mpg.de

Aste, Tomaso Dynamique des Fluides Complexes, Institut de Physique - Univ. Louis Pasteur, F-67084 Strasbourg, France, tomaso@ldfc.u-strasbg.fr

Axel, Françoise Physique des Solides, Bât. 510, F-91405 Orsay Cedex, France, axel@lps.u-psud.fr

Balian, Roger Service de Physique Théorique, Orme des Merisiers, F-91191 Gif-sur-Yvette Cedex, France

Ben-Abraham, Shlomo I. Dept. of Physics, Ben Gurion University, Il-84105 Beer-Sheba, Israël, benabr@bgumail.bgu.ac.il

Benkadda, Sadruddin Équipe Turbulence Plasma, UMR 6633 CNRS - Univ. de Provence, F-13397 Marseille Cedex 20, France, benkadda@newsup.univ-mrs.fr

Berger, Claire LEPES-CNRS, 25 avenue des Martyrs, F-38042 Grenoble Cedex 05, France, berger@lepes.polycnrs-gre.fr

Bouchaud, Jean-Philippe SPEC, Orme des Merisiers, F-91191 Gif-sur-Yvette Cedex, France, bouchau@amoco.saclay.cea.fr

Campbell, Ian A. Physique des Solides, Bât. 510, F-91405 Orsay Cedex, France, campbell@lps.u-psud.fr

Cockayne, Eric J. Ceramics Division, Materials Science and Engineering Laboratory, National Institute of Standards and Technology, Gaithersburg, MA 20899-8520, USA, cockayne@nist.gov

Cont, Rama CMAP, École Polytechnique, F-91128 Palaiseau Cedex, France, cont@cmapx.polytechnique.fr

Courtens, Éric Laboratoire des verres, Univ. Montpellier II, F-34095 Montpellier Cedex 5, France, courtens@ldv.univ-montp2.fr

Cuniberti, Gianaurelio Dipartimento di Fisica, room 607, Università di Genova, I-16146 Genova, Italie, cunibert@fisica.unige.it

Dekking, Michel Techn. Hogeschool Delft, Dept. of Mathematics, NL-2600 AJ Delft, Pays Bas, f.m.dekking@twi.tudelft.nl

Dénoyer, Françoise Physique des Solides, Bât. 510, F-91405 Orsay Cedex, France, denoyer@lps.u-psud.fr

Ezzine De Blas, Ana-Zahra Physique des Solides, Bât. 510, F-91405 Orsay Cedex, France, ezzine@lps.u-psud.fr

Family, Fereydoon Physics Dept., Emory University, GA 30322, USA, phyff@emory.edu

Filoche, Marcel Physique de la Matière condensée, École Polytechnique, F-91128 Palaiseau Cedex, France, marcel.filoche@polytechnique.fr

Florescu, Mihai Modélisation Stochastique et statistique, Univ. Paris-Sud, F-91405 Orsay, France, florescu@stats.math.u-psud.fr

Gachok, Volodymyr Syndergetics Dept., Bogolyubov Inst. for Theoretical Physics, Kyiv 143, Russie, vgachok@gluk.apc.org

Gazeau, Jean-Pierre LPTMC, Univ. Paris 7 - Denis Diderot, F-75251 Paris Cedex 05, France, gazeau@ccr.jussieu.fr

Hellekalek, Peter Institut für Mathematik, Universität Salzburg, A-5020 Salzburg, Autriche, peter.hellekalek@sbg.ac.at

Janssen, Ted Dept. of Theoretical Physics, Katholieke Universiteit, NL-6525 ED Nijmegen, Pays Bas, edj@baserv.uci.kun.nl ou ted@sci.kun.nl

Jones, Peter Dept. of Mathematics, Yale University, New Haven CT 06520, USA, jones@math.yale.edu

Kockelkoren, Julien Dept. of Theoretical Physics, Univ. of Nijmegen, NL-6525 ED Nijmegen, Pays Bas, julienk@sci.kun.nl

Kramer, Peter Inst. für Theor. Phys. der Universität, Auf der Morgenstelle 14, D-72076 Tübingen, Allemagne, peter.kramer@uni-tuebingen.de

Krejcar, Rudolph Univ. Paris 7 - Denis Diderot, Physique Théorique Matière Condensée, F-75251 Paris Cedex 05, France, krejcar@ccr.jussieu.fr

Lamb, Jeroen Mathematics Dept. , University of Houston, Texas 77204-3476, USA, lamb@maths.warwick.ac.uk

Liardet, Pierre Mathématiques et Informatique, 36 rue Joliot Curie, F-13453 Marseille Cedex 13, France, liardet@gyptis.univ-mrs.fr

Mari, Pierre-Olivier Physique des Solides, Bât. 510, F-91405 Orsay Cedex, France, mari@lps.u-psud.fr

Menotti, Chiara Scuola Normale Superiore, I-56126 Pisa, Italie, menotti@cibs.sns.it

Moody, Robert V. Dept. of Mathematical Sciences, Univ. of Alberta, Edmonton, AB, T6G 2G1, Canada, rvm@miles.math.alberta.ca

Pando Lambruschini, C.L. Instituto de Fisica, Univ. Autonoma de Puebla, Puebla, Mexique, carlos@sirio.ifuap.buap.mx

Parisi, Giorgio Dipartimento di Fisica, Univ. degli Studi di Roma "La Sapienza", I-00185 Roma, Italie

Patera, Jiri Centre de Recherches Mathématiques, Univ. de Montréal, Montréal (Québec) H3C 3J7, Canada, patera@CRM.Umontreal.CA

Peyrière, Jacques Dept. de Mathématiques, Univ. de Paris-Sud, F-91405 Orsay Cedex, France, Jacques.Peyriere@anh.math.u-psud.fr

Porrà, Josep M. Dept. Fisica Fonamental - Divisio de Ciencies, Experimentals i Matematiques - Univ. de Barcelona, E-08028 Barcelona, Espagne, jporra@hermes.ffn.ub.es

Rivier, Nicolas Dynamique des Fluides complexes, Inst. de Physique, Univ. Louis Pasteur, F-67084 Strasbourg Cedex, France, nick@fresnel.u-strasbg.fr

Rubinstein, Boaz Dept. of Physics, Ben Gurion Univ., IL-84105 Beer-Sheba, Israël, boazrb@bgumail.bgu.ac.il

Sapoval, Bernard Physique de la Matière condensée, École Polytechnique, F-91128 Palaiseau Cedex, France, bernard.sapoval@polytechnique.fr

Shlesinger, Michael Office of Naval Research, Physical Sciences Division, Arlington VA 22217-5660, USA, shlesim@onr.navy.mil

Solomon, Sorin Racah Institute of Physics, The Hebrew Univ. of Jerusalem, Jerusalem, Israël, sorin@vms.hujs.ac.il

Twarock, Reidun Arnold Sommerfeld Institut, TU Clausthal, D-38678 Clausthal-Zellerfeld, Allemagne, reidun.twarock@tu-clausthal.de

Verger-Gaugry, Jean-Louis Institut Fourier, CNRS UMR 5582, UFR de Mathématiques, 38402 St-Martin d'Hères, France, lverger@ltpcm.inpg.fr

Wen, Zhi-Xiong Dept. of Mathematics, Wuhan University, 430072 Wuhan, République Populaire de Chine, zhxwen@whu.edu.cn

Zaslavsky, George Courant Institute for Mathematical Sciences, 251, Mercer Street, New York, NY 10012, USA, zaslav@cims.nyu.edu

CONTENTS

COURSE 3

Electronic Properties of Quasicrystals. A Comparison with Approximant Phases and Disordered Systems

given by C. Berger

written by C. Berger and T. Grenet

COURSE 4

Exact Electron States in 1D (Quasi-) Periodic Arrays of Delta-Potentials

given by P. Kramer

written by P. Kramer and T. Kramer

COURSE 5
Random Tiling Models for Quasicrystals
by E. Cockayne

COURSE 6
Model Sets: A Survey
by R.V. Moody

COURSE 7

Acceptance Windows Compatible with a Quasicrystal Fragment

given by J. Patera

written by Z. Masáková, J. Patera and E. Pelantová

COURSE 8

Counting Systems with Irrational Basis for Quasicrystals

given by J.P. Gazeau

written by J.P. Gazeau and R. Krejcar

COURSE 9

Acoustic-Like Excitations in Strongly Disordered Media

given by E. Courtens

written by E. Courtens and R. Vacher

COURSE 10

Intermittent Dynamics and Ageing in Glassy Systems

by J.-Ph. Bouchaud

COURSE 11

A Short Introduction to Ergodic Theory and Its Applications

by F.M. Dekking

COURSE 12

Fractality and the Kinetics of Chaos

by G.M. Zaslavsky

COURSE 13

Long-Tailed Distributions in Physics

given by M.F. Shlesinger

written by M.F. Shlesinger, J. Klafter and G. Zumofen

COURSE 14

Distribution of Galaxies: Scaling *vs.* Fractality

by R. Balian

Course N° 1

Dynamics and Transport Properties of Aperiodic Crystals

T. Janssen

Institute for Theoretical Physics,
University of Nijmegen,
Toernooiveld, 6525 ED Nijmegen, The Netherlands

Abstract. The physical properties of aperiodic, but well ordered systems as incommensurate crystal phases and quasicrystals, are in many respects different from those of lattice periodic structures. The usual techniques based largely on the use of the Brillouin zone are no longer applicable. Here a number of approaches of the study of structure, lattice dynamics, electronic structure and transport properties are discussed.

1. STRUCTURE AND SYMMETRY

By aperiodic crystal we mean a structure with Bragg peaks in the diffraction spectrum but without lattice periodicity [1,16]. Lattice periodicity would mean that the diffraction peaks are Bragg peaks on a reciprocal lattice. A more general situation is found in quasiperiodic crystals, where the diffraction peaks are Bragg peaks on a Fourier module M^*, *i.e.* the peaks are located on positions

$$\vec{H} = \sum_{i=1}^{n} h_i \vec{a}_i^*$$

with integer indices h_i, and where the vectors $\vec{a}_i^*$ form a (minimal) basis of rank n. If n is equal to the dimension of physical space and the vectors $\vec{a}_i^*$ are linearly independent then the crystal has lattice periodicity and the Fourier module is just the reciprocal lattice. If n is larger than the dimension of physical space the structure is aperiodic. Examples of such structures are incommensurately modulated (IC) phases, incommensurate composite or misfit structures, and quasicrystals.

These aperiodic crystals are sometimes considered to be between lattice periodic and disordered systems. The order, however, of an ideal quasiperiodic system is just as high as that of an ideal lattice periodic system. In the following we shall discuss whether one can say that the physical properties are between those for periodic and amorphous systems. Another point is that, as in all solids, there is disorder present which can cause a diffuse component in the diffraction. It is not immediately clear whether this natural disorder has the same effect in an aperiodic crystal as it has in a lattice periodic crystal.

The system can be described by a density function ρ that has a discrete component ρ_o for which the support of the Fourier transform is a sum of delta peaks:

$$\rho_o(\vec{r}) \; = \; \sum_{\vec{k} \in M^*} \hat{\rho}(\vec{k}) \; \exp(-i\vec{k}\vec{r}).$$

Besides this component there is usually a diffuse component which we shall disregard for the moment. The symmetry of the diffraction pattern is the point group that leaves the intensity function $I(\vec{k})$ invariant. This function is the modulus squared of the structure factor $F(\vec{k})$. Because the peaks in $I(\vec{k})$ above a certain threshold are discrete the symmetry group (the Laue group) is finite. For a 3-dimensional physical space this means that the diffraction symmetry group is one of the finite subgroups of O(3). These are isomorphic to a cyclic group C_n, a dihedral group D_n, the tetrahedral group T, the octahedral group O, the icosahedral group I or a direct product of one of these with the group generated by the central inversion. Only for lattice periodic systems the group is necessarily crystallographic. For a general quasiperiodic system there is no such crystallographic condition.

There are three large classes of quasiperiodic crystal structures.

The first is that of incommensurately modulated phases. For such a structure there is a lattice periodic basic structure with space group symmetry. The positions of the atoms are $\vec{n} + \vec{r}_j$, where $\vec{n}$ is a lattice vector, and $\vec{r}_j$ the position in the unit cell of the j-th atom. For a displacive modulation the positions of the atoms are shifted to

$$\vec{n} + \vec{r}_j + \vec{f}_j(\vec{n}), \tag{1}$$

where $\vec{f}_j(\vec{r})$ is a periodic function. A simple example is a sinusoidal one-dimensional modulation:

$$\vec{n} + \vec{r}_j \; + \; \vec{f}_j \sin(\vec{Q}\vec{n} \; + \; \phi), \tag{2}$$

where $\vec{f}_j$ is the modulation polarisation vector and $\vec{Q}$ the modulation wave vector. The diffraction pattern shows besides the spots belonging to the reciprocal lattice of the basic structure (main reflections) satellites separated from the main reflections by multiples of $\vec{Q}$. In this case the vectors of the Fourier module are

$$\vec{H} = h\vec{a}^* + k\vec{b}^* + \ell\vec{c}^* + m\vec{Q}$$

and the rank is four.

A second class consists of incommensurate composites. In this case there are two or more subsystems having a lattice periodic basic structure. Besides there is a displacive modulation, caused by the fact that the atoms in a certain subsystem have different surroundings. The interaction with the other subsystems gives rise to this modulation. If $\vec{a}^*_{\nu i}$ form a basis for the reciprocal lattice of the basic structure of the ν-th subsystem the diffraction vectors are

$$\vec{H} = \sum_{\nu} \sum_{i=1}^{3} h_{\nu i} \vec{a}^*_{\nu i} = \sum_{i=1}^{n} h_i \vec{a}^*_i, \tag{3}$$

where all $\vec{a}^*_{\nu i}$ belong to the Fourier module spanned by $\vec{a}^*_i$, and $h_{\nu i}$ and h_i are integers. In general the modulation wave vectors of one subsystem belong to the span of the reciprocal lattice basis vectors of the other subsystems. Usually the rank n is less than three times the number of subsystems. Examples are systems consisting of a host lattice in the channels of which other systems (like mercury in channels of AsF_6) diffuse, inclusion compounds like urea-alcane, intercalates and layers of noble gases adsorbed at a crystalline surface.

A third class consists of the quasicrystals [14]. These are systems where one cannot distinguish a basic structure, even not an incommensurate one. Often they can be described as a, possibly modulated, quasiperiodic space filling by means of a finite number of building blocks, like in a Penrose tiling. Because there is no crystallographic restriction, which still exists for IC phases, they may have non-crystallographic symmetries, and that is often, but not always, the case. Examples are intermetallic alloys with icosahedral symmetry, or consisting of three-dimensional stacks of quasiperiodic layers with octagonal, decagonal or dodecagonal symmetries.

The perfect ordering of an ideal quasiperiodic crystal can be used to apply the usual crystallographic techniques if one embeds the 3-dimensional structure into a space of dimension equal to the rank of the Fourier module [15, 40]. To that end the basis vectors $\vec{a}^*_i$ are considered as projection of linearly independent vectors $(\vec{a}^*_i, \vec{b}^*_i)$ in n-dimensional space. These span a reciprocal lattice Σ^*. In this superspace a metric can be introduced as follows. The elements R of the symmetry group of the diffraction pattern transform basis vectors into vectors of the Fourier module:

$$R\vec{a}^*_i = \sum_{j=1}^{n} \Gamma^*(R)_{ji} \vec{a}^*_j. \tag{4}$$

The matrices $\Gamma^*(R)$ form an integer representation of the finite symmetry group K. Therefore, there is an invariant metric in superspace. Using this metric one may construct the direct lattice Σ. If the crystal has density function ρ with

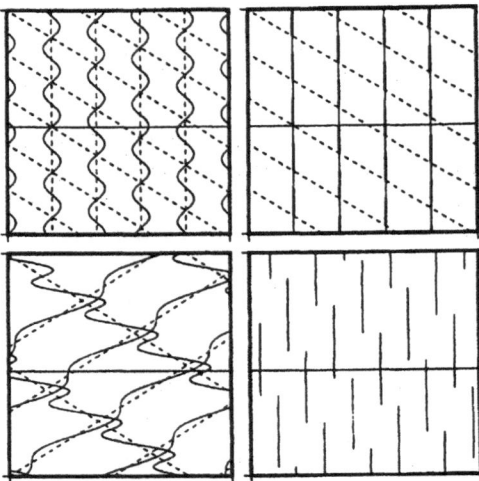

Fig. 1. — Embeddings into higher-dimensional space of a number of quasiperiodic structures. Left top: sinusoidally modulated chain, left bottom: IC composite, right top: composition modulated phase, right bottom: Fibonacci chain.

Fourier components $\hat{\rho}(\vec{k})$ the function with n variables given by

$$\rho_s(\vec{r}, \vec{r}_I) \;=\; \sum_{\vec{k} \in M^*} \hat{\rho}(\vec{k}) \exp(-i(\vec{k}\vec{r} + \vec{k}_I \vec{r}_I)), \tag{5}$$

(where the vector $(\vec{k}, \vec{k}_I)$ belongs to the reciprocal lattice Σ^* in superspace) is lattice periodic with the lattice Σ. Moreover,

$$\rho(\vec{r}) \;=\; \rho_s(\vec{r}, 0).$$

Therefore, the density function in physical space is the restriction of the lattice periodic function ρ_s and the projection of the Fourier transform of ρ_s is the function $\hat{\rho}(\vec{k})$. The function ρ_s is fully determined by its value in the unit cell of Σ and the function ρ is the restriction of ρ_s to physical space. The value of the Fourier transform of ρ_s in $(\vec{k}, \vec{k}_I)$ is equal to the value of $\hat{\rho}$ in its projection $\vec{k}$.

The quasiperiodic crystal can in this way be embedded in n-dimensional space. Examples are given in Figure 1.

The symmetry of the quasiperiodic crystal can be considered to be that of its embedding, and that is an n-dimensional crystallographic space group. An element $g \;=\; \{(R, R_I) | (\vec{a}, \vec{a}_I)\}$ is a symmetry operator for ρ_s if

$$\rho_s(\vec{r}, \vec{r}_I) \;=\; \rho_s(R\vec{r} + \vec{a}, R_I \vec{r}_I + \vec{a}_I). \tag{6}$$

The Fourier transform of this symmetry relation is

$$\hat{\rho}(\vec{k}) \ = \ \hat{\rho}(R\vec{k})\exp(i\vec{k}\vec{a})\exp(i\vec{k}_I\vec{a}_I). \tag{7}$$

This relation contains only functions on physical space. If $\vec{a}_I = 0$ it is just the condition for having $\{R|\vec{a}\}$ as symmetry operator for ρ. In the present case there may be an additional phase factor, which can be considered as a gauge transformation.

2. PHONONS

Because there is no lattice periodicity, and hence no Brillouin zone the usual approach to lattice excitations cannot be applied [5, 30], and it is not clear whether there are propagating phonon modes in quasiperiodic crystals. If $\vec{u}_j$ is the displacement of the j-th particle from its equilibrium position, the equations of motion in the harmonic approximation still read

$$m_j\ddot{\vec{u}}_j \ = \ -\sum_m \Phi_{jm}\sqrt{m_j m_m}\vec{u}_m \ = \ -\sum_m \frac{\partial^2 V}{\partial\vec{u}_j\partial\vec{u}_m}\Big|_{\vec{u}_j=\vec{u}_m=0} \ \vec{u}_m \ . \tag{8}$$

We can put $\vec{u}_j(t) = \vec{v}_j\exp(i\omega t)/\sqrt{m_j}$. This leads to the eigenvalue equation

$$\omega^2 v_j = \sum_m \Phi_{jm} v_m. \tag{9}$$

To study the solutions we consider some simple models. A simple model for the excitations in an IC phase is the modulated spring model [21], a chain of particles with harmonic nearest-neighbour interaction. If the displacement of particle n is u_n the potential energy is

$$V \ = \ \frac{1}{2}\sum_n \alpha_n(u_n \ - \ u_{n-1})^2,$$

where the spring constants are given by a periodic function, e.g.

$$\alpha_n \ = \ \alpha\{1 + \Delta\cos(Qn + \phi)\}$$

for an irrational value of $Q/2\pi$. The equations of motion for this chain are

$$\omega^2 v_n \ = \ [\alpha_n(v_n - v_{n-1}) + \alpha_{n+1}(v_n - v_{n+1})]/m.$$

For $\Delta = 0$ this gives the dispersion curves

$$\omega^2 \ = \ (4\alpha/m)\sin^2(k/2).$$

For $0 < \Delta < 1$ there is coupling of modes, and for $\Delta = 1$ the stability limit is reached. For small values of Δ the spectrum can be calculated by

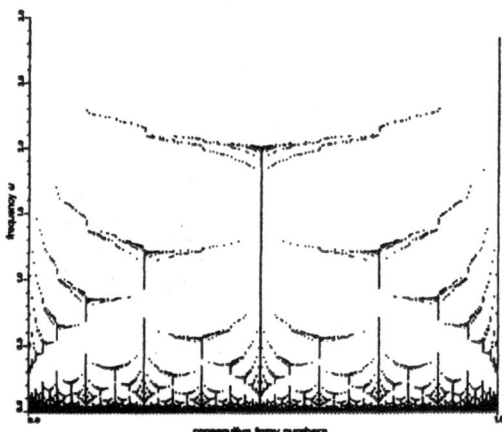

Fig. 2. — The spectra of the modulated spring model for $\Delta = 1$, plotted as function of the wave vector $Q/2\pi$. The latter is chosen to be rational (M/N). The values of M/N represent the Farey numbers of the first 9 generations.

perturbation methods, for larger values the usual approach is via commensurate approximants. The wave vector $Q/2\pi$ is replaced by the rational truncation of its continued fraction expansion. For $Q = 2\pi M/N$ the modulation is commensurate and there are N phonon branches. The spectra in the (Q-ω)-plane show a strong hierarchy which shows that the spectra for incommensurate values of Q have self-similarity and are singular continuous for $\Delta = 1$. In Figure 2 this hierarchy is shown for the spectra for all values of $Q = 2\pi M/N$ occurring in the first 9 generations of the Farey tree. Here the n–th generation consists of the fractions M/N which are obtained from all pairs of consecutive fractions $M_1/N_1, M_2/N_2$ from the $(n-1)$–th generation via $M/N = (M_1 + M_2)/(N_1 + N_2)$. See also the contribution of Dekking in this volume.

A model for the origin of an IC phase is the DIFFOUR model [17], a linear chain with variables x_n and a anharmonic potential energy:

$$V = \sum_n \left(\frac{1}{2}\alpha x_n^2 + \frac{1}{4}x_n^4 + \beta x_n x_{n-1} + \delta x_n x_{n-2} \right). \tag{10}$$

The ground state of this chain shows within certain ranges of the parameters (α/δ and β/δ) commensurate and incommensurate structures. The phonons in this model are the eigenvectors of the dynamical matrix. Next to the phonons which have their origin in coupled vibrations of the unmodulated chain ($x_n = 0$), there are modes which are qualitatively different. These are related

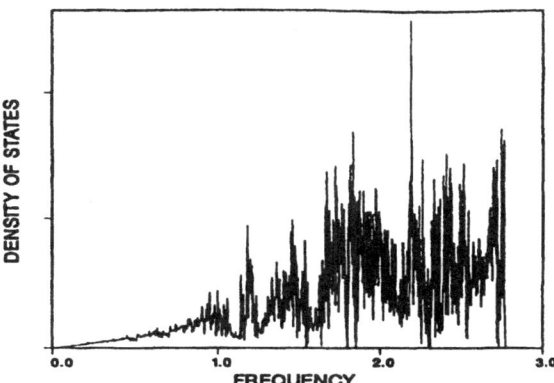

Fig. 3. — Phonon density of states for the octagonal tiling (Los *et al.* [23]).

to the modulation function f which represents the displacements $x_n = f(Qn)$, where f is periodic. The additional modes are spatially varying oscillations in the amplitude and the phase of the modulation function. They are called amplitudons and phasons, respectively. The wave vector Q may vary continuously or discontinuously as a function of the parameters.

The incommensurate composites are sometimes modelled by the Frenkel-Kontorova model [7], consisting of a harmonic chain with lattice constant a on a periodic substrate with period b. Its potential energy is

$$V = \sum_n \left[\alpha(x_n - x_{n-1} - a)^2/2 + \lambda\cos(2\pi x_n/b + \nu) \right]. \tag{11}$$

The ground state is determined by $\partial V/\partial x_n = 0$ and the spectrum of lattice vibrations by the eigenvalues of the dynamical matrix $\partial^2 V/\partial x_n \partial x_m$. The unrealistic feature of this model is that the substrate is rigid, whereas in many composites all subsystems become modulated. This can be studied in a double chain model with dynamical variables x_n for one chain and y_m for another chain. The potential energy then is

$$V = \sum_n f(x_n - x_{n-1}) + \sum_m g(y_m - y_{m-1}) + \sum_{nm} v(x_n - y_m). \tag{12}$$

A simple version is that with harmonic intrachain couplings

$$f(x) = \alpha_1(x - a_1)^2/2, \quad g(y) = \alpha_2(y - a_2)^2$$

and a rational approximant for the ratio of the two lattice constants such that a_1 and a_2 have a smallest common multiple a. The interchain interaction then can be taken to be a periodic function with period a, whereas the positions x_n and y_m are limited to the unit cell of the lattice generated by a (Radulescu and Janssen 1998).

Numerically one can solve the ground state equations for $pa_1 = qa_2 = a$. The solutions can be described as a modulation of the equidistant arrays $n_1 a_1$ and $n_2 a_2$.

$$x_n = n\, a_1 + f_1(n\, a_1), \quad y_m = m\, a_2 + f_2(m\, a_2) \tag{13}$$

with

$$f_1(x\, +\, a_2) = f_1(x), \quad f_2(y\, +\, a_1) = f_2(y).$$

For fixed interchain interaction, the solutions show two regimes as function of the intrachain spring constants. For small values of $1/\alpha_1$ and $1/\alpha_2$ the modulation functions are smooth. For large values the modulation functions show discontinuities, and in the $1/\alpha_1 - 1/\alpha_2$ plane there is a line where the discontinuities open. This is comparable to the behaviour in models for incommensurately modulated phases due to frustration. There also appears a sharp transition inside the IC phase where the modulation function becomes discontinuous. Usually this is called the transition to the discommensuration regime. Also in the Frenkel-Kontorova model such a transition has been found, coined "breaking of analyticity" by Aubry [2]. Without interchain coupling the double chain model has two sets of dispersion curves, and two zero frequency modes. After switching on the interaction the modes become coupled. For small strength of the interaction perturbation calculation shows that discontinuities open up at the new zone boundary (π/a), and in the center of the zone and that crossings of the original branches are lifted. This means that for incommensurate lattice constants, there is an infinite number of jumps in the dispersion curves. For small values of the interchain interaction the two zero frequency modes persist, in a symmetric combination as the zero frequency acoustic phonon mode for the whole crystal, and in an antisymmetric combination as sliding mode or phason where the two chains move with respect to each other. For increasing interaction the frequency of this sliding mode is no longer zero: a phason gap opens up exactly along the line in the $1/\alpha_1$ and $1/\alpha_2$ plane where the discontinuity in the modulation opens up. Therefore, the mode for which the frequency goes to zero for decreasing interaction is a soft mode for a transition from the regime with a smooth modulation to one with a discontinuous modulation. For still higher strength of the interaction the phason frequency increases, but its characterization becomes more complicated.

The breaking of analyticity has been brought in connection with the breaking up of KAM tori in dynamical systems. An example is given by the DIFFOUR model for incommensurate crystal phases. It considers a linear chain of particles at the positions of an equidistant array. Each particle has a degree of freedom x_n, and the potential energy for the system is given by

$$V = \sum_n \left(\frac{\alpha}{2} x_n^2 + \frac{1}{4} x_n^4 + \beta x_n x_{n-1} + \delta x_n x_{n-2} \right).$$

The ground state is given by a solution of the equation

$$\frac{\partial}{\partial x_n} V = 0,$$

or

$$\alpha x_n + x_n^3 + \beta(x_{n-1} + x_{n+1}) + \delta(x_{n-2} + x_{n+2}) = 0.$$

This formula can be rewritten as a discrete mapping in four-dimensional space:

$$\begin{pmatrix} x_{n+2} \\ x_{n+1} \\ x_n \\ x_{n-1} \end{pmatrix} = \begin{pmatrix} (-\beta x_{n+1} - \alpha x_n - x_n^3 - \beta x_{n-1})/\delta - x_{n-2} \\ x_{n+1} \\ x_n \\ x_{n-1} \end{pmatrix}$$

$$\Leftrightarrow v_{n+1} = S(v_n).$$

For values of α between $-2\beta - 2\delta$ and $2\beta + 2\delta$ the origin is an elliptic fixed point. In its neighbourhood there are periodic orbits as well as orbits with an irrational winding number. Close to the origin these fill smooth manifolds, but going further from the origin these orbits fall on discontinuous manifolds. This corresponds to the breaking of smooth KAM tori, and for the crystal configuration it corresponds to a transition from smooth to non-analytic modulation functions. See also [6].

Studies of the lattice vibrations of quasicrystals go along the same lines. Simple models consider a 2- or 3-dimensional tiling decorated with atoms, e.g. in the vertices of the tilings. Short range springs are taken into account only. An example is an Amman-Beenker 2-dimensional octagonal tiling or a 3-dimensional icosahedral tiling with springs along the edges of the tiles [23, 24]. For the numerical calculations one considers commensurate approximants which are lattice periodic and can be obtained from a quasiperiodic tiling by affecting a linear strain in the additional space to the embedding of the quasicrystal model in 4 or 6 dimensions. For a series of approximants the spectrum must be calculated as the eigenvalues of the Fourier transform of the dynamical matrix

$$D_{ij} = (m_i m_j)^{-1/2} \frac{\partial^2}{\partial u_i \partial u_j} V.$$

A typical choice for the dynamical matrix D is

$$D_{ij} = (\alpha_i + \alpha_{i+1})\delta_{ij} - \alpha_i \delta_{ij+1} - \alpha_i \delta_{ij-1}, \tag{14}$$

for a linear chain with a quasiperiodic binary series of interatomic spring constants α_i (e.g. a Fibonacci chain with spring constant A over a long interval and B over a short interval). Other models are 2D or 3D tilings with atoms on the vertices and potential energy

$$V = \frac{1}{4} \sum_{ij} (\alpha_{ij} u_{ij\parallel}^2 + \beta_{ij} u_{ij\perp}^2), \tag{15}$$

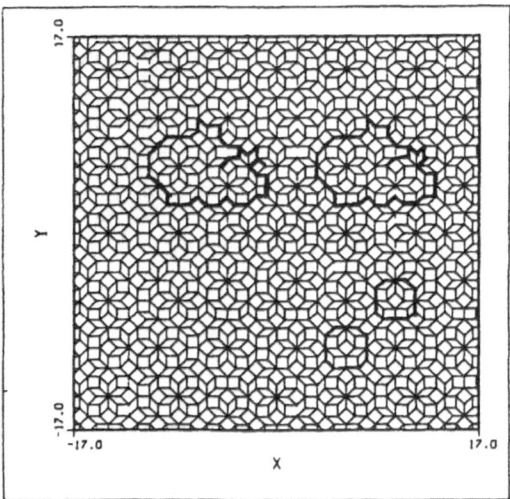

Fig. 4. — Bounded configurations in tilings occur infinitely often, but without periodicity. The wave functions on equal patches are similar, but with different amplitudes.

for pairs ij connected by an edge. The coefficients α and β are here the stretching and bending spring constants. More sophisticated models use model structures for specific compounds and more realistic Lennard-Jones, Morse or pseudo-potentials [10,11]. Details differ for these models, but there are several common properties which are different from those for lattice periodic crystals:

1. the phonon density of state is very spiky. In two-dimensional models there are often narrow gaps, in 3 dimensions pseudo-gaps (Fig. 3).

2. The DOS becomes smoother for long wave lengths. On the average the behaviour is as ω^{d-1}, but there are pseudo-gaps for arbitrarily low frequencies.

3. For 1-dimensional models it has been proven that the eigenvectors are neither extended as in lattice periodic systems nor localized as in disordered systems, but critical. There are indications for such eigenvectors in higher dimensions as well.

4. Nevertheless long wave length excitations behave as propagating modes. For higher frequencies the eigenvectors are rather localized in a unit cell of an approximant. In a quasiperiodic tiling there is no lattice periodicity but every patch of the tiling occurs infinitely many times in other positions. The eigenvectors show in identical patches often the same shape, but with different amplitude (Fig. 4).

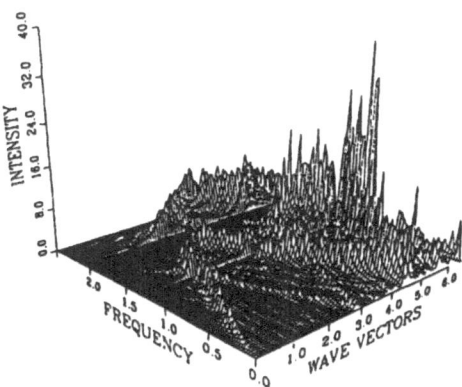

Fig. 5. — Dynamical structure factor: cross section as function of wave vector and frequency for the Fibonacci chain. One may observe the linear dispersion for low frequencies, the broadening at higher frequencies, the gaps in the spectrum and the pseudo-zone boundary between the origin and the strong Bragg peak near 4.5.

5. These features can be found in the differential cross section for inelastic neutron scattering (Fig. 5). The dynamic structure factor $S(\vec{q}, \omega)$ is for low frequencies concentrated on a narrow region around straight lines in the $(\vec{q}, \omega)$-plane. These lines correspond to dispersion curves. For higher frequencies the regions in which the $S(\vec{q}, \omega)$ does not vanish broaden. This is related to the more localized character of the critical states.

6. The quasi-dispersion curves bend down and reach a maximum in a point of a perpendicular bisecting plane between two strong Bragg peaks. These planes (or lines in 2D) correspond to pseudo-Brillouin zone boundaries.

7. Dispersion curves of subsequent commensurate approximants show a scaling behaviour: the lower frequency part of one approximant is a scaled version of that for the next approximant. This is an indication that the limit for the incommensurate structure has self-similarity properties.

8. The fine features of the phonon density of states, like the pseudo-gaps, disappear when phason disorder is present in the model.

3. DOMAIN WALL MOTION

The discontinuities in the modulation function found for certain parameters in the DIFFOUR, Frenkel-Kontorova and double chain models can be viewed as discommensurations. These discommensurations make that a commensurate chain becomes incommensurate. For example in a spin model with spin up and down the period-4 system with unit cell ↑↑↓↓ becomes an incommensurate chain by introducing regular defects. For example a fraction $\sqrt{5} - 2$ of cells of length 5 makes the wave vector of the modulation $\sqrt{5} - 2$ instead of $1/4$. Such a chain looks like ↑↑↓↓↑↑↓↓↑↑↓↓↑↑↓↓ ... where the extra spin down in position 13 is an example of a discommensuration, which is very sharp in this case. The spin chain could be the ground state for the ANNNI (anisotropic, next-nearest-neighbour Ising) model, a spin system on a tetragonal lattice with Hamiltonian

$$H = \sum_n \left(J_0 \sum_{m \text{ neighbrs } xy \text{ plane}} S_n S_m + J_1 S_n S_{n'} + J_2 S_n S_{n''} \right), \qquad (16)$$

where n' and n'' are first and second neighbours in the z-direction [33]. For the DIFFOUR model, which is a soft ANNNI model, the Hamiltonian

$$H = \sum_n \left(\frac{p_n^2}{2m} + \frac{1}{2}\alpha x_n^2 + \frac{1}{4}x_n^4 + \beta x_n x_{n-1} + \delta x_n x_{n-2} \right) \qquad (17)$$

has a ground state for which the positions x_n satisfy

$$\alpha x_n + x_n^3 + \beta(x_{n-1} + x_{n+1}) + \delta(x_{n-2} + x_{n+2}) = 0. \qquad (18)$$

The ground state is the solution with minimal energy, and this depends on the parameters. For a specific, fixed choice of β and δ the solution is $x_n = 0$ for $\alpha > \alpha_i$ and it is a N-cycle (a commensurate structure with N particles in the unit cell) for $\alpha < \alpha_c$. For values of α between α_c and α_i the character of the solution changes qualitatively. Close to α_i the projection of the solution on the $x_n - x_{n-1}$ plane fill out a smooth curve, corresponding to a smooth modulation function. Near α_c the points are concentrated in clouds around the N points of the solution below α_c. In between the transition mentioned above occurs.

These discommensurations are quasiperiodically arranged and differ in that from domain walls in lattice periodic crystals. They can be seen as non-linear excitations on a periodic structure (the stable commensurate solution below α_c). Their dynamics is governed by:

$$\ddot{x}_n = -\alpha x_n - x_n^3 - \beta(x_{n-1} + x_{n+1}) - \delta(x_{n-2} + x_{n+2}). \qquad (19)$$

Splitting off the fast fluctuating part by the substitution

$$x_n = \exp(ikn)P(n),$$

where k is the commensurate wave vector of the periodic solution below α_c, the equation of motion may be approximated by a continuum approximation, which can be solved exactly:

$$P(r) = \sqrt{-\alpha + 2\beta} \tanh \left[(r - r_o - vt)\sqrt{\frac{-\alpha + 2\beta}{4\beta - v^2}} \right]. \qquad (20)$$

The solution of the continuum approximation may be used for a starting configuration in a numerical integration of the original equations of motion. For low velocities ($v < v_1$) the continuum solution remains stable in the discrete system. For $v_1 < v < v_2$ the solitary wave looses energy by phonon emission. For still higher velocities ($v > v_2$) the solitary wave falls apart in independent solitary waves moving both to the left and to the right. The velocity v_1 corresponds with the phason velocity. Such domain wall motion is specific for incommensurate phases and can also be expected to be present in incommensurate composites.

4. ELECTRONS

The study of electronic properties in quasiperiodic crystals faces the same problems as that of the phonons. There is no lattice periodicity, hence no Brillouin zone. Therefore, generally the electron wave functions will not be Bloch waves and the spectrum is expected to be more complicated than one with bands.

There are several approaches to the problem. Analytically there are only rigorous proofs for simple one-dimensional models. Some results have been obtained using renormalization methods. A related method is the trace map approach for one-dimensional chains. Numerically one may calculate the band structure and wave functions for commensurate approximants of the quasiperiodic system or for clusters of increasing size.

In one dimension one has studied the Kronig-Penney chain, a 1D Schrödinger problem with potential

$$V(x) = a \sum_n \delta(x - x_n),$$

where the x_n denote the positions of (identical) atoms in a quasiperiodic structure, such as the points in a Fibonacci chain or in a modulated chain $na + \Delta \cos(Qn)$ [21]. The results are very similar to those of tight-binding models

$$H = \sum_{nm} |n > t_{nm} < m| + \sum_n |n > \epsilon_n < n|. \tag{21}$$

For a one-dimensional chain with nearest neighbour hopping the latter leads to the Schrödinger equation

$$t_n c_{n-1} + t_{n+1} c_{n+1} + \epsilon_n c_n = E c_n.$$

The hopping frequencies t_n and/or the site energies ϵ_n form a quasiperiodic sequence. This Schrödinger equation may be rewritten as a map in R^2.

$$\begin{pmatrix} c_{n+1} \\ c_n \end{pmatrix} = T_n \begin{pmatrix} c_n \\ c_{n-1} \end{pmatrix} = \begin{pmatrix} (E - \epsilon_n)/t_{n+1} & -t_n/t_{n+1} \\ 1 & 0 \end{pmatrix} \begin{pmatrix} c_n \\ c_{n-1} \end{pmatrix}. \tag{22}$$

Using the inflation symmetry of certain quasiperiodic chains [39] one can write a map in three-dimensional space $(x_p, x_{p-1}, x_{p-2}) \rightarrow (x_{p+1}, x_p, x_{p-1})$, where $x_n = \frac{1}{2}\mathrm{Tr}M_p$ and M_p is the product of T_n matrices [19, 26]. A bounded orbit of this trace map corresponds to an electron state of energy E. For further references to trace maps see Peyrière [27] and Süto [36].

The results for these one-dimensional models give the following picture.

1. For the modulated Kronig-Penney model with a continuous range of positions or of strengths of the delta peaks the eigenfunctions are not extended Bloch waves, but either localized, pseudo-localized, or extended. This can very well be seen in superspace (Fig. 6).

2. For the Harper equation, which is a one-dimensional equation for a tight-binding electron in an external magnetic field, of the form

$$c_{n-1} + c_{n+1} + \lambda\cos(Qn + \theta)c_n = Ec_n \qquad (23)$$

   the spectrum is absolutely continuous for $\lambda < 2$, discrete for $\lambda > 2$ and singular continuous for $\lambda = 2$. The eigenfunctions are for these cases extended, localized or critical, respectively.

3. The tight-binding Hamiltonians for chains obtained by substitution rules, such as the Fibonacci chain, have generically singular continuous spectra. The Hamiltonian is determined by the hopping frequencies between neighbours, which take a finite number of values. The eigenfunctions are critical, they have an envelope that decays with a power law [20]: $|c_n| \sim |n|^{-\gamma}$. The value of γ may be quite big, and then such a state behaves for all practical purposes as a localized state (Fig. 7).

4. Eigenfunctions often are self-similar. These and the spectral properties may be studied using a multifractal analysis. If the spectrum of the p-th approximant to a quasiperiodic chain consists of N_p bands of width $\Delta_i^{(p)}$ one may calculate the distribution of exponents ϵ with which the band widths tend to zero. For the function

$$F_p(\beta) = \frac{1}{p}\log\sum_i^{N_p}\left(\Delta_i^{(p)}\right)^\beta \qquad (24)$$

   the distribution function is

$$S(\epsilon) = \lim_{p\to\infty} S_p(\epsilon); \qquad S_p(\epsilon) = F_p(\beta) - \epsilon\frac{\mathrm{d}F_p}{\mathrm{d}\beta}. \qquad (25)$$

   A discussion of the behaviour of spectra and wave functions using this technique can be found in Hiramoto and Kohmoto (1992).

5. The spectrum for the tight-binding model on substitution chains is a Cantor set, which can be approximated by the band spectra of approximants. The splitting of the bands when one approximates the quasiperiodic chain by periodic chains corresponding to the subsequent truncation

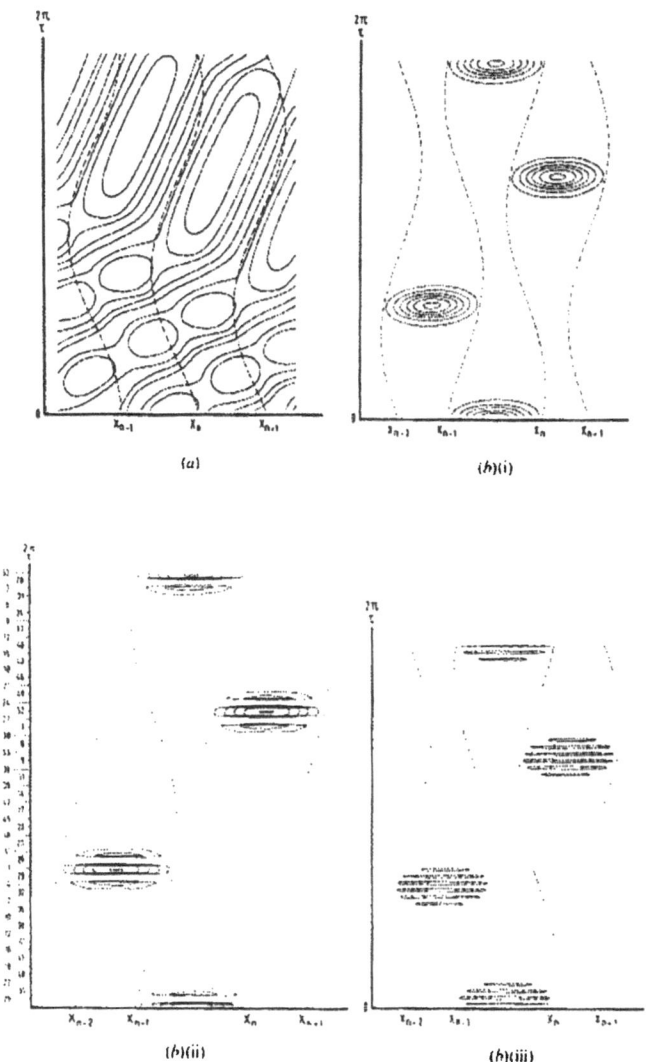

Fig. 6. — Electron wave functions in superspace for the modulated Kronig-Penney model. Extended state left top, pseudo-localized state in approximants in the three other parts (de Lange and Janssen 1984). Reprinted from Physica A, Vol. 127, pp. 125-140 (1994) with permission from Elsevier Science.

of the continued fraction expansion of the irrational number, gives a hierarchical structure. The integrated density of states for these chains is a Cantor function.

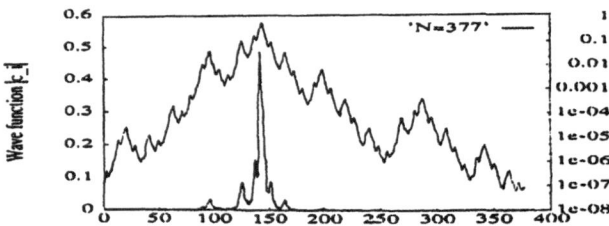

Fig. 7. — Modulus of an electron wave function for an approximant to a quasiperiodic chain. Although the wave function looks strongly localized in the unit cell, the
corresponding wave function for the quasiperiodic chain is critical. Narrow figure:
modulus, other figure: logarithm of the modulus.

Besides there are numerical calculations of the electronic structure of tiling
models or models based on structure analysis of specific quasicrystals. One
has studied tight-binding models on 2-dimensional tilings, like the Penrose
or Ammann-Beenker tiling [18, 34], or on the standard 3-dimensional icosahedral tiling, obtained by intersection of 3D physical space with a 6D periodic
array of 3D triacontahedra (the projection of the 6D unit cell on the internal space). More specific models use structural models, as presented by de
Boissieu *et al.*, Burkov or Elser *et al.* Fujiwara *et al.* and Hafner and Krajči
have used Schrödinger equations on more or less realistic structural models
using pseudo-potentials [8, 9, 11]. The common features are the following.

1. The electron density of state shows a very spiky character.

2. Near the Fermi energy the density of states has a pseudogap.

3. There are numerical indications that there are critical wave functions in
   3D as well.

Using the moments method one finds a less spiky DOS [29].

5.   TENSORIAL PROPERTIES

Physical properties often are described using tensors. Tensors are multilinear
functions of vectors, and the independent coefficients (also called free parameters) are restricted by symmetry. In lattice periodic crystals the arguments are
3-dimensional vectors, and the symmetry groups are point groups for tensors
which are not spatially varying. The number of free parameters of a tensor of
certain rank and given permutation symmetry of the coefficients is the multiplicity of the trivial representation in the representation of the symmetry group
carried by the space of these tensors. As example, the elasticity tensor in 3D

is a rank four tensor T_{ijkm} such that $T_{ijkm} = T_{jikm} = T_{ijmk} = T_{kmij}$. The space of tensors is 21-dimensional and carries a representation of a point group which acts on a tensor according to

$$(RT)_{i_1 j_1 k_1 m_1} = \sum_{i_2 j_2 k_2 m_2} R_{i_1 i_2} R_{j_1 j_2} R_{k_1 k_2} R_{m_1 m_2} T_{i_2 j_2 k_2 m_2}.$$

For example, the cubic point group 432 leaves a 3-dimensional subspace pointwise invariant. Therefore, the number of free parameters in the elasticity tensor for a crystal with 432 symmetry is three.

For quasiperiodic crystals the symmetry group is an n-dimensional crystallographic point group, where n is the rank of the Fourier module. These point groups have the property that they do not mix coordinates in physical space and coordinates in perpendicular space. Therefore, the point group leaves the physical space invariant, and this means that it acts also on the space of tensors in 3D space. For example, the point group of icosahedral quasicrystals is the 6-dimensional group $5\bar{3}m(5^2\bar{3}m)$. It acts on physical space according to the non-crystallographic icosahedral point group $5\bar{3}m$, and also on the 21-dimensional space of rank 4 tensors with the symmetry of the elasticity tensor. This space contains a 2-dimensional pointwise invariant subspace. Therefore, the elasticity tensor in the usual sense for an icosahedral quasicrystal has 2 free parameters [3].

This tensor describes the elasticity tensor with 3D strain. However, a quasicrystal may also have a strain with components in perpendicular space. This corresponds to a linear phason strain. Such a strain corresponds to a tensor $f_{ik} = \partial_i u_k$, where $i = 1, 2, 3$ and $k = 4, \ldots, n$. The total elastic energy then is given by

$$U = \frac{1}{2} \sum_{ijkl} c^{E}_{ijkl} e_{ij} e_{kl} + \frac{1}{2} \sum_{ijkl} c^{I}_{ijkl} f_{ij} f_{kl} + \sum_{ijkl} c^{EI}_{ijkl} e_{ij} f_{kl},$$

where e is the usual elastic tensor, and the c's are the generalized elastic constants. There are 3 subspaces each carrying a representation of the n-dimensional point group and corresponding to, respectively, the phonon degrees of freedom in physical space, the phason degrees of freedom in internal space, and a coupling. For example, the 3 spaces carry each a pointwise invariant subspace under the icosahedral point group. The dimension of these subspaces is 2, 2 and 1, respectively. Therefore, there are 2 free parameters for phonon elasticity, 2 for phason elasticity, and one for the coupling.

6. SURFACE EFFECTS

In lattice periodic crystals electronic and vibrational states can change their character by the presence of a surface. In quasiperiodic crystals this phenomenon also occurs but there the variety of states is already bigger from the start. To study the influence of a surface on physical properties it is important to be able to characterize the surface. The first problem encountered here is that the situation is experimentally not so clear. Surfaces with neat, flat terraces as well as rough surfaces showing clusters have been observed.

For a lattice periodic crystal flat surfaces are net planes. The net planes of the embedding of a quasiperiodic structure in higher dimensions can be characterized by reciprocal lattice vectors as well. If the nD lattice is spanned by e_i and the reciprocal lattice by e_i^* a net plane corresponding to $H_s = \sum_i h_i e_i^*$ is given by $\sum_i h_i x_i = N$. Lattice points in such a net plane are $r_s = \sum_i n_i e_i$ with $\sum_i h_i n_i = N$. The net planes intersect the physical space along 2D planes: Points in the intersection of the physical space with the net plane $\sum_i h_i x_i = N$ satisfy $H_s r_s = H_{sE} r_E = Q.r = N$. The 3-vector Q is the projection of H_s on the physical space V_E. The planes $Q.n = N$ form a family of parallel planes with distance $d(H_s) = 1/|Q| = 1/|H_{sE}|$. The positions of the atoms are the intersections of atomic surfaces with V_E. These atomic positions are distributed around the net planes $Q.n = N$.

For each H_s there is a unique Q, but in contrast to 3D crystals there is no vector with minimal length perpendicular to the net planes. For example, for a 6D icosahedral lattice all 6D vectors $[h\ell\ell\ell\ell\ell]$ give Q-vectors parallel to $(1, \Phi, 0)$, where $\Phi = (\sqrt{5}+1)/2$. The lengths of these vectors are $\sqrt{2 + \Phi}(H + L\Phi)$ and fill densely the positive real axis. This can be used to characterize a slab with a fixed chosen thickness. To characterize atoms near a surface one can distinguish slabs (all atoms with the same $\sum_i h_i n_i$) and planes (all atoms with the same $Q.r$).

The symmetry of the surface is the symmetry of the characterizing reciprocal lattice vector. For example, for the icosahedral lattice the plane [100000] has 5-fold, the plane [0001$\bar{1}$0] has 2-fold and the plane [110001] 3-fold symmetry. This however is only true for flat surfaces. There are experimental indications that the surface can be rough, and follow the surface of the clusters which form the quasicrystal. Even for flat surfaces the microscopic structure is unknown. For example, surface roughening and reconstruction may be present, just as they occur for lattice periodic crystals.

There are some calculations of the influence of a surface in 1D models. Truncation of a Fibonacci chain in a tight-binding model shows, that next to the critical states of the bulk, there are surface induced gap states. These gap states decay exponentially from the surface, but the decay rate may be so slow that the localization is not within the first atomic layers and the wave function penetrates deeply into the bulk.

Numerical calculations on 3D icosahedral models show the existence of strictly localized states, concentrated at the surface. Also the density of states varies as function of the distance to the surface. If one defines the slab density of states

$$SDOS(\omega, i) = \sum_{\text{states} k} w_{ki} \, \delta(\omega - \omega_k), \qquad (26)$$

where w_{ki} is the participation i

$$w_{ki} = \sum_{n \in \text{layer } i} |\Psi_k(n)|^2$$

of layer i to state k, there is a variation with i. In this model the pseudogap that appears in the bulk spectrum at the Fermi energy persists till the surface layer.

7. TRANSPORT PROPERTIES

Quasicrystals have interesting transport properties (see the contribution by Berger in this volume). In the first place the electric conductivity in the icosa-hedral phase is low (between 100 and 300 $(\Omega \text{ cm})^{-1}$ for AlMnPd), and decreases with improved quality. The conductivity as function of temperature shows an increase according to $\sigma(T) = \sigma_o + \delta\sigma(T)$, where the second term is sample independent. Many quasicrystals have approximants, structure with locally almost the same order as the quasicrystal, but globally having lattice period-icity. For example, AlCuFe and AlMnPd have approximants. The behaviour for approximants is quite similar to that of quasicrystals, except for the low-est approximants. This holds for the low conductivity and for the positive derivative of $\delta\sigma(T)$. For low approximants the conductivity is much higher and decreases with temperature. This suggests that the local order becomes the determining factor if the unit cell of an approximant becomes big enough. On the other hand, closely related structures, like the icosahedral phases of AlMnPd and AlMnRe may behave quantitatively very differently. For exam-ple, the conductivity of AlMnRe is 2 orders of magnitude lower then that of AlMnPd [13,28]. The temperature dependence reminds of semiconductors, but on a closer look there are important differences. The increase in conductivity follows approximately a law $\sigma(T) \sim T^\beta$ with $1 < \beta < 1.5$.

It is believed that the anomalous behaviour (with respect to lattice periodic crystals) is due to the critical character of the wave functions, and to the existence of a pseudogap at the Fermi level. The wave functions are often pseudo-localized, and this makes hopping possible between regions with such states. Another mechanism is that of anomalous diffusion. The propagation of a wave packet has been found in models to satisfy a law $L(t)t^\beta$ (with $\beta < 1$). This implies

$$\sigma(\tau) \sim A\tau^{2\beta-1}. \tag{27}$$

This means that for $\beta < \frac{1}{2}$ the conductivity decreases with τ. Work in this direction has been done by Sire and Bellissard [4, 35].

The transmission has been calculated by Tsunetsugu [37,38] for the Fibonacci chain and the Penrose tiling using the Landauer formula for resistance. They find

$$\frac{\sigma}{L} = g(L) \sim L^\alpha, \qquad (\alpha \approx -0.3). \tag{28}$$

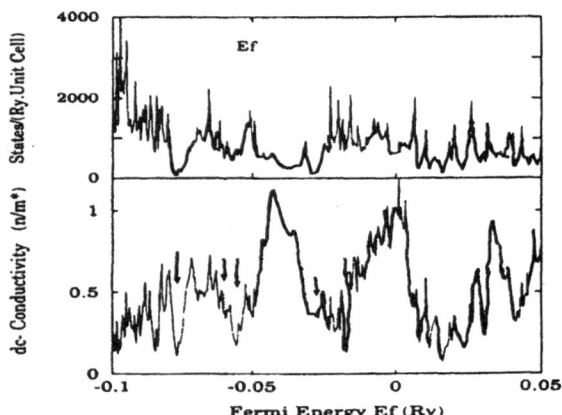

Fig. 8. — Dependence of the conductivity as function of energy. It varies strongly (After Fujiwara *et al.* 1993)

Linear combinations of Muffin Tin Orbitals (LMTO)-calculations on a model for an AlMn alloy have been performed by Fujiwara *et al.* [8]. They find effects of the band structure. Using the Boltzmann formalism they find a conductivity which depends strongly on the Fermi energy (Fig. 8). The conductivity appears to depend strongly on the disorder [25, 32].

Measurements of the thermal conductivity of AlCuFe show a plateau between 10 K and 100 K. Phonon wave functions have the same critical character as the electron wave functions, but neither their role nor the role of phason hopping is clear at the moment.

8.   CONCLUDING REMARKS

Quasiperiodic crystals form an interesting family of compounds having changed our view on material properties. For IC modulated and composite structures there is an understanding of their structure, although there remain many questions. In particular, the existence and the role of discommensurations is not clear. For quasicrystals there are more problems. Although there is a global understanding, most models fail to explain the weak diffraction spots. The possible role of randomness remains a source of debate.

Concerning the dynamics of IC modulated phases and composites the question of the existence of a phason gap and of a sliding mode has been answered in very few cases only. In the case that the phason gap vanishes the phason behaves like a nonlinear excitation. The consequences of that have not yet been investigated. In quasicrystals the dynamics show properties of periodic crystals and of amorphous systems. The question here is to what extent the dynamics is determined by the local structure or by the global ordering.

With the improvement of the quality of quasicrystals more information has become available concerning electronic properties. These have been put into relation with the critical character of the wave functions. However, this character could easily got lost if one takes defects and disorder into account.

In a certain sense one may defend the statement that quasiperiodic structures are between lattice periodic crystals and amorphous materials. IC phases have much in common with lattice periodic crystals, apart from phenomena as phasons and the role of metastability. Quasicrystals behave differently, but here the disorder can become important. In all cases it seems that boundary conditions, for example the existence of surfaces, play a more important role in quasiperiodic systems than in the periodic ones.

REFERENCES

[1] *Beyond Quasicrystals*, edited by F. Axel and D. Gratias (Springer/Les Éditions de Physique, Heidelberg/Les Ulis, 1995).

[2] S. Aubry and P.Y. Le Daeron, *Physica D* **8** (1983) 381-422.

[3] P. Bak, *Phys. Rev. B* **32** (1985) 5764-5772.

[4] J. Bellissard and H. Schulz-Baldes, in *Quasicrystals, Proceedings of ICQ5*, edited by C. Janot and R. Mosseri (1996) 439.

[5] R. Currat and T. Janssen, *Solid State Physics* **41**, edited by H. Ehrenreich and D. Turnbull (Academic Press, 1988) 201-302.

[6] F.M. Dekking, in *Beyond Quasicrystals*, edited by F. Axel and D. Gratias (Springer/Les Éditions de Physique, 1995) 415-432.

[7] Y.I. Frenkel and T. Kontorova, *Zh. Eksp. Teor. Fiz.* **8** (1938) 1340.

[8] T. Fujiwara, S. Yamamoto and G. Trambly de Laissardière, *Phys. Rev. Lett.* **71** (1993) 4166.

[9] J. Hafner and M. Krajči, *Phys. Rev. Lett.* **68** (1992) 2321.

[10] J. Hafner and M. Krajčí, *J. Phys. Cond. Mat.* **5** (1993) 2489.

[11] J. Hafner and M. Krajčí, *Phys. Rev. B* **47** (1993) 11795.

[12] H. Hiramoto and M. Kohmoto, *Int. J. Mod. Phys.* **B6** (1992) 281-320.

[13] Y. Honda, K. Edagawa, A. Yoshioka, T. Hashimoto and S. Takeuchi, *Jpn. J. Appl. Phys.* **33** (1994) 4929-4935.

[14] C. Janot, *Quasicrystals, A primer* (Oxford UP, Oxford, 1996).

[15] A. Janner and T. Janssen, *Phys. Rev. B* **15** (1977) 643-658.

[16] T. Janssen and A. Janner, *Adv. Phys.* **36** (1987) 519-624.

[17] T. Janssen and J.A. Tjon, *J. Phys. C* **16** (1983) 4789-4810.

[18] M. Kohmoto and J.R. Banavar, *Phys. Rev. B* **34** (1986) 563.

[19] M. Kohmoto, L.P. Kadanoff and C. Tang, *Phys. Rev. Lett.* **50** (1983) 1870.

[20] M. Kohmoto and B. Sutherland, *Phys. Rev. B* **34** (1986) 3849-3853.

[21] C. de Lange and T. Janssen, *J. Phys. C* **14** (1981) 5269-5292.

[22] C. de Lange and T. Janssen, *Physica A* **127** (1984) 125-140.

[23] J. Los, T. Janssen and F. Gähler, *Int. J. Mod. Phys. B* **7** (1993) 1505-1525.

[24] J. Los, T. Janssen and F. Gähler, *J. Phys. France* **3** (1993) 1431-1461.

[25] K. Moulopoulos and S. Roche, *Phys. Rev. B* **53** (1996) 212.

[26] S. Ostlund, R. Pandit, D. Rand, H.J. Schellnhuber and E.D. Siggia, *Phys. Rev. Lett.* **50** (1983) 1873.

[27] J. Peyriere, in *Beyond Quasicrystals*, edited by F. Axel and D. Gratias (Springer/Les Éditions de Physique, 1995) 563-584.

[28] F.S. Pierce, Q. Guo and S.J. Poon, *Phys. Rev. Lett.* **73** (1994) 2220.

[29] G. Poussigue, C. Benoit, M. de Boissieu and R. Currat, *J. Phys. Cond. Mat.* **6** (1994) 659.

[30] M. Quilichini and T. Janssen, *Rev. Mod. Phys.* **69** (1997) 277-314.

[31] O. Radulescu and T. Janssen, Proceedings Aperiodic 97 (World Scientific, Singapore, 1998).

[32] S. Roche and D. Mayou, *Phys. Rev. Lett.* **79** (1997) 2518.

[33] W. Selke, in *Phase transitions and critical phenomena*, Vol. 15, edited by C. Domb and J.L. Lebowitz (1992) 1-72.

[34] C. Sire, *Europhys. Lett.* **10** (1989) 483-488.

[35] C. Sire, in *Quasicrystals, Proceedings of ICQ5*, edited by C. Janot and R. Mosseri (1996) 415.

[36] A. Süto, in *Beyond Quasicrystals*, edited by F. Axel and D. Gratias (Springer/Les Éditions de Physique, 1995) 483-550.

[37] H. Tsunetsugu, T. Fujiwara, K. Ueda and T. Tokihiro, *J. Phys. Soc. Jpn.* **55** (1986) 1420.

[38] H. Tsunetsugu and K. Ueda, *Phys. Rev. B* **38** (1988) 10109.

[39] Wen Zhi-Ying, in *Beyond Quasicrystals*, edited by F. Axel and D. Gratias (Springer/Les Éditions de Physique, 1995) 433-440.

[40] P.M. de Wolff, *Acta Cryst.* **A30** (1974) 777-785.

Diffraction Experiments on Quasicrystals and Related Phases

F. Dénoyer

Laboratoire de Physique des Solides, Associé au CNRS, bâtiment 510, Université Paris-Sud, 91405 Orsay Cedex, France

1. INTRODUCTION

Structural studies are essential for understanding formation, growth stability and physical properties of quasicrystals and related phases. Diffraction is a powerful tool for studying order and disorder in these materials. Electron diffraction has played a key role in the discovery of icosahedral [1, 2], decagonal [3] and dodecagonal [4] quasicrystals and their characterization. In many quasicrystalline alloys, the size of quasicrystalline grains being small (typically 1 μm), only electron diffraction and X-ray or neutron powder diffraction methods were feasible. Inconvenient with electron diffraction is the importance of multiple Bragg scattering making intensity calculations generally hard to handle. The disadvantage of powder methods is that the three-dimensional diffraction patterns of all individual small single grains are angularly averaged and reduced to one-dimensional informations. It is one of the reasons to go further in the structure determination of these materials, since the beginning, that efforts were undertaken to prepare large single grains. Of course, this rests on the assumption that the quasicrystal is stable at least in some composition and temperature range. With the elaboration of large single grains with triacontahedral shapes, Al_6Li_3Cu was the first reported stable icosahedral quasicrystal [5]. This opened the way for structure determination by X-ray or neutron diffraction. The use of X-ray synchrotron beam combining high-flux and high-resolution is particularly well-suited for studying very weak diffraction peaks that allow to distinguish between quasiperiodic order and complex domain arrangements of approximant phases. For a long time, quasicrystals have been described as quasiperiodic structures with a crystallographically disallowed orientational symmetry (existence of 5-fold, 8-fold, 10-fold, 12-fold axes, ...). These restrictions based on the notion of forbidden symmetry are not mandatory to build quasiperiodic order. Furthermore, very recently P. Donnadieu et al., studying a rapidly solidified Al-Mg alloy by electron diffraction, discovered a quasicrystal characterized by a cubic point group and an inflation symmetry [6]. At the present time, it is the only example of quasicrystal which does not exhibit a disallowed orientational symmetry. Therefore, it will be preferred to define a quasicrystal as a material without translational periodicity along any direction of the three dimensional space (in the case of three dimensional quasicrystals) or of the two dimensional space (in the case of bidimensional quasicrystals), but with Bragg peaks in diffraction patterns. The basis for describing such peculiar but ideally perfect structures lies in higher dimensional crystallography [7–10].

In this lecture we will limit ourselves to the study of bidimensional quasicrystals and their related phases in order to give a pedagogical review of their main characteristic features.

2. DECAGONAL QUASICRYSTALS

2.1. Generalities

Decagonal quasicrystalline phases are interesting in the sense that they are periodic along their unique tenfold axis $\vec{c}$ and quasiperiodic in the planes perpendicular to it (Fig. 1). Simplifying aspects of these bidimensional quasicrystals make their study attractive, experiments and theoretical models easier.

Decagonal symmetry has been reported in many alloys, using selected area electron diffraction technique (a list of decagonal phases was given in the review paper of Kelton [11]). So far only a few decagonal quasicrystals have been grown as "large" single grains with decaprism morphology and high "crystallographic" quality, mainly in the ternary alloy systems Al-Co-Ni and Al-Co-Cu and, only very recently, in the Al-Pd-Mn system [12], allowing conventional and synchrotron X-ray diffraction using single-crystal methods. Most stable decagonal quasicrystals are stable only at high temperature and transform to crystalline approximant phases at lower temperature. Most often the approximant phases have a complicated crystalline microdomain structure, the size of these microdomains typically varying from 20 nm to over 500 nm.

2.2. X-ray structure determination of decagonal quasicrystals

2.2.1. Metrics and space groups

All crystallographic determinations were based upon the five-dimensional description of ideal decagonal quasiperiodic structures introduced by Janssen [13]. Basic concepts on structure and symmetry are given in this volume

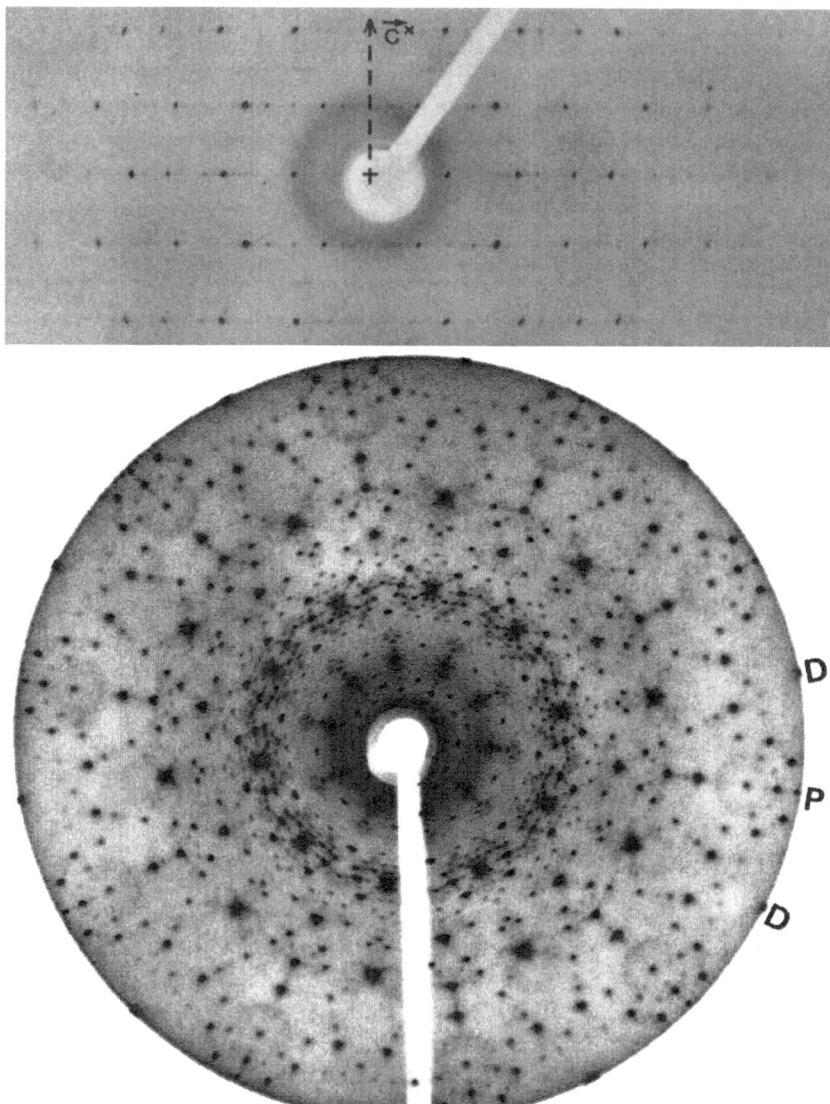

Fig. 1. — Monochromatic X-ray diffraction patterns of $Al_{63}Co_{17.5}Cu_{17.5}Si_2$. A rotating-crystal pattern showing equidistant layers of Bragg peaks along the ten-fold axis (on the top) and a precession pattern showing the decagonal symmetry in the plane perpendicular to the periodic reciprocal axis (on the bottom).

(*cf.* the course of Janssen). The five-dimensional periodic space $E = (E^{\parallel}, E^{\perp})$ decomposes into two orthogonal subspaces, the three-dimensional physical or parallel space $E^{\parallel}$ and the two-dimensional perpendicular space $E^{\perp}$. The periodic lattice is decorated with objects called atomic surfaces lying in the perpendicular space and the physical structure is obtained by a cut through the periodic decorated lattice. Details on the method are given in the course of Cockayne in this volume. The observed diffraction pattern is the projection of the five-dimensional periodic reciprocal space upon the parallel reciprocal space.

Any Bragg reflection is indexed by five integers $(h_1 h_2 h_3 h_4 h_5)$ from which the physical-space reciprocal lattice vectors are defined in a unique way:

$$\overrightarrow{Q^{\parallel}} = h_1 \overrightarrow{a_1^*} + h_2 \overrightarrow{a_2^*} + h_3 \overrightarrow{a_3^*} + h_4 \overrightarrow{a_4^*} + h_5 \overrightarrow{a_5^*}$$

$$\text{with} \quad \overrightarrow{a_i^*} = a_0^* \left(\cos\left(\frac{2\pi i}{5}\right) \overrightarrow{e_1} + \sin\left(\frac{2\pi i}{5}\right) \overrightarrow{e_2} \right) \quad \text{for } i = 1, .., 4$$

where a_0^* is the reciprocal quasilattice parameter and $\overrightarrow{e_1}, \overrightarrow{e_2}$ are two orthogonal unit vectors respectively lying along the P and D directions of the decagonal reciprocal plane

$$\text{and} \quad \overrightarrow{a_5^*} = c^* \overrightarrow{e_3}$$

with $\overrightarrow{e_3}$ lying along the tenfold periodic direction and $c^* = 1/c$, where c is the lattice constant along the tenfold direction.

Each diffraction peak $\overrightarrow{Q^{\parallel}}$ has associated with it a value of $\overrightarrow{Q^{\perp}}$ defined by:

$$\overrightarrow{Q^{\perp}} = h_1 \overrightarrow{a_1^{*\perp}} + h_2 \overrightarrow{a_2^{*\perp}} + h_3 \overrightarrow{a_3^{*\perp}} + h_4 \overrightarrow{a_4^{*\perp}}$$

$$\overrightarrow{a_i^{*\perp}} = a_0^* \left(\cos\left(\frac{4\pi i}{5}\right) \overrightarrow{e_4} + \sin\left(\frac{4\pi i}{5}\right) \overrightarrow{e_5} \right), i = 1, .., 4$$

where $\overrightarrow{e_4}$ and $\overrightarrow{e_5}$ are two orthogonal unit vectors (Fig. 2).

Single-crystal X-ray structure determinations have been published for $Al_{78}Mn_{22}$ [14], $Al_{65}Co_{20}Cu_{15}$ [15], $Al_{70}Ni_{15}Co_{15}$ [16] and $Al_{70.5}Mn_{16.5}Pd_{13}$ [17]. Generally based upon a detailed analysis of X-ray diffraction patterns the possible space group(s) and the metrics are first determined. In all the mentioned above decagonal quasicrystals, the centrosymmetric superspace group $10_5/mmc$ was attributed and the basis vectors are given in Table I.

It is worth noting that the periodicities along the 10-fold axis $c = 1/a_5^*$ are all a multiple of the fundamental periodic length 0.41 nm ($c \simeq N * 0.41$ nm, $N = 1, ..., 4$).

2.2.2. Structure refinement

Once determined the symmetry and metrics, the next step consists in determining the 5-D Patterson functions. They are obtained by a Fourier transformation

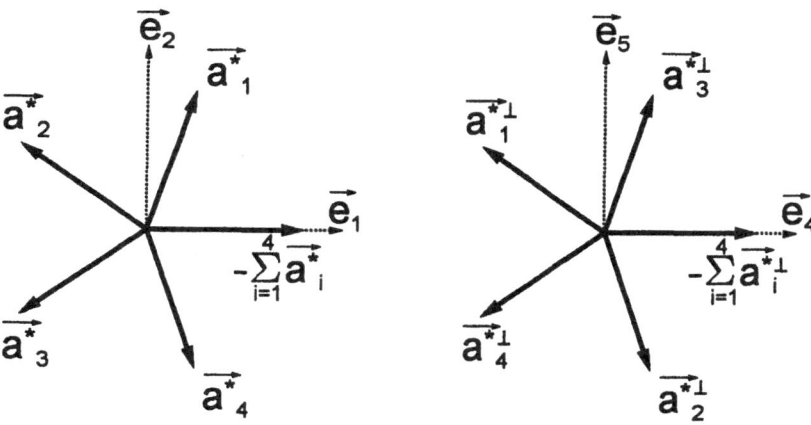

Fig. 2. — reciprocal basis vectors for a decagonal quasicrystal: the decagonal reciprocal plane of the 3D physical space (on the left) and the perpendicular reciprocal space (on the right).

Table I. — Metrics.

Alloys	$a_i^*(\mathrm{nm}^{-1})$, $i = 1, .., 4$	$a_5^*(\mathrm{nm}^{-1})$	reference
$\mathrm{Al}_{65}\mathrm{Co}_{20}\mathrm{Cu}_{15}$	2.656(2)	2.4107(3)	[15]
$\mathrm{Al}_{70}\mathrm{Ni}_{15}\mathrm{Co}_{15}$	2.636(1)	2.4506(3)	[16]
$\mathrm{Al}_{78}\mathrm{Mn}_{22}$	2.556(1)	0.8065(5)	[14]
$\mathrm{Al}_{70.5}\mathrm{Mn}_{16.5}\mathrm{Pd}_{13}$	2.570(1)	0.7964(1)	[17]

of Bragg peak intensities (square of structure factor amplitudes) using X-ray single crystal diffraction experiments (data collection measured with a four-circle diffractometer). In all alloys quoted above, the Patterson analyses have revealed that all atomic surfaces or "5-D atoms" occupy special Wyckoff positions. Geometrical shape, size and chemical fine structure of atomic surfaces, temperature factors in physical and perpendicular space have then to be refined. The number of variables to be refined is generally too large with respect to the number of observed reflections. Therefore, in order to reduce the number of variable to be refined, instead of fine chemical structures of atomic surfaces, site occupancy factors assuming a statistical chemical distribution over each "5-D atom" were introduced and temperature factors were suppressed in the perpendicular space. The least-square refinement gives the best set of phases in structure factors and the Fourier transform of structure factors (based on calculated phases and observed amplitudes) gives density maps. In the cases of Al-Ni-Co and Al-Pd-Mn, subsequent 3-D physical space maximum-entropy method (MEM) was applied in the course of structure factor phasing.

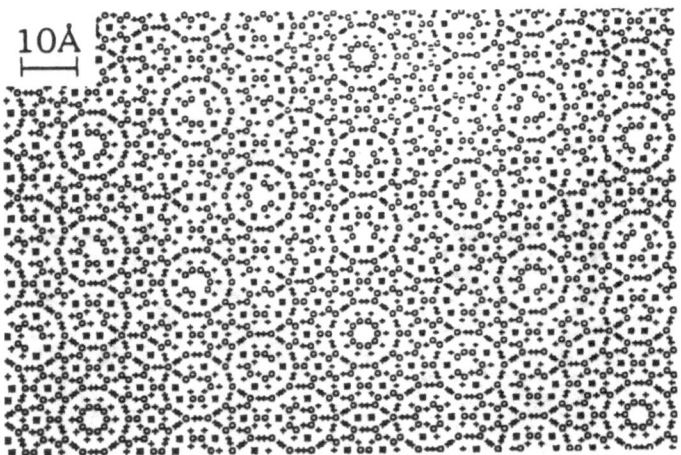

Fig. 3. — The quasicrystalline structural model of Steurer and Kuo [15]. Atomic positions are projected along the periodic direction on the decagonal plane. (o) correspond to atomic positions with an occupancy factor of 100%, where 92% are Cu/Co atoms and 8% Al atoms, (+) and (x) correspond to Al atomic positions with 86% and 25% respective occupancy factors. From [18].

Structure results are given in [14–17] and presented in the form of selected perpendicular and parallel sections of the 5-D Fourier functions. Perpendicular sections gives the density distribution of the "5-D atoms" in the unit cell and their perpendicular shape and parallel sections give the electron density in the physical space *i.e.* the quasiperiodic structure. An example of a calculated projected structure taken from [18] is illustrated in Figure 3.

The quality of the least-square fit is estimated using the reliability factor

$$R = \frac{\sum || F_{\text{obs}} | - | F_{\text{calc}} ||}{\sum | F_{\text{obs}} |},$$

where F_{obs} is the measured structure factor of a diffraction peak, F_{calc} the calculated one, the sum running over all observed independent peaks. The smaller the reliability factor, the better the structure refinement is.

2.2.3. Quasicrystals or not?

The R values obtained for these quasicrystals are high compared to those obtained for the recent structure refinements of their related approximant crystalline phases such as for example μ-Al$_{4.12}$Mn [19], Al$_3$Mn [20] and Al$_{13}$Fe$_4$ [21]. Deficiencies in the refined model is a possible explanation. These are illustrated by Fourier difference plots (the difference between the electron density resulting from the observed structure amplitude and that of the refined model) which indicate deviations from homogeneity for the "5-D atoms" and important residual electron density in the physical space [14–17]. For this purpose, it is worth noting the amount of disorder present in the structure; in cases of $Al_{70}Ni_{15}Co_{15}$ and $Al_{70.5}Pd_{13}Mn_{16.5}$, the intensity of the diffuse scattering being of the same order as that of Bragg peaks, the real structure is necessarily highly disordered and obviously only an "average" structure can be derived if diffuse scattering is neglected. Intrinsic experimental problems possibly play a role in the bad quality of refinements; most of decagonal quasicrystals being only stable at high temperature, we have to be certain that the thermal treatment in sample preparation yields the quasicrystal at room temperature. Resolution of conventional X-ray experiments (typically a few 10^{-2} nm^{-1}) are sometimes not sufficient to decide in favor or not of a quasicrystal. In fact, within these resolutions the sharpness of observed Bragg reflections is consistent with either diffraction from a quasicrystal or a high-order approximant decagonally twinned sample (see Sect. 3).

2.2.4. "Average" structure

In spite of these approximations and possible experimental problems, it is interesting to give the main characteristics of these "average" structures. The details of structural informations can be found in [14–17].

From electron density maps, these structures may be described as periodic stackings of quasiperiodic layers. Two planar layers "A" and "a" related by the tenfold screw axis have been observed in $Al_{65}Co_{20}Cu_{15}$ and $Al_{70}Ni_{15}Co_{15}$ while a periodic stacking of two different quasiperiodic layers "A" (puckered layer) and "B" (planar layer) with sequence ABAaba have been observed in $Al_{78}Mn_{22}$ and $Al_{70.5}Mn_{16.5}Pd_{13}$ (a and b denote the layers respectively A and B rotated by 36°).

Basic structure-building elements corresponding to columnar clusters of diameter $\simeq 2$ nm have been identified. Sections of these clusters form a decoration of an underlying Penrose basic quasilattice with edge length of the unit rhombus $\simeq 0.25$ nm. Furthermore, it has been possible to determine how these columnar clusters are linked to each others.

It is also worth noting that slightly distorted characteristic structure elements forming some intermetallic crystalline phase of $Al_{13}Fe_4$-type [22,23] (i.e. monoclinic $Al_{13}Fe_4$ [21], $Al_{13}Co_4$ [24] and orthorhombic Al_3Mn [20]) have been locally observed in the quasiperiodic layers.

2.2.5. What can we do to improve structure determinations?

- We have to make sure that the sample is in a quasicrystalline state;

- Recently it has been demonstrated that a combination of Patterson method with the maximum-entropy method allows to obtain a better description of atomic surfaces, i.e. the structure [25];

- Another way to go further into a quantitative description would be to get additional data. For instance structure of icosahedral $Al_{73}Mn_{21}Si_6$ [26,27] has been performed in coupling X-ray and neutron data, "large" differences existing between X-ray scattering factors and neutron scattering lengths. Another possibility would consist in measuring partial structure factors that could be derived from contrast variation experiments. In X-ray experiment, this last method consists in changing the scattering factors of transition metals near their absorption edges while in neutron experiment it consists in changing scattering lengths using either isotopes or isomorphous substitution. Structure determinations of icosahedral phases $Al_{74}Mn_{21}Si_5$ [28], $Al_{63}Cu_{25}Fe_{12}$ [29], Al_6Li_3Cu [30] and $Al_{70.5}Pd_{21}Mn_{8.5}$ [31–33] have been performed using these contrast methods. However, despite all these efforts, no structure determination has achieved agreement with data comparable with the agreement obtained in classical crystallography.

3.   DECAGONAL SYMMETRY AND TWINNING

3.1.   Tenfold twinning of $Al_{13}Fe_4$ and $Al_{13}Fe_4$-type structure

Decagonal point symmetry of diffraction patterns have been reported in many intermetallic phases and attributed to tenfold twinning (Fig. 4).

We define decagonally twin samples as samples formed by an assembly of micro (or nano) domains of the same phase with orientations which differ by $2\pi n/10$ angles, n being an integer (tenfold twinning). Note that the arrangement of these micro (or nano) domains is in general complex.

Because of close resemblances in local order with decagonal structure of quasicrystals, the case of $Al_{13}Fe_4$ in which twinning was reported as early as 1955 [22, 23] seems particularly interesting.

$Al_{13}Fe_4$ crystallizes in a monoclinic phase (space group C2/m) with lattice constant: $a = 1.5489$ nm, $b = 0.8083$ nm, $c = 1.2476$ nm and $\beta = 107.71°$ [22, 23]. Note that the ratio of lattice spacings $d_{(200)}/d_{(001)}$ is close to the golden mean ($\tau = 2\cos 36°$). Combining transmission electron microscopy (TEM) images and electron diffraction patterns, microtwinning was identified ([34] and Fig. 4) and atomistic model of $Al_{13}Fe_4$-domains, completely linked by simple twinning was recently proposed [35].

The case of the crystalline τ^2-$Al_{13}Co_4$ is even more interesting. This phase was first reported in 1992 by Ma and Kuo [36] and described by a unit cell with lattice constant: $A = 3.984$ nm $\simeq \tau^2 * a$, $B = 0.8148$ nm $\simeq b$, $C = 0.3223$ nm $\simeq \tau^2 * c$ and $\beta = 107.97°$ where a, b and c are the lattice parameters of the monoclinic $Al_{13}Co_4$ (space group Cm, $a = 1.5183$ nm, $b = 0.8122$ nm, $c = 1.234$ nm, $\beta = 107.9°$ [24]). Various and complex microdomain structures were observed in a recent high-resolution electron microscopy (HREM) investigation [37] (Fig. 5).

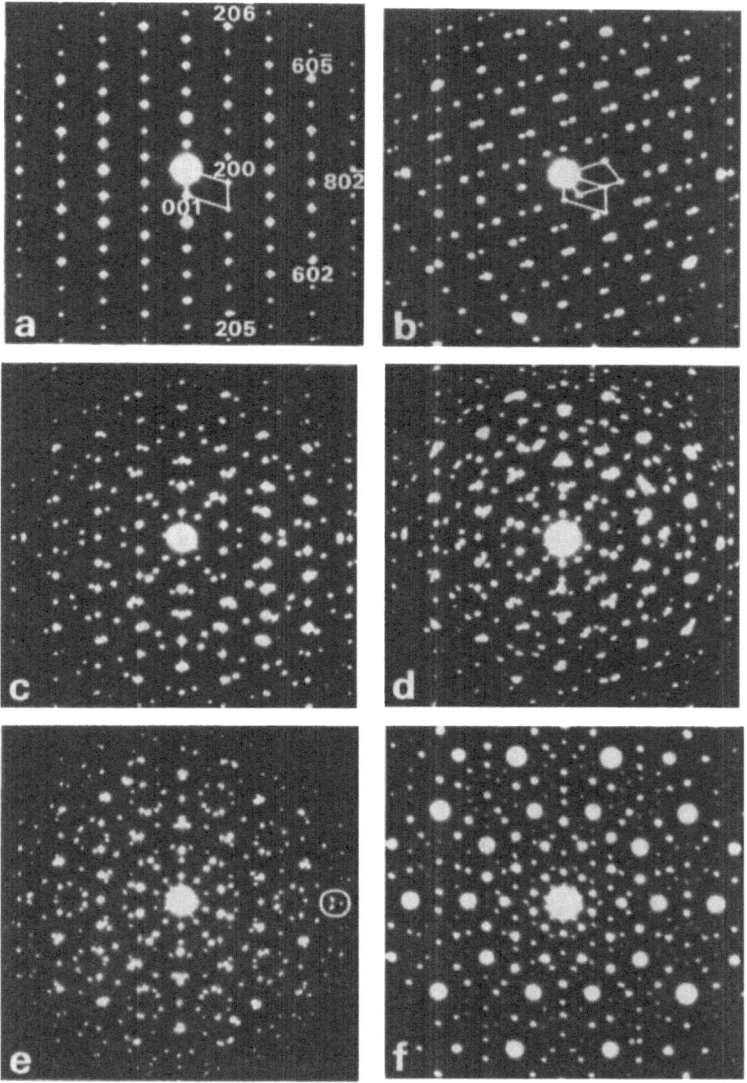

Fig. 4. — Tenfold diffraction pattern from tenfold decagonally twinned sample of the monoclinic phase of Al$_{13}$Fe$_4$ (e) and decagonal phase (f). The development of the pseudo-tenfold pattern due to the micro (or nano) domains is illustrated in (a–e), where everything happens as if a single crystal had been rotated around the b direction by $2n\pi/10$ angles ($n = 1, .., 5$). For a comparison, the tenfold pattern of the Al-Fe decagonal quasicrystalline phase is shown in (f). From [34] at http://www.tandf.co.uk/journals

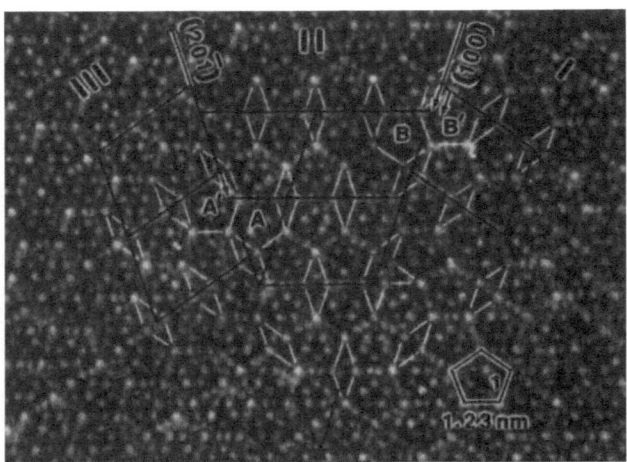

Fig. 5. — HREM image illustrating rotational microdomains with the simultaneous presence of (100), (20$\bar{1}$) glide twins I, II, III. Repeated (100), (20$\bar{1}$) glide twinning gives a set of tenfold twins. From [37] at http://www.tandf.co.uk/journals

3.2. Twinning in the structure of decagonal phases and fine structure of diffraction peaks

Recently, single-crystal X-ray diffraction studies of as-cast decaprisms of $Al_{63}Cu_{17.5}Co_{17.5}Si_2$ [18], $Al_{62}Cu_{20}Co_{15}Si_3$ [38], $Al_{70}Co_{15}Ni_{15}$ [39] and $Al_{70}Pd_{18}Mn_{12}$ [12] have emphasized the importance of high resolution. In these experiments, perfect decagonal symmetry is still observed but, instead of single diffraction peaks arising from a decagonal quasilattice, multicomponent peaks forming typical geometric arrangements along the "P" and "D" directions have been observed (Fig. 6) and attributed to microcrystalline or nanocrystalline domains (tiled with large unit cell) whose orientations in decagonal plane differ by 72° or 36°.

3.3. Twinning and main characteristic features of diffraction patterns

The set of diffraction peaks positions in diffraction patterns of all the mentioned-above decagonal phases, twinned $Al_{13}Fe_4$ and $Al_{13}Fe_4$-type structure can be schematically described by a superposition of five rotated crystalline reciprocal lattices tiled with rhombic unit cell (Figs. 7, 8). The peaks arising from the five contributing twin domains are exemplified in the two cases, where domains are tiled with rhombuses of identical edge but different angles, *i.e.* 108° and 144°.

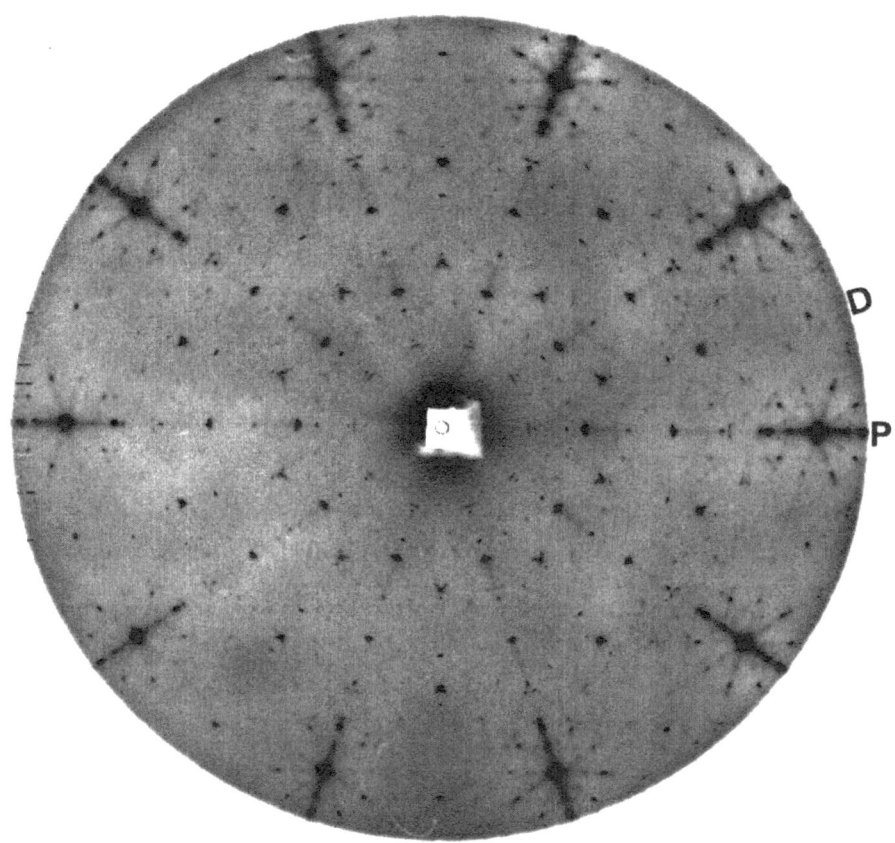

Fig. 6. — High-resolution X-ray synchrotron precession diffraction pattern of $Al_{63}Cu_{17.5}Co_{17.5}Si_2$.

Note that the descriptions of twinning in Figures 4 and 7 (on the bottom) are completely equivalent provided that one has the rhombus edge length equal to the d-spacing $d_{(200)}$ of the monoclinic cell of $Al_{13}Fe_4$ ($d_{(200)} = 0.7377$ nm). The model on the top of Figure 7 fits the set of experimental peak positions of $Al_{63}Cu_{17.5}Co_{17.5}Si_2$ [18], $Al_{62}Cu_{20}Co_{15}Si_3$ [38] and $Al_{70}Co_{15}Ni_{15}$ [39], the domains being tiled by rhombuses of edge length respectively equal to 5.1515 nm, 5.153 nm and 5.1909 nm. The model on the bottom of Figure 7 fits the set of experimental peak positions of $Al_{70}Pd_{18}Mn_{12}$ [12] and τ^2-$Al_{13}Co_4$ [36] provided that the rhombus edge lengths be respectively equal to 3.2813 nm and 1.8948 nm.

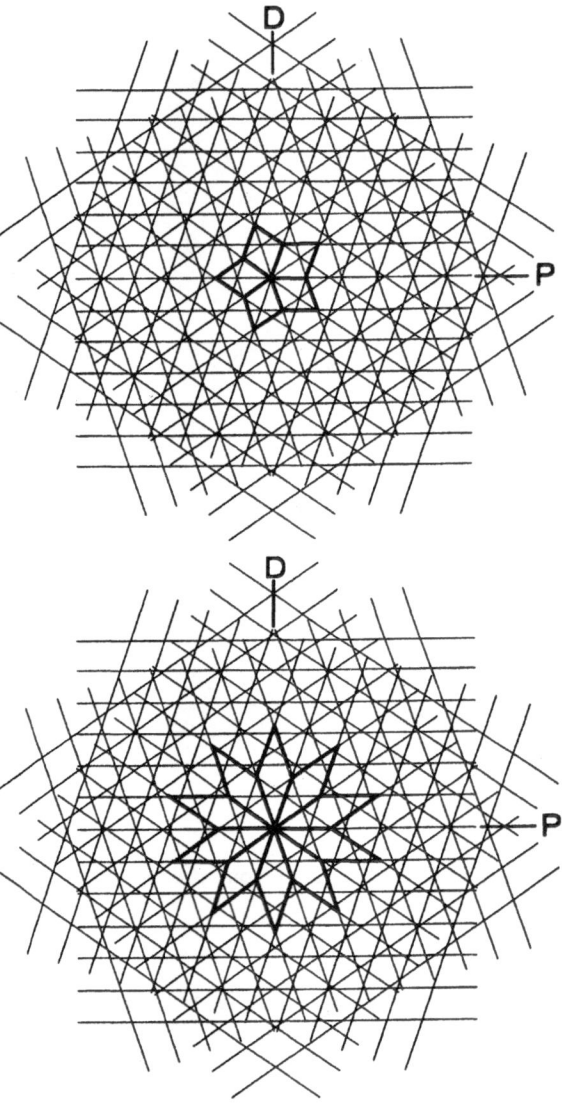

Fig. 7. — Illustration of "twinning" where the decagonal symmetry is reproduced by superposing five rotated reciprocal lattices (thin lines) tiled with reciprocal rhombic cell (thick lines). The reciprocal rhombic cell has an angle of 72° (on the top), an edge τ times larger and an angle of 36° (on the bottom), corresponding to rhombic tiles of same edge length but of angle 108° and 144° respectively.

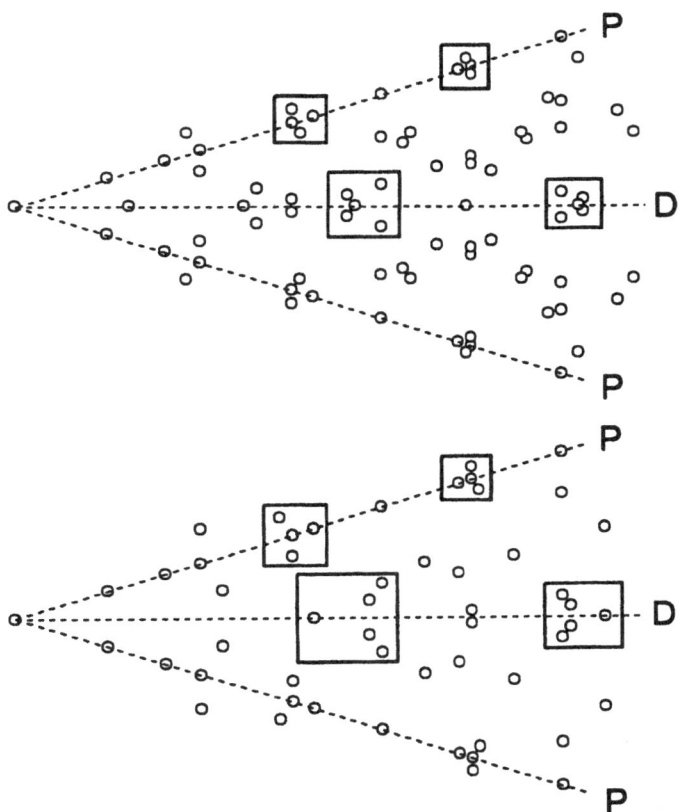

Fig. 8. — Diffraction peaks (o) corresponding to Figure 7. Domains are tiled with rhombuses of 108° angle (on the top) and with rhombuses of same edge length but of 144° angle (on the bottom).

Construction of reciprocal space is straighforward and certain scaling rules can be deduced from Figures 7 and 8. Rectangle areas, with inside several peaks close to each others and arising from different domains, select characteristic geometrical "motifs" along the "P" and "D" directions in Figure 8. The form of "motifs" does not change along a "P" or a "D" direction, but two consecutive "motifs" defined by their centre corresponding to wavector $\overrightarrow{Q_1}$ and $\overrightarrow{Q_2}$ and with $|\overrightarrow{Q_2}| / |\overrightarrow{Q_1}| \cong \tau$ alternate with sizes in a ratio equal to τ.

3.4. Description of microstructures in terms of phason-strain quasicrystals

The close resemblance of diffraction patterns of phases resulting of micro (or nano)-crystalline domains with diffraction patterns of decagonal quasicrystals, especially when unit cell parameters are large, incite us to do a link *via* the phason-strain concept.

The linear phason theory of quasicrystals, first introduced by Lubensky *et al.* [40] and Horn *et al.* [41] has been developed by Zhang and Kuo [42] to describe approximants of decagonal quasicrystals.

Consider a quasicrystal embedded in a higher dimensional periodic lattice.

The electron density $\rho(\vec{r})$ can be expanded in a Fourier series in superspace by:

$$\rho(\vec{r}) = \sum_{\{\vec{Q}\}} \rho_{\vec{Q}} \exp 2\pi i(\vec{Q} \cdot \vec{r} + \Phi_{\vec{Q}})$$

where the sum runs over the 5D reciprocal lattice vectors $\vec{Q} = \overrightarrow{Q^{\parallel}} + \overrightarrow{Q^{\perp}}$.

The phases $\Phi_{\vec{Q}}$ can be decomposed into two components:

$$\Phi_{\vec{Q}} = \vec{u}(\vec{r}) \cdot \overrightarrow{Q^{\parallel}} + \vec{w}(\vec{r}) \cdot \overrightarrow{Q^{\perp}}$$

where by definition the strain fields $\vec{u}(\vec{r})$ and $\vec{w}(\vec{r})$ are respectively the phonon strain in the parallel or physical space and the phason strain in the perpendicular space.

If the spatial variations in $\vec{w}(\vec{r})$ are linear in a small region of a quasicrystal, then we can write $\vec{w}(\vec{r}) = \overline{\overline{M}} \cdot \vec{r}$ where $\overline{\overline{M}}$ is a second rank tensor

$$\overline{\overline{M}} = \begin{pmatrix} m_{11} & m_{12} \\ m_{21} & m_{22} \end{pmatrix}.$$

In the case of a uniform compression or shear (linear phonon-strain) in a crystal or a quasicrystal, there is no change in the atomic arrangements. These manifest themselves in diffraction peaks which shift. The shifts being proportional to $\overrightarrow{Q^{\parallel}}$, a compression or a shear of the entire reciprocal lattice or quasilattice will be observed.

By the action of linear phason-strain in $\vec{w}(\vec{r})$ which produces changes in the atomic arrangements, distorsions of the reciprocal quasilattice also occur and the new set of diffraction peaks can be described by:

$$\overrightarrow{Q^{\parallel}}' = \overrightarrow{Q^{\parallel}} + \overline{\overline{M}} \cdot \overrightarrow{Q^{\perp}}. \tag{1}$$

Shifts of diffraction peaks will be along the strain direction, the magnitude of shifts will be proportional to $\overrightarrow{Q^{\perp}}$ and since the intensity of peaks roughly decreases with $\overrightarrow{Q^{\perp}}$, the weaker the peaks, the farther the peaks shift. As already mentioned in [40] and [43], addition of non-linear components in the same direction will broaden the peaks, with a broadening in the same direction as the shifts.

Following the pioneering work of Entin-Wohlman *et al.* [44], Zhang and Kuo [42] have shown that by introducing phason-strain independently along two orthogonal directions "P" and "D" in the quasiperiodic plane perpendicular to the periodic tenfold direction and approximating the golden mean τ by two consecutive Fibonacci numbers F_{n+1}/F_n the Penrose tiling becomes periodic with large unit cell parameters, consisting of finite pieces of Penrose tiling.

In the case where:

$$m_{12} = m_{21} = 0$$
$$m_{11} = (-1)^{n+1}\tau^{-2(n-2)}$$
$$m_{22} = (-1)^{n}\tau^{-2(n-2)}/\tau$$

the periodicities a_P and a_D along the "P" and "D" directions are:

$$a_P = \sqrt{5}a_R\tau^{n-2}$$
$$\text{and } a_D = \sqrt{5}a_R\tau^{n-1}/\sqrt{1+\tau^2}$$

where a_R is the edge length of the Penrose rhombus. Combination of these a_P and a_D give a series of orthorhombic **Penrose-tiling approximants** (*cf.* Tab. II). For example, the crystalline C-centered orthorhombic approximant with lattice parameters:

$$a_P = 6.0560 \text{ nm and } a_D = 8.3354 \text{ nm}$$

can be described as a phason-strain quasicrystal with respective tensor values:

$$m_{11} = -1/\tau^{12} \text{ and } m_{22} = -1/\tau^{15}.$$

It is worth noting that this unit cell is equivalent to a rhombic tile of edge 5.1515 nm and rhombic angle 108° as found in microcrystalline $Al_{63}Cu_{17.5}Co_{17.5}Si_2$ [18].

It is this same approximant which is still found in microcrystalline $Al_{62}Cu_{20}Co_{15}Si_3$ [38] and $Al_{70}Co_{15}Ni_{15}$ [39] providing that we take a_R equal to 0.15097 nm and 0.15208 nm respectively.

A phason-strain quasicrystal with:

$$m_{11} = -1/\tau^{12} \text{ and } m_{22} = 1/\tau^{9}$$

would describe another crystalline C-centered orthorhombic approximant. With $a_R = 0.15555$ nm the lattice parameters a_P and a_D becomes:

$$a_P = 6.2414 \text{ nm and } a_D = 2.0280 \text{ nm},$$

they correspond to an equivalent rhombic tile of edge 3.2813 nm and rhombic angle 144° as found in microcrystalline $Al_{70}Pd_{18}Mn_{12}$ [12].

The same analysis can be done for the crystalline phase of $Al_{13}Fe_4$ and $Al_{13}Co_4$ that can be described as an approximant crystalline phase or a phason-strain quasicrystal with tensor values $m_{11} = 1/\tau^6$, $m_{22} = -1/\tau^3$ and an edge length Penrose rhombus $a_R = 0.14814$ nm ($Al_{13}Fe_4$) and $a_R = 0.14507$ nm

Table II. — Phason strain tensor and discrete allowed periodicities along the "P" and "D" directions (assuming a Penrose rhombus edge = 0.15093 nm). In bold: the values m_{11} and m_{22} lead to the lattice parameters of the approximant phase of $Al_{63}Cu_{17.5}Co_{17.5}Si_2$.

n	m_{11}	a_P (nm)	m_{22}	a_D (nm)
1	τ^2	0.2086	$-\tau$	0.1774
2	-1	0.3375	$1/\tau$	0.2871
3	$1/\tau^2$	0.5461	$-1/\tau^3$	0.4645
4	$-1/\tau^4$	0.8836	$1/\tau^5$	0.7516
5	$1/\tau^6$	1.4300	$-1/\tau^7$	1.2161
6	$-1/\tau^8$	2.3132	$1/\tau^9$	1.9677
7	$1/\tau^{10}$	3.7428	$-1/\tau^{11}$	3.1838
8	$-1/\tau^{12}$	**6.0560**	$1/\tau^{13}$	5.1515
9	$1/\tau^{14}$	9.7988	$-1/\tau^{15}$	**8.3354**
$\vdots$	$\vdots$	$\vdots$	$\vdots$	$\vdots$
$\infty(= DQC)$	0	∞	0	∞

($Al_{13}Co_4$). In the case of τ^2-$Al_{13}Co_4$ phase, $m_{11} = 1/\tau^{10}$ and $m_{22} = -1/\tau^7$ and $a_R = 0.14534$ nm.

In summary phason-strain quasicrystals with tensor ratio values $m_{22}/m_{11} = \tau^3$ and $m_{22}/m_{11} = -\tau^3$ describe crystalline centered approximants, their unit cell being equivalent to rhombic tiles with rhombic angle equal to 108° and 144° respectively.

Within this description, it is worth noting that a one-dimensional quasicrystal, quasiperiodic along the "P" (or "D") direction and periodic along the "D" (or "P") direction should correspond to tensor values $m_{11} = 0$ (or $m_{22} = 0$), m_{22} (or m_{11}) being equal to one of the allowed discrete values (cf. Tab. II and [45, 46]).

Now coming back to diffraction patterns of approximant phases. Using expression (1), we are able to calculate peak positions of diffraction patterns.

Figures 9 and 10 give an illustration of peak-shifts caused by application of linear phason-strain in the cases where

$$\overline{\overline{M}} = \left(\begin{array}{cc} m_{11} = -1/\tau^{12} & m_{12} = 0 \\ m_{21} = 0 & m_{22} = -1/\tau^{15} \end{array} \right)$$

$$\text{and } \overline{\overline{M}} = \left(\begin{array}{cc} m_{11} = -1/\tau^{12} & m_{12} = 0 \\ m_{21} = 0 & m_{22} = 0 \end{array} \right) \text{ respectively.}$$

Note that the two resulting approximant phases have the same periodicity along the "P" direction and that the last approximant is quasiperiodic along the "D" direction.

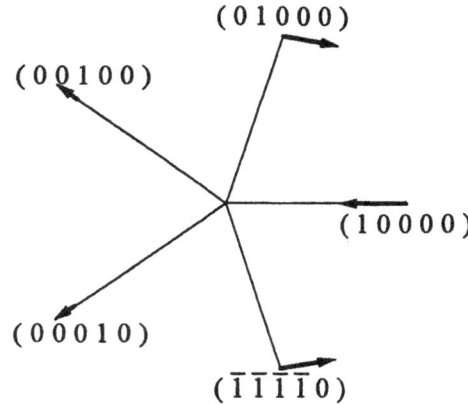

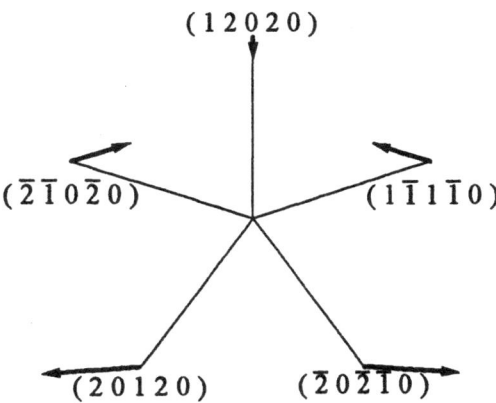

Fig. 9. — Schematic illustration of peak-shifts caused by application of a linear phason-strain with $m_{11} = -1/\tau^{12}$, $m_{22} = -1/\tau^{15}$ – on quasicrystal reflection (10000) and its four symmetry equivalent reflections lying along "P" directions (on the top) – on quasicrystal reflection (12020) and its four symmetry equivalent reflections lying along "D" directions (on the bottom). Bold arrows denote shifts of peak positions (note that peak shifts are enlarged compared to the initial positions of equivalent reflections).

Since there is no *a priori* reason to distinguish between the symmetry related directions in a decagonal quasicrystal, all these directions should be treated in the same manner. This is done by subsequent fivefold twinning of diffraction

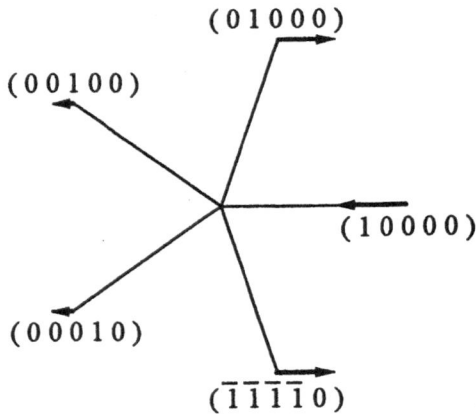

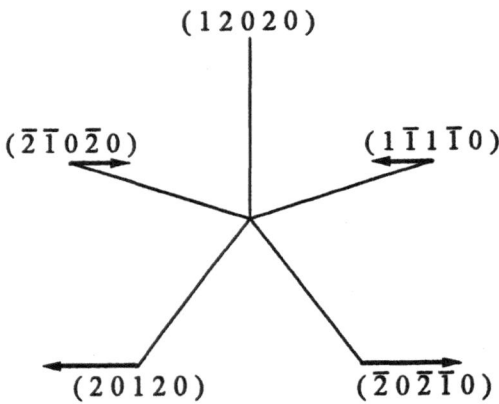

Fig. 10. — Schematic illustration of peak-shifts caused by application of a linear phason-strain with $m_{11} = -1/\tau^{12}$, $m_{22} = 0$ on quasicrystal reflections (10000) and (12020) and their four symmetry equivalent reflections.

pattern. The resulting diffraction pattern exhibits a full decagonal symmetry but now, instead of quasicrystalline peaks, multicomponent peaks are observed (Figs. 11 and 12). On the right-hand side of Figure 11, obviously we find again the form of "motifs" previously observed on the top of Figure 8. With the same periodicity along "P" direction and quasiperiodicity along "D" direction, peaks positions on the left-hand side of Figure 11 are slightly differents. If we take a quasilattice parameter $a_0^* = 2.650$ nm^{-1}, as deduced from [18] in the case of $Al_{63}Cu_{17.5}Co_{17.5}Si_2$, sizes of squares in Figure 11 correspond to areas equal to $(15.9 \times 10^{-3}$ nm$^{-1})^2$ on the top and are three times larger on the bottom. With such values, it is obvious that distinction between for example top right-hand and left-hand side in Figure 11 is not an easy task, experimental resolution better than 10^{-3} nm^{-1} being absolutely necessary.

Very recently, a twinned one-dimensional quasicrystal has been reported in a high-resolution X-ray synchrotron diffraction experiments on $Al_{70}Co_{15}Ni_{15}$ [46]. Using a resolution better than 10^{-3} nm^{-1}, Kalning et $al.$ have been able to make the difference between a twinned crystalline orthorhombic approximant of lattice parameters $a_P = 6.1022$ nm and $a_D = 8.3981$ nm and its twinned related one-dimensional quasicrystal ($a_P = 6.1022$ nm and $a_D = \infty$).

Separation of peaks is easier in the example of $Al_{70}Pd_{18}Mn_{12}$ (Fig. 12) where the splitting exhibits the characteristic "motifs" of Figure 8 (bottom). The distance between peaks being $\simeq 3 \times 10^{-2}$ nm^{-1} in Figure 12 (left-hand side), a resolution of 10^{-2} nm^{-1} is adequate in this case.

In summary, distinction between a decagonal quasicrystal, the micro (or nano) twinned structures of crystalline approximant with large lattice parameters a_P and a_D and the micro(or nano) twinned structures of a one-dimensional quasicrystal is not experimentally an easy task. Analyses in term of phason-strain quasicrystals allow us to satisfactorily explain peak positions of diffraction patterns and to make a link between decagonal quasicrystals and their related phases. Studies of peak widths (rocking curves) indicates nearly perfect ordering with correlation length of 6000 nm in $Al_{70}Co_{15}Ni_{15}$ [46] whereas the size of domains observed by electron microscopy is typically varying between a few ten and a few hundred nanometers. These results can be understood in assuming that spatially "separated" domains with identical orientation coherently scatter.

4. QUASICRYSTAL TRANSFORMATIONS

Another way to make a link between these microstructures and the decagonal quasicrystal should be to study these alloys as a function of temperature. This has been performed in $Al_{63}Cu_{17.5}Co_{17.5}Si_2$ [47]. Figure 13 gives the high-resolution precession patterns obtained in a temperature in-situ X-ray synchrotron experiment at 700 °C and 800 °C [47]. Peak splitting observed in as-cast sample [18] and attributed to a microstructure of the orthorhombic approximant ($a_P = 6.0560$ nm and $a_D = 8.3354$ nm) is still present at 700 °C. This splitting has disappeared at 800 °C and peak positions now correspond to a decagonal quasicrystal.

Song and Ryba [48] propose for such a transformation a mechanism based on the reduction of the domain size ($i.e.$ an increase in twinning or an increase of disorder). When the size of domains would become of the size of the cell, the transformation to the quasicrystal should occur. It is nevertheless more

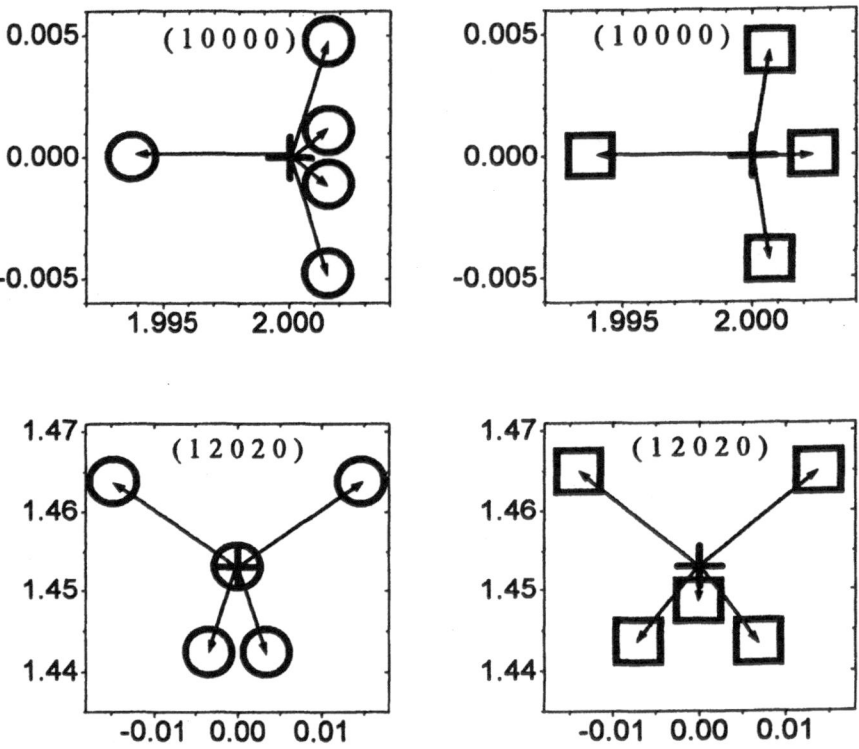

Fig. 11. — Effect of application of linear phason-strain and subsequent fivefofd twinning on decagonal quasicrystal reflections (10000) and(12020). Calculated peak positions: (+) perfect quasicrystal ($m_{11} = m_{22} = m_{12} = m_{21} = 0$), ($\square$) Fivefold twinned orthorhombic approximant ($m_{11} = -1/\tau^{12}$ and $m_{22} = -1/\tau^{15}$), (O) Fivefold twinned one dimensional quasicrystal ($m_{11} = -1/\tau^{12}$ and $m_{22} = 0$).

difficult to imagine a mechanism that should force the reverse transformation, that has been observed with decreasing temperature [47].

5. ICOSAHEDRAL SHORT RANGE ORDER IN GLASSES

Basic complex cluster of atoms of icosahedral symmetry have been observed in quasicrystals and in their related phases. At the present time, all known icosahedral quasicrystalline phases can be described from either the Mackay or the Bergman icosahedral cluster. The simplest crystalline arrangements of these clusters are found in the cubic structure of α-(Al,Si)$_{57}$Mn$_{12}$ [49] with two

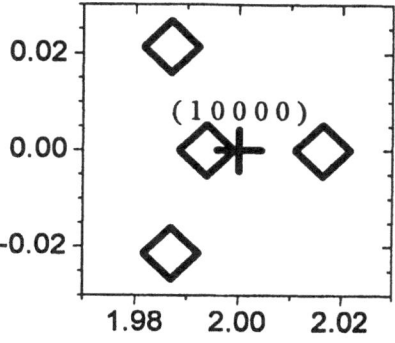 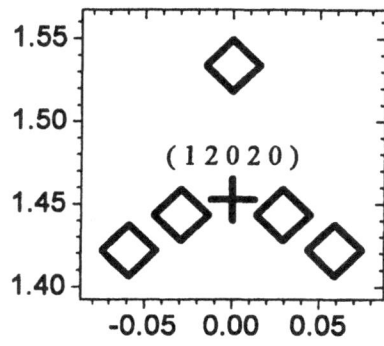

Fig. 12. — Effect of application of linear phason-strain and subsequent fivefofd twinning on decagonal quasicrystal reflections (10000) and (12020) in the case of $Al_{70}Pd_{18}Mn_{12}$. Calculated peak positions: (+) perfect quasicrystal ($m_{11} = m_{22} = m_{12} = m_{21} = 0$), ($\square$) Fivefold twinned orthorhombic approximant ($m_{11} = -1/\tau^{12}$ and $m_{22} = 1/\tau^9$). With a quasilattice constant $a_0^* = 2.5715$ nm^{-1}, the square on the left-hand side has an area of $(77.145 \times 10^{-3}$ nm$^{-1})^2$ whereas the square on the right-hand side is three times larger.

Mackay clusters per unit cell and in the cubic T-phase $(Al,Zn)_{49}Mg_{32}$ [50] or in the cubic R-phase Al_5Li_3Cu [51] with here two Bergman clusters per unit cell. In an analogous fashion, decagonal structures are formed by changes in the basic stacking sequence of similar icosahedral clusters forming decagonal columnar building blocks with different periodicities along the tenfold direction [14–17, 52]. The question which naturally arises is the following: is there a correlation in structure between liquid, glasses or amorphous systems related to quasicrystal phases?

It was in the early 1950's that Frank [53] pointed out that an icosahedral arrangement of 13 Lennard-Jones atoms should have a binding energy larger by 8.4% that FCC or HCP arrangemements and he proposed that icosahedra should be prevalent in supercooled metals. He interpreted the experimentally observed ability to supercool simple liquid metals well-below the equilibrium melting temperature as being due to this prevalence of icosahedral clusters. Evidence of a large undercoolability of melts of Al-Cu-Co and Al-Cu-Fe alloys forming quasicrystals were very recently observed by Holland-Moritz et al. [54].

Molecular dynamics simulations of undercooled Lennard-Jones liquids were performed by several groups and confirmed that upon supercooling about 10% below the melting temperature, predominantly icosahedral orientational order develops [55]. Simulating the cooling of a two-component Lennard-Jones system (an alloy $A_{20}B_{80}$) from liquid temperatures to below the glass transition, Jónsson and Andersen [56] have found a structure which is predominantly

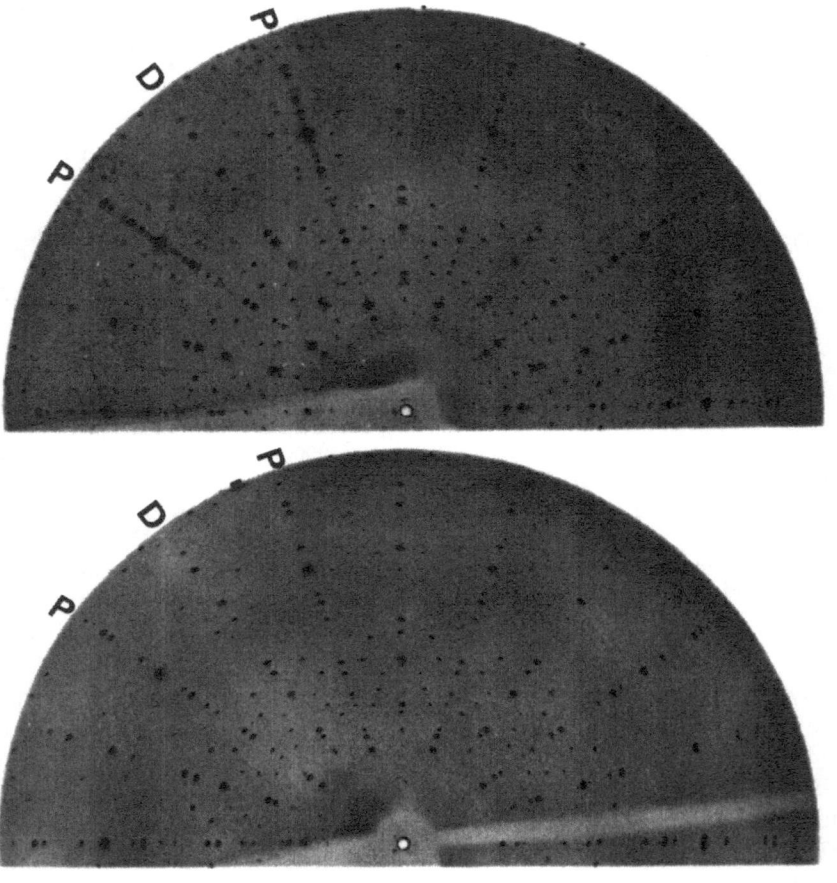

Fig. 13. — $Al_{63}Cu_{17.5}Co_{17.5}Si_2$: High-resolution X-ray synchrotron precession diffraction patterns at $T = 700\ ^\circ$C (on the top) and $T = 800\ ^\circ$C (on the bottom).

icosahedral with on average each icosahedron interpenetrating 1.9 icosahedra and sharing a face with 1.7 other icosahedra.

Liquid alloys $Al_{1-x}Mn_x$, in the composition range forming quasicrystals and a few ten degrees above the liquidus temperature, has been studied by neutron scattering-contrast variation method in order to obtain the structure $i.e.$ together the topological and the chemical short range order (SRO) [57]. After some mathematical manipulations (see for example chapter 3 in [58]), total and partial radial distribution functions (RDF) may be obtained. In fact the RDF is a one-dimensional representation of the 3-dimensional structure that can only give a limited amount of structural informations on average pair distances (from peak positions), average coordination number (from peak area), average fluctuations in pair distances (from peak width) and coherence length (from the limit of density oscillations). From their experimental results and combined to a subsequent modelling, Maret *et al.* [57] concluded in favor of topological order with a moderate chemical order. Strong similarities were found between the local orders of the liquid phase and of the α-$(Al,Si)_{57}Mn_{12}$ crystalline phase.

X-ray diffraction spectra of amorphous and quasicrystalline films of sputtered $Al_{72}Mn_{22}Si_6$ were compared by Robertson *et al.* [59]. The amorphous film shows a typical metallic-glass structure factor with the main characteristic features of icosahedral ordering and is very close to the results obtained in [57] on liquid $Al_{1-x}Mn_x$ alloy. Combining particle-size and phason strain broadening to the normalized and reduced structure factor of the quasicrystalline film result in a spectrum which coincide with the amorphous spectrum. These authors conclude that the glass (or amorphous) phase represents a defect limit of the quasicrystal. The importance of disorder in icosahedral splat-cooled Al-Mn alloy was emphasized by Laridjani [60] in X-ray diffraction experiments where the intensity of diffuse scattering and diffraction peaks are of the same order of magnitude. The analysis of diffuse scattering indicates the relationships between disorder and quasicrystal.

6. CONCLUSION

In this course, we discussed the features of quasicrystalline decagonal phases structural models in five-dimensional space.

Usually, these models do not fit sufficiently well diffraction data, in particular the peaks having weak intensity.

In the framework of n-dimensional crystallography, these diffraction peaks having weak intensity are associated with large components of the diffraction scattering vector in perpendicular space. Therefore, they are very sensitive to "phason-type" disorder.

In spite of many efforts, no structure determination has approached experimental diffraction with an agreement comparable to what is obtained in classical cristallography.

It is important to remark that the common feature found in all structure determinations of quasicrystalline materials made up to now, is the existence of building blocks based on clusters of a few nanometers (1–2 nm).

In this course, we related the quasicrystals with their approximants and rational approximants. We emphasized the existence of tenfold or fivefold twinning of numerous intermetallic phases.

On the basis of diffraction experiments, we also traced the limits reached by our capabilities to clearly differenciate between a "true" quasicrystal and a complex assembly of micro (or nano) domains of some approximant phase with large unit cell parameters.

In certain cases even, the intensity diffused outside of Bragg reflections can reach the same order of magnitude than the total Bragg peaks intensity itself: this proves how very important disorder is. Such is the case in particular for "quasicrystalline" alloys obtained by splat-cooling, but also for many decagonal single grains prepared by Bridgman techniques. This disorder could contribute to the weak values observed in electrical conductivity measurements.

Acknowledgments

I am grateful to M. Laridjani and R. Reich for helpful discussions.

REFERENCES

[1] Schechtman D. and Blech I., *Metall. Trans. A* **16** (1984) 1005.
[2] Schechtman D., Blech I., Gratias D. and Cahn J.W., *Phys. Rev. Lett.* **53** (1984) 1951.
[3] Bendersky L., *Phys. Rev. Lett.* **55** (1985) 1461.
[4] Ishimasa T., Nissen H.-U. and Fukano Y., *Phys. Rev. Lett.* **55** (1985) 511.
[5] Dubost B., Lang J.M., Tanaka M., Sainford P. and Audier M., *Nature* **324** (1986) 48.
[6] Donnadieu P., Su H.L., Proult A., Harmelin M., Effenberg G. and Aldinger F., *J. Phys. I France* **6** (1996) 1153.
[7] Duneau M. and Katz A., *Phys. Rev. Lett.* **54** (1985) 2688.
[8] Kalugin P.A., Kitaev A.Y. and Levitov L.S., *JETP Lett.* **41** (1985) 145.
[9] Elser V., *Phys. Rev. B* **32** (1985) 4892.
[10] Janssen T., Beyond Quasicrystals, edited by Axel F. and Gratias D. (Les Éditions de Physique - Springer, 1994), 75.
[11] Kelton K.F., *Intern. Material Rev.* **38** (1993) 105.
[12] Matsuo Y., Yamamoto K. and Ishii Y., *J. Phys. Cond. Matt.* **10** (1998) 983.
[13] Janssen T., *Acta Cryst.* **A42** (1986) 261.
[14] Steurer W., *J. Phys. Cond. Matt.* **3** (1991) 339.
[15] Steurer W. and Kuo K.H., *Acta Cryst. B* **46** (1990) 703.
[16] Steurer W., Haibach T., Zhang B., Kek K. and Lück R., *Acta Cryst. B* **49** (1993) 66.
[17] Steurer W., Haibach T., Zhang B., Beeli C. and Nissen H.U., *J. Phys. Cond. Matt.* **6** (1994) 613.
[18] Fettweis M., Launois P., Dénoyer F., Reich R. and Lambert M., *Phys. Rev. B* **49** (1994) 15573.
[19] Shoemaker C.B., Keszler D.A. and Shoemaker D.P., *Acta Cryst. B* **45** (1989) 13.
[20] Hiraga K., Kaneko M., Matsuo Y. and Hashimoto S., *Philos. Mag. B* **67** (1993) 193.

[21] Grin J., Burkhardt U., Ellner M. and Peters K., *Z. für Kristallographie* **209** (1994) 479.
[22] Black P.J., *Acta Cryst.* **8** (1955) 43.
[23] Black P.J., *Acta Cryst.* **8** (1955) 175.
[24] Hudd R.C. and Taylor W.H., *Acta Cryst.* **15** (1962) 441.
[25] Haibach T. and Steurer W., *Acta Cryst. A* **52** (1996) 277.
[26] Cahn J.W., Gratias D. and Mozer B., *Phys. Rev. B* **38** (1988) 1938.
[27] Gratias D., Cahn J.W. and Mozer B., *Phys. Rev. B* **38** (1988) 1943.
[28] Janot C., de Boissieu M., Dubois J.M. and Pannetier J., *J. Phys. Cond. Matt.* **1** (1989) 1029.
[29] Cornier-Quicandon M., Bellissent R., Calvayrac Y., Cahn J.W., Gratias D. and Mozer B., *J. Non-Cryst. Solids* **153-154** (1993) 10.
[30] de Boissieu M., Janot C., Dubois J.M., Audier M. and Dubost B., *J. Phys. Cond. Matt.* **3** (1991) 1.
[31] Boudard M., de Boissieu M., Janot C., Dubois J.M and Dong C., *Philos. Mag. B* **64** (1991) 197.
[32] Boudard M., de Boissieu M., Janot C., Heger G., Beeli C., Nissen H.U., Vincent H., Ibberson R., Audier M. and Dubois J.M., *J. Phys. Cond. Matt.* **4** (1992) 10149.
[33] de Boissieu M., Stephens P., Boudard M., Janot C., Chapman D.L. and Audier M., *J. Phys. Cond. Matt.* **6** (1994) 10725.
[34] Fung K.K., Zou X.D. and Yang C.Y., *Philos. Mag. Lett.* **55** (1987) 27.
[35] Ellner M. and Burkhardt U., *Mat. Sc. Forum* **150-151** (1994) 97.
[36] Ma X.L. and Kuo K.H., *Met. Mat. Trans. A* **23** (1992) 1121.
[37] Ma X.L. and Kuo K.H., *Philos. Mag. A* **71** (1995) 687.
[38] Wittmann R., Fettweis M., Launois P., Reich R. and Dénoyer F., *Philos. Mag. Lett.* **71** (1995) 147.
[39] Kalning M., Kek S., Burandt B., Press W. and Steurer W., *J. Phys. Cond. Matt.* **6** (1994) 6177.
[40] Lubensky T.C., Socolar J.E.S., Steinhardt P.J., Bancel P.A. and Heiney P.A., *Phys. Rev. Lett.* **57** (1986) 1440.
[41] Horn P.M., Menzfeldt W., Di Vicenzo D.P., Toner J. and Gambino R., *Phys. Rev. Lett.* **57** (1986) 1944.
[42] Zhang H. and Kuo K.H., *Phys. Rev. B* **42** (1990) 8907.
[43] Bancel P.A. and Heiney P.A., *J. Phys. Colloq. C3* **47** (1986) 341.
[44] Entin-Wohlman G., Kléman M. and Pavlovitch A., *J. Phys. France* **49** (1988) 587.
[45] He L.X., Lograsso T. and Goldman A.I., *Phys. Rev. B* **46** (1992) 115.
[46] Kalning M., Kek S., Krane H.G., Dorna V., Press W. and Steurer W., *Phys. Rev. B* **55** (1997) 187.
[47] Fettweis M., Launois P., Reich R., Wittmann R. and Dénoyer F., *Phys. Rev. B* **51** (1995) 6700.
[48] Song S. and Ryba E.R., *Philos. Mag. Lett.* **65** (1992) 85.
[49] Cooper M. and Robinson K., *Acta Cryst.* **20** (1966) 614.
[50] Bergman G., Waugh J.L.T. and Pauling L., *Acta Cryst.* **10** (1957) 254.

[51] Audier M., Pannetier J., Leblanc M., Janot C., Lang J.M. and Dubost B., *Physica B* **153** (1988) 136.

[52] Daulton T.L., Kelton K.F. and Gibbons P.G., *J. Non-Cryst. Solids* **153-154** (1993) 15.

[53] Frank F.C., *Proc. Roy. Soc.* **215A** (1952) 43.

[54] Holland-Moritz D., Schroers J., Herlach D.M., Grushko B. and Urban K., Proceedings of the 5th Intern. Conf. on Quasicrystals, edited by C. Janot and R. Mosseri (World Scientific, 1996) 636.

[55] Steinhardt P.J., Nelson D.R. and Ronchetti M., *Phys. Rev. B* **28** (1983) 784.

[56] Jónsson H. and Andersen H.C., *Phys. Rev. Lett.* **60** (1988) 2295.

[57] Maret M., Chieux P., Dubois J.M. and Pasturel A., *J. Phys. Cond. Matt.* **3** (1991) 2801.

[58] Elliot S.R., Physics of amorphous materials (Longman, London, 1984) 54.

[59] Robertson J.L., Moss S.C. and Kreider K.G., *Phys. Rev. Lett.* **60** (1988) 2062.

[60] Laridjani M., *J. Phys. I France* **6** (1996) 1347.

Electronic Properties of Quasicrystals. A Comparison with Approximant Phases and Disordered Systems

Given by C. Berger

Written by C. Berger and T. Grenet

Laboratoire d'Études des Propriétés Électroniques des Solides, CNRS, BP. 166, 38042 Grenoble Cedex 9, France

1. INTRODUCTION

The fascination with the quasicrystalline phase has started from its discovery [1] when it was understood to be a new state of matter. Quasicrystalline alloys are highly ordered, but they exhibit a non periodic long range atomic order in contrast with ordinary crystals [2, 3]. The interest was renewed by their unusual properties. Indeed, the icosahedral phases of high structural quality have low electrical and thermal conductivities, although there are composed only of atoms of metallic elements. This is exactly the contrary to the well known metallic behaviour. The conventional way of treating electronic properties in metallic alloys is based on periodicity, or alternatively on disorder. These classical treatments cannot fully explain the behaviour of quasicrystalline phases, and alternative approaches have been suggested so as to express their specific structure [2].

The scope of this article is the study of electronic properties of the quasicrystalline phases. The reader can usefully refer to previous review articles [4–6]. We wish here to relate selected properties (electrical conductivity and electronic density of states) to structural considerations: the effect of long range order (quasiperiodicity in quasicrystals *versus* periodicity in approximants – Sect. 4), the role of the symmetry of the orientational order, described by the icosahedral point group, and of the icosahedral atomic clusters – Section 5, and finally the role of random disorder – Section 6. For this, we first present the electrical conductivity observed in quasicrystalline materials (Sect. 2). This allows to point out the different classes of alloy behaviour and compare them with known systems (Sect. 3).

1.1. Quasiperiodic order

The originality of the QC structure (see also Dénoyer this book) comes from the presence of long range order (sharp diffraction spots) having a rotational symmetry incompatible with translational periodicity (five fold symmetry for the icosahedral phase). This structure is neither periodic, nor amorphous for which halos of diffraction are observed.

This complex order can be modeled by quasiperiodicity. A quasiperiodic function can be defined as the projection onto our physical space of a periodic function belonging to a higher dimensional space (the latter is defined by the two complementary subspaces: the physical and the "perpendicular" ones). An introduction to this description can be found for instance in reference [3]. The easiest illustrative example is the projection of the 2d square lattice

$$g(x,y) = \sum_{n,m} \delta(x - na).\delta(y - ma)$$

onto the 1d line $x = \alpha y$, so to obtained a 1d $f_\alpha(x)$ series. Note that if one wishes to avoid unphysically short distances between $f_\alpha(x)$ points, only $g(x,y)$ points selected in a strip are projected (see for instance Ref. [3]). If α is a rational number $\alpha = p/q$, the 1d $f_\alpha(x)$ series is periodic. But if α is an irrational number the series is quasiperiodic. Note that the Fourier transform of $f_\alpha(x)$, which is a cut in the Fourier space, is calculated from the Fourier transform of a 2d periodic series convoluted by the strip selection function. In consequence the Fourier transform is also discrete. So it is understandable that a quasiperiodic atomic structure possess sharp diffraction peaks although being non periodic. However, since α is an irrational number, the multiple of α modulo one cover the unit interval in a dense manner. Consequently the density in reciprocal space is very large, which means that in principle diffraction peaks are present everywhere (see Refs. [3,8] for instance).

One can consider a series of rational numbers α_n converging to the irrational α. It is then easily seen that the $f_{\alpha_n}(x)$ series approaches the quasiperiodic $f_\alpha(x)$. This is a simple definition of crystalline approximants (see Dénoyer, this book):

$$(\alpha_n)^{-1} = \frac{1}{1+\frac{1}{1+...}} = 1/1, \ 2/1, \ 3/2, \ 5/3, \ 8/5...$$

As n increases, the local environments of the approximant ressemble more closely that of the i-phase. These common local atomic environments have been described by large clusters of icosahedral symmetry, which are displayed on quasiperiodic or periodic lattices.

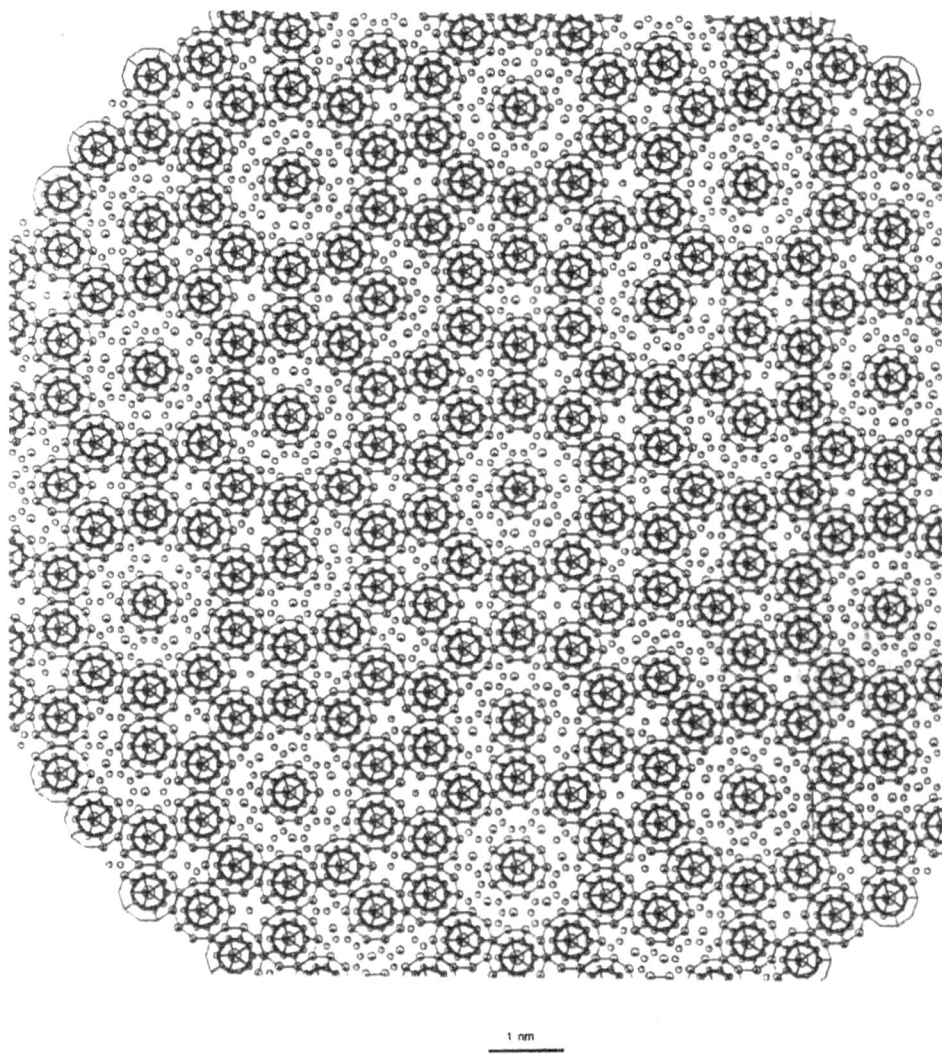

1 nm

Fig. 1. — 2-dimensional cut of an atomic structure model for the 3-dimensional icosahedral i-AlCuFe phase (courtesy of Gratias).

The icosahedral quasicrystalline phases can be described by 6d periodic (hypercubic) lattices, with the use of the golden mean $\tau = (1 + \sqrt{5})/2$ as the irrational number. Also some crystalline phases have been recognized as approximants of the icosahedral i-phase.

Table I. — Scheme of some structural relation between quasicrystalline, amorphous and crystalline phases.

crystal	quasicrystal	amorphous
long range order - periodic -	long range order - quasiperiodic -	long range disorder
short range order	short range order	short range order

periodicity breaking higher orientational symmetry	loss of long range order loss of orientational symmetry (isotropy)

1.2. Quasicrystals, crystals and amorphous phases

An example of such a model is given in Figure 1 for the *i*-phase. We note in particular that all the rings, which are slices of the icosahedral clusters, have the same orientation. Around each local environment of typical size L the same environment can be found at a typical distance $2L$ (Conway theorem).

This property of homogeneous distribution of the local environments at any scale L is obviously obeyed in a crystal, but it is in contrast with amorphous phases. This almost-periodicity would tend to delocalize the electronic wave function (similarly to a metal), whereas the non-periodicity would tend to localize it (like in disordered systems). In this respect it is significant to recall that the very first models for electronic properties have predicted either a zero or a finite conductivity in perfect quasicrystals [9]. Also subtle and specific features are predicted (see for instance Kramer and Janssen, this book).

As summarized in Table I, quasicrystals can be seen as intermediate between crystals and amorphous phases, in so far as translation order is concerned. But since similar local environments can be found in these different structures, our first question is: to what extent are electronic structures and derived properties sensitive to a particular long range order type rather than to local structural environments. In other words, can electronic properties distinguish between periodic, quasiperiodic or disordered packing of similar building units? The second question is how sensitive are the properties to the orientational symmetry. In the icosahedral phases indeed, there is on the one hand an especially large number of equivalent Bragg peaks, and on the other hand the real space structure may be depicted by means of an especially large number of icosahedral clusters.

1.3. Samples of high structural quality in ternary alloys

In order to probe the intrinsic properties of the QC state, high quality samples are required which can now be produced in the icosahedral phase. In the ternary alloys AlLiCu, AlCu(Fe,Ru), AlPd(Mn,Re), GaMgZn, MgZnRE (RE = rare earth), within a restricted range of composition and temperature, the *i*-phase belongs to the *equilibrium* phase diagram. Long time annealing treatment can thus be processed to improve the structural state by removing casting defects.

The first discovered stable *i*-phase, in AlLiCu, can be obtained as nicely facetted single grains. But it still contains a lot of defects. It is indeed difficult to control the stochiometry during the thermal treatment. The *i*-AlCuFe can be produced in the microcrystalline state (grain size typically a few μm), as well as in quasicrystalline single grains (about 1 millimeter size). In perfect QC, diffraction peak widths are experimental resolution limited. The profile of the peaks as a function of the diffraction vector Q (more precisely its component is the perpendicular direction) allows to characterize the defects (phasons, see Dénoyer, this book). The promising *i*-AlCuRu and *i*-AlPdRe are yet much less studied. One of the reasons is that samples are often non homogeneous, because of the difficult mixture of a refractory metal with Al of low melting point. Studies on the GaMgZn and MgZnRE *i*-phases are just at their beginning, although a large effort is put on the preparation of *i*-MgZnRE for magnetic properties.

This is in the AlPdMn system that the best icosahedral phase can actually be grown. Single grains up to centimeter size [11], which are pulled out from the melt and subsequently annealed, present a high degree of structural perfection [2]. The diffraction peak widths for such a material is lowered down to the instrumental resolution of the best diffractometers and the mosaicity is as good as for the best silicon crystals. Studies requiring large single grains can now be performed, such as angle-resolved diffraction (or photoemission), etc.

It is of interest to mention the first discovered quasicrystal [1]. In contrast with the previous phases, the AlMn icosahedral phase is metastable. It is easily produced from a rapidly quenched melt, but it transforms to crytalline phase(s) upon annealing at moderate temperature. Therefore, "quenched" defects cannot be removed from the sample. The intensity of diffuse scattering outside the Bragg reflexions was observed to be of the same order of magnitude than the total Bragg peaks intensity itself [10]. The structure might be described as well by an icosahedral glass, *i.e.* an assembly of clusters all oriented in the same direction, but with random positions.

Many other phases have been identified as quasicrystals in Al or Ga based systems, but also in alloys of transition metals (TiZr for instance). Eight-fold or twelve-fold symmetry axis have been observed, as well as low dimension quasicrystalline stacking (1d and 2d) (for instance see [3] and references therein). Because of the very few possible studies of electronic properties on these phases, often composed of mixture of phases, these will not been discussed in this article. However, we will be concerned by the AlCoCu and AlCoNi decagonal phase [12] (ten-fold axis). It consists of quasicrystalline 2d planes stacked periodically. Single grains can be grown in the form of tiny needles, with the periodic axis along the needle. These allow the observation of anisotropic properties between the periodic and the quasiperiodic directions.

1.4. Unexpected physical properties

The high resistivity of the high structural quality *i*-phases is one of their most salient properties. Considering these alloys are made entirely of metallic elements, with about 60–70 at.% Al, rather low resistivity values would have been expected on the contrary.

As seen in Figure 2, the electrical resistivity of icosahedral phases is at least two orders of magnitude higher than the constitutive metals. In high structural quality *i*-phases, like *i*-AlCuFe or *i*-AlPdMn the resistivity is at least 10 times larger, and even more than 10^3 times for *i*-AlPdRe, than the resistivity of a typical amorphous metal. So far, the highest reported resistivity value at 4 Kelvin reaches more than 1 Ωcm in *i*-AlPdRe [13–16]. This unprecedented value in a structurally ordered metal alloy is well into the range of doped semiconductors.

By contrast, the electrical resistivity of metastable *i*-AlMn phases is comparable to the one of the amorphous counterpart (see Fig. 2). The resistivity value is of the order of 200–600 10^{-6} Ωcm and varies slightly (a few percent) with temperature. Note that the lowest resistivity values are to be attributed to the presence of a secondary Al rich phase in the samples. As outlined in the previous section, these metastable *i*-phases are of poor structural quality. The intrinsic property of quasicrystals could be hidden by whatever disorder is present in the samples.

Other quite unexpected properties have focused attention. Indeed, these make the quasicrystals very attractive for applications [17]. The surface of quasicrystals is hardly wetted, contrary to aluminum. In consequence food is claimed to stick less on it than on other metallic surfaces, and quasicrystals are already used as cookware coatings. Friction coefficients of quasicrystal surfaces can be much lower than that of steel. Together with the hardness of the material this makes quasicrystals useful for wear resistance coatings. As quasicrystals are bad electrical conductors, they also are bad thermal conductors. With thermal insulating properties which compare to those of zirconium oxide, they may be used as thermal barriers. Finally, they also are present as precipitates in some alloy matrices where they are reinforcing the mechanical behaviour.

2. CONDUCTIVITY AND DENSITY OF STATES IN QUASICRYSTALS

2.1. Low electrical conductivity values in *i*-phases

We now turn to the electrical conductivity in the high quality *i*-AlCu(Fe,Ru), *i*-AlPd(Mn,Re). The conductivity values are low even at room temperature,

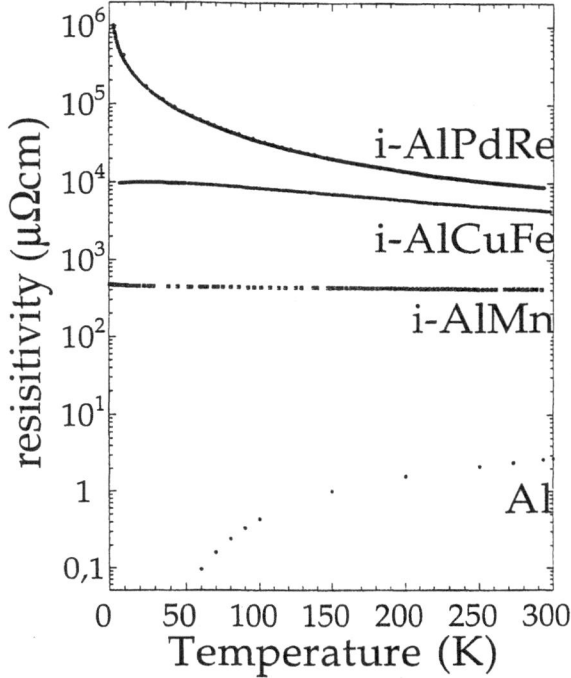

Fig. 2. — Resistivity scale for selected icosahedral phases.

typically $\sigma = 1/\rho \leq 400\,\Omega^{-1}\,\mathrm{cm}^{-1}$, and decrease as the structural quality improves, for instance after defect removing by annealing treatments [18]. This feature is just the contrary of the well known behaviour of a metal. By comparison, the metastable i-AlMn, i-AlMgZn or i-AlCuMg [4] phases where defects cannot be removed, have significantly higher σ values (see Fig. 2). These are comparable to metallic glasses in the range $\sigma \sim 2\ 10^3$ to $10\ 10^3\ \Omega^{-1}\,\mathrm{cm}^{-1}$.

In all the high quality i-phases, i-AlCu(Fe,Ru), i-AlPd(Mn,Re), the temperature dependence $\sigma(T)$ is much stronger than in amorphous metals. These have $R = \sigma_{300\,\mathrm{K}}/\sigma_{4\,\mathrm{K}} \sim 2$ compared to R close to one with a few percent increase with temperature in metallic glasses. In contrast with amorphous metals, no saturation of $\sigma(T)$ is observed up to 1000 K, and power laws of the temperature are observed. Figure 3 shows that the so-called inverse Matthiessen rule is present in the conductivity of these i-phases: $\sigma(T) = \sigma_0 + \delta\sigma(T)$. Only the σ_0 term varies from one sample to another, as a function of the structural quality and composition. The temperature dependence $\delta\sigma(T)$ is the same for all samples and in the low temperature range, it is similar to amorphous

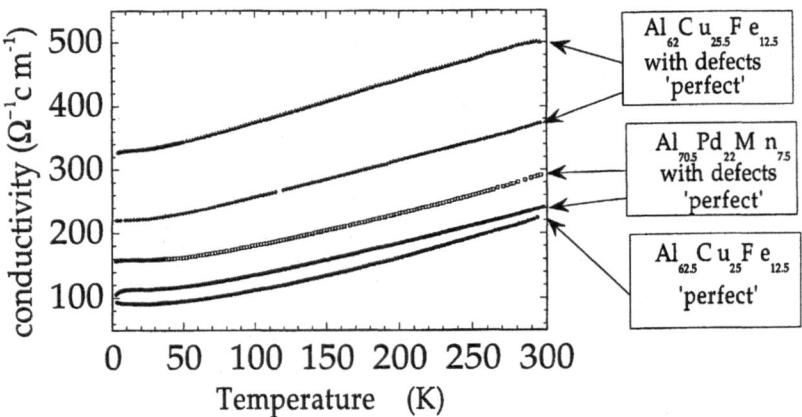

Fig. 3. — Inverse Matthiessen rule in i-AlCuFe, for samples of different composition and structural states.

metals [19]. These systems behave inversely to the normal metal case where $\rho(T) = 1/\sigma(T) = \rho_0 + \delta\rho(T)$. In that case, ρ_0 results from scattering on defects or impurities, and $\delta\rho(T)$ has essentially the same value as the one of the pure matrix.

Even more stricking is the very low temperature dependence that we present in Figure 4 for i-AlPdRe samples of low conductivity and strong temperature dependence ($\sigma_{4\,K} \leq 1\ \Omega^{-1}\,cm^{-1}$, $R = \sigma_{300\,K}/\sigma_{4\,K} \sim 100$) [20]. The conductivity sharply drops at very low temperature below ~ 600 mK, as temperature tends to zero. The same conductivity law was observed in disordered systems, containing insulators or semiconductors like InO_x, SiP or SiY for instance [21], very close to the metal-insulator transition and on the *insulating* side of the transition. That is to say in i-AlPdRe samples the extrapolation of this law down to $T = 0$ gives a zero conductivity as in insulators.

At this point it is worth wondering whether these transport properties are really related to any structural specificity of the quasicrystals. An important step was provided by the observation of anisotropy in the conductivity of single grains of the decagonal phase (see Fig. 5). Along the periodic axis, the conductivity $\sigma_{//}$ is of metallic type with a high value $\sigma_{//}(4\,K) = 2\ 10^4\ \Omega^{-1}\,cm^{-1}$ and a negative temperature dependence. On the contrary, in the quasicrystalline planes σ is lower by a factor 10 to 20 ($\sigma_\perp(4\,K) = 1500$ to $2800\ \Omega^{-1}\,cm^{-1}$) with positive $\sigma(T)$ [22, 23].

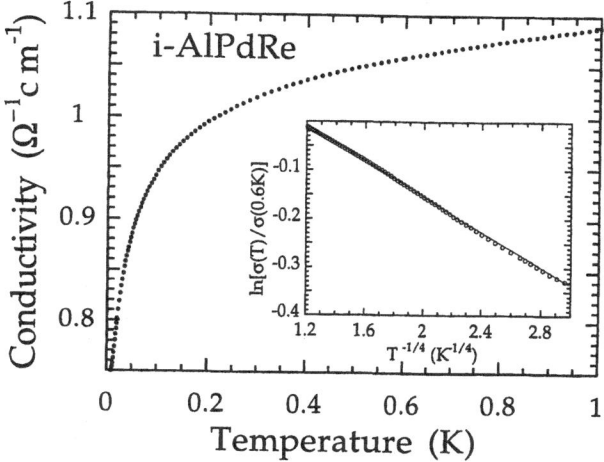

Fig. 4. — Very low temperature conductivity in i-AlPdRe. Insert, $\ln\sigma(T)$ *versus* $T^{-1/4}$ for highly resistive i-AlPdRe ($R = \sigma_{300\,K}/\sigma_{4\,K} = 117$) in the 20 mK – 600 mK range. The $\sigma(T)$ data are normalized at 600 mK for clarity. From reference [20].

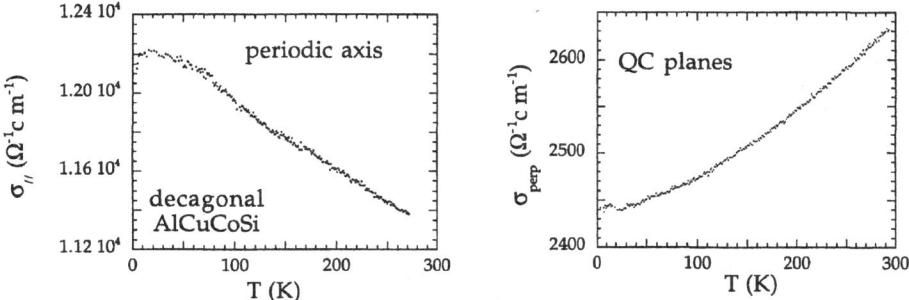

Fig. 5. — Anisotropy of the electrical conductivity in a single grain of the decagonal AlCuCo phase.

2.2. Low electronic density of states in quasicrystals

Since quasicrystalline phases can have conductivity values in the range of doped semiconductors, the existence of a gap was searched in the electronic density of states. The electronic density of states at the Fermi level $N(E_{\mathrm{F}})$ can be evaluated by specific heat $C_{\mathrm{p}}(T) = \gamma T$ at low temperature, where γ is proportional

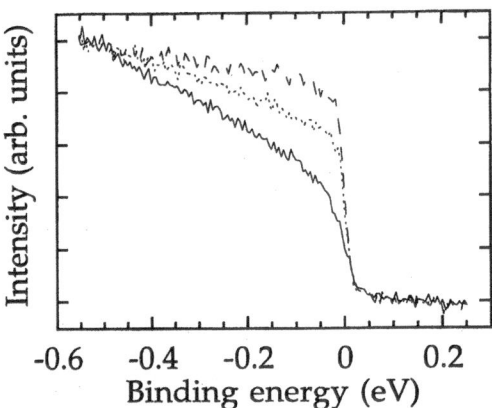

Fig. 6. — Ultra-violet Photoemission Spectroscopy of a *i*-AlPdMn single grain for two surface states: from bottom to top quasicrystalline surface, disordered surface and molybdenum metallic surface for comparison. From reference [29].

to $N(E_F)$. In the stable *i*-AlCu(Fe,Ru) and AlPd(Mn,Re) phases, the γ term is as low as $\frac{1}{3}$ to $\frac{1}{10}$ of Al which is the main constituent [4,5,13,14]. A low $N(E_F)$ for a metal is confirmed by Nuclear Magnetic Resonance data [24–27], and by the total diamagnetic susceptiblity of *i*-AlCuFe and *i*-AlPd(Mn,Re): $\chi_{Total} < 0$, revealing a weak Pauli paramagnetism χ_P ($\chi_P \sim N(E_F) \leq -\chi_{Total}$) [18].

Direct measurements of the electronic structure were performed. Soft X-ray spectroscopy experiments indicate that a pseudo-gap deepens in the Al band [28] from Al to the defective *i*-Al$_{86}$Mn$_{14}$ and to the high quality *i*-Al$_{63}$Cu$_{25}$Fe$_{12}$. Photoemission experiments are sensitive to the total $N(E_F)$ of a few atomic layers at the surface of the samples. As presented in Figure 6, $N(E)$ close to E_F is reduced for a clean surface of *i*-AlPdMn compared to a reference metal, and becomes comparable to a metal by imposing disordering onto the surface [29]. The general trend is observed that the lower the conductivity the lower $N(E_F)$. Such a correlation was noticed also between the conductivity and the γ term of specific heat [6,30].

However $N(E_F)$ is of metallic type, and there is no indication for a semiconducting gap. It should be noted here that a gap opens in the Al$_2$Ru phase, which is also an Al-Transition Metal compound, albeit of non related atomic structure. The latter gap results in an activated temperature dependence of conductivity, instead of the power laws generally observed for *i*-phases [13,14].

Table II. — Electronic transport scale for alloys containing metals, with σ in $\Omega^{-1}\,cm^{-1}$ the typical conductivity at room temperature, τ_e and τ_{ie} the elastic and inelatic scattering time, respectively, ℓ_e the mean free path. a is the interatomic distance.

Metals	weak scattering amorphous simple metals	strong scattering amorphous transition metals	strong disorder metal-semiconductor alloy
$\sigma > 10^5\text{-}10^6$	$10^4 \leq \sigma \leq 2\,10^4$	$2000 \leq \sigma \leq 6500$	$\sigma \leq 50$
$\ell_e \gg a$	$\ell_e > a$	$\ell_e \sim a$	localized wave-
$\tau_{ie} \ll \tau_e$	$\tau_{ie} \leq \tau_e$	$\tau_{ie} \gg \tau_e$	function
nearly free electron	Faber-Ziman diffraction	Quantum Interference effects	Anderson localization

$\longleftarrow$quasicrystalline phases$\text{-----}\rightarrow$

3. COMPARISON WITH OTHER METALLIC ALLOYS

3.1. Scale of conductivity in metallic alloys

We now compare the quasicrystalline phases with other well known systems in the same conductivity range for which transport mechanism have been well studied. We present in Table II a conductivity scale for alloys containing metals. Here the underlying physical picture is that the conductivity decreases with increasing disorder, as measured by the lowering of the electronic mean free path ℓ_e and of the scattering time τ_e (average distance and time respectively between two scattering events) [31].

Starting from the conducting side, an ideal metal at zero Kelvin is perfectly conductor due to the lattice periodicity. The finite conductivity comes from the scattering of electrons by lattice imperfections or phonons: in the Drude approximation $\sigma = ne^2\tau/m$, with n the electronic concentration and m the electron mass. At room temperature, the conductivity reaches a few 10^6 $\Omega^{-1}\,cm^{-1}$. By increasing static disorder, the electronic mean free path ℓ_e is limited in amorphous alloys to a few interatomic distances. Conductivity values are then typically 2 to 10 $10^4\,\Omega^{-1}\,cm^{-1}$ and increase slightly (a few percent) with increasing temperature [32]. We recall this is typically the behaviour of the metastable and defective i-AlMn, i-PdUSi or i-AlMgZn phases. The crossover between negative (metallic) and positive temperature dependences of conductivity occurs around $\sigma = 6500\,\Omega^{-1}\,cm^{-1}$ for all systems. This universal correlation is known as the Mooij correlation. It is thus no surprise to find positive $\sigma(T)$ dependence in i-phases of low conductivity values. With a strong enough disorder potential, the wave function itself experiences localization: this is the Anderson regime [33, 34].

3.2. Effect of diffraction

Transport mechanisms were proposed in these systems to account for their conductivities. In the free electron model for metals, the ion-electron interaction is neglected. A more realistic picture is to look at the collective effect of ion potentials on electrons. This can be described by a diffraction effect.

We define the Brillouin zone boundary (Bragg plane), associated to the vector K of the reciprocal lattice, by the set of points in the plane perpendicular to K at the distance $K/2$. For a plane wave of vector k pointing in the Bragg plane, the Bragg condition is fulfilled (in the K direction this corresponds $k = K/2$). The two waves k and $K - k$ have the same kinetic energy, and the states can be mixed by the pseudo potential V_K. The stronger $|V_K|$ the stronger the mixing. This results in the opening of a gap in the dispersion relation $E(k)$ for energies at which $k \sim K/2$. As shown in Figure 7, the Fermi surface is no more spherical, and the total density of states is reduced for $E(k = K/2)$. For an alloy of a given composition, *i.e.* of a given number of electrons, the filling of the electron band will favor a structure for which the Fermi energy E_F lies within the pseudogap. Indeed in that case, more states are repelled to an energy lower compared to the free electron case (no diffraction effect), and the total electronic band energy is lower. Crystalline phases for which the stability is dominated by such electron band term are known as Hume-Rothery phases [35].

Note that this diffraction effect is the same as for X-ray diffraction. Thus the electronic diffraction is strong for Bragg planes K associated to intense peaks of the diffraction pattern. This effect can be generalized in the non periodic case. In a quasicrystal, diffraction peaks can be associated to K vectors in reciprocal space thus defining a pseudo-Brillouin zone, and the same diffraction effects can be defined.

A diffraction picture also was proposed by Faber-Ziman [36] to describe the scattering of electronic waves in disordered simple metals and liquids metals in order to account for the low conductivity and for the positive temperature dependences of conductivity.

The resistivity can be seen in this case as arising from the interferences of waves scattered by all the scattering centers. This information is contained in the structure factor $S(k)$ (see Dénoyer, this book). The strength of the diffraction is expressed by the potential $|V_K|$ as above. As $S(k)$ is related to the diffraction, one can relate the wide peak feature of X-ray diffraction to the resistive behaviour (in the approximation of weakly scattered free electrons, for large mean free path, $\ell_e > a$, and neglecting multiple scattering).

In Figure 8, the experimental $S(k)$ is presented in liquid lead for different temperatures [37]. Considering electrons at the Fermi level, the stronger effects of diffraction occur when $S(k)$ is maximum for $k/2 = k_F$. In simple liquid metals, the temperature dependence of $S(k)$ is in general agreement with the temperature dependence of resistivity. For divalent metals (dashed line at $q \sim 2.2$ Å in Fig. 8), both temperature coefficients are negative, whereas

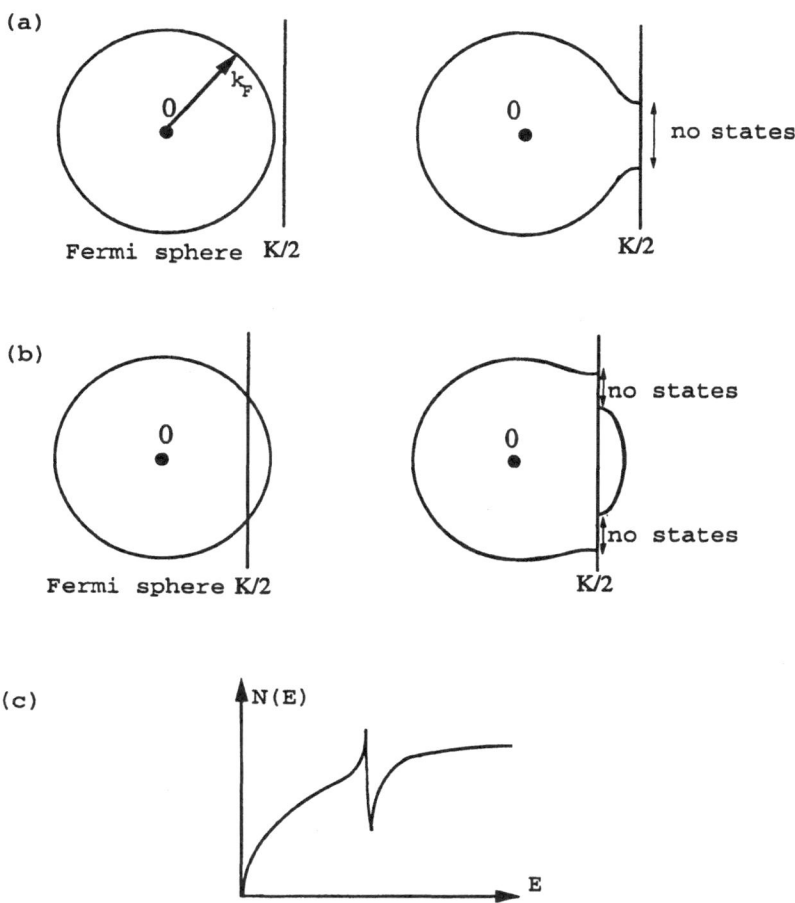

Fig. 7. — Effect of diffraction by the Bragg planes. Effect on the Fermi sphere of radius k_F: (a) $k_F \leq K/2$, (b) $k_F \geq K/2$, (c) effect on the total density of states $N(E)$.

for monovalent metal (dashed line at $q \sim 1.9$ Å) they are positive. In fact, the presence of phonons lower the diffraction peak intensity, and decrease the effective potential seen by the electrons (Debye-Waller effect).

3.3. Quantum interference effects in disordered systems

As presented above, the simple diffraction picture could only account for weak pseudo potential. The actual origin of negative temperature dependences of resistivity in amorphous metals (most of them containing transition metal) was

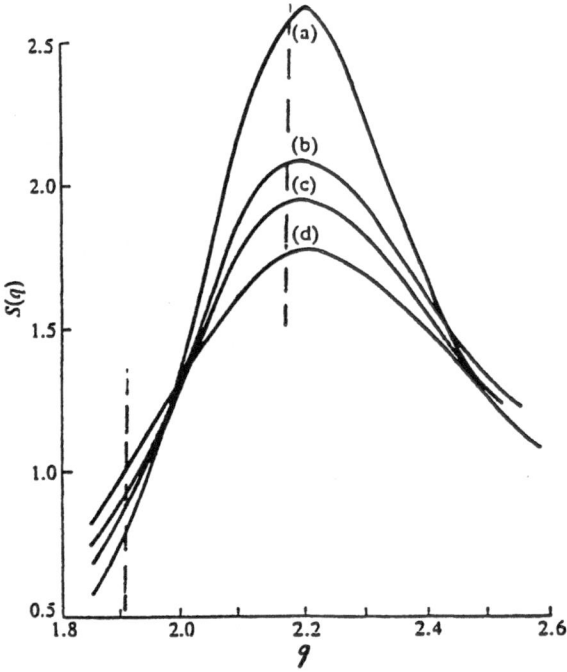

Fig. 8. — Experimental structure factor $S(k)$ in liquid lead for different temperatures: (a) 340 °C, (b) 600 °C, (c) 780 °C, (d) 1100 °C. Dashed lines: $k \sim 2k_F$ for monovalent and divalent simple metals ($q \sim 1.9$ Å^{-1}) and $q \sim 2.2$ Å^{-1} respectively). From reference [37].

much debated until the formulation [38] of the quantum interference effects (QIE) theories taking into account the effect of multiple scattering. In disordered metals, wave functions are extended and the electronic propagation is diffusive. This induces a non zero probability of loop paths for electrons. A constructive interference then can arise between wave functions for electrons travelling on the same loop but in opposite directions (see Fig. 9). This results in a backscattering probability twice the classical case. The net consequence is a so called weak localization of electrons, and thus a higher resistivity compared to the classical diffusion. However, any inelastic scattering event on the loop (essentially due to phonons but also to electron-electron collision at low temperature) destroys the interference coherence at the origin of the enhanced resistivity.

An applied magnetic field also strongly affects the interference pattern, since the two paths in opposite directions are no more symmetrical. The net result

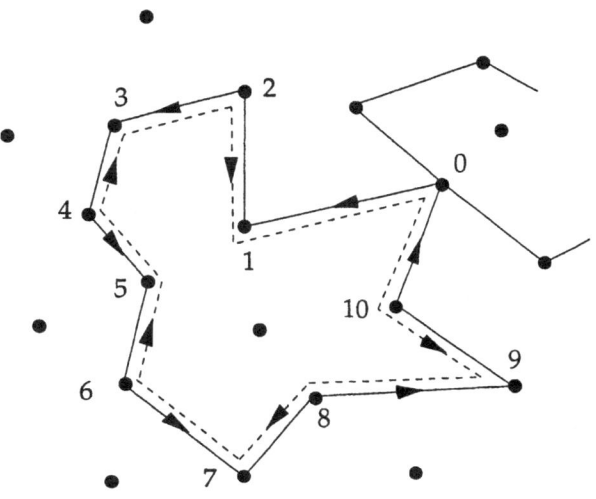

Fig. 9. — Schematic close path in a diffusive system leading to the constructive quantum interference effect.

is to reduce the interference effect. A further complication comes from the complex phenomena of electron-electron interaction effects [39], which can be physically pictured as arising from the enhanced charge pattern produced by the interference function. An external electron can probe this spatially inhomogeneous charge distribution, thus leading to electron-electron interaction.

Both weak localization and electron-electron interaction are small corrections to the Boltzmann conductivity. Precise temperature and field dependences are predicted, which depend on a number of parameters such as the diffusivity D, the temperature dependent inelastic scattering time $\tau_{ie}(T)$, the spin-orbit coupling strength, the electron screening term.

In disordered metals and in amorphous alloys, QIE could depict quite precisely the temperature and magnetic field B dependences of the conductivity, in particular the positive temperature coefficient (the conductivity increases as QIE are destroyed by temperature), the occurrence of $\sqrt{T}$ contributions at low temperature, the characteristic $\sqrt{B}$ magnetoconductance... QIE proves to be also a powerful tool to estimate transport parameters, like $\tau_{ie}(T)$, which are otherwise not accessible because the usual Boltzmann conductivity doesn't account for the observed behaviour.

Surprisingly enough the characteristic features of QIE are also observed in the quasicrystalline phases, such as i-AlLiCu, i-AlCu(Fe,Ru), i-AlPdMn, and in the not too resistive i-AlPdRe. This was not expected since these i-phases firstly are well ordered phases, which call into question mark on the origin of

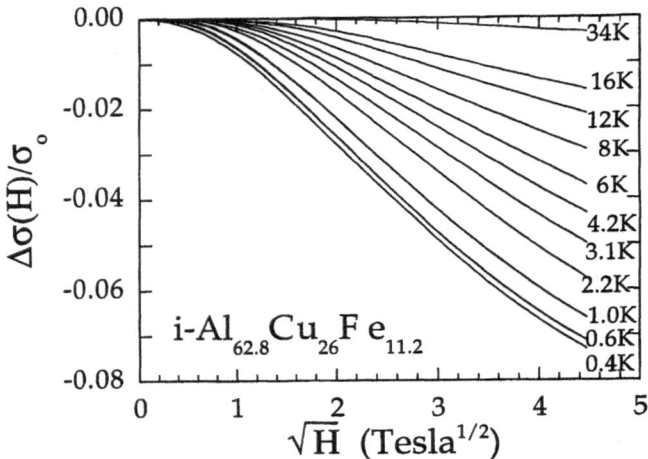

Fig. 10. — Quantum interference effect in i-AlCuFe. Note the $\sqrt{B}$ magnetoconductance $\Delta\sigma/\sigma = (\sigma(B) - \sigma(B = 0))/\sigma(B = 0)$. The negative term arises from the combined effect of spin orbit coupling and electron electron interaction. The line is a fit to the QIE theory.

back-scattering. Secondly they have conductivity values at least one order of magnitude lower than the amorphous metallic alloys, albeit the temperature and field dependence $\delta\sigma(T, B)$ are of similar magnitude.

In i-AlCuFe, by attributing the temperature dependence entirely to QIE effects, one can roughly estimate an order of magnitude for the electronic mean path ℓ_e [19]. To this end, the diffusion constant $D = 0.1–0.3\,\mathrm{cm}^2/\mathrm{s}$ was estimated from the Einstein formula $\sigma = e^2 DN(E_F)$ taking the electronic density of states $N(E_F)$ from the specific heat. From the fitting of the $\delta\sigma(B)$ curve, one can estimate QIE are destroyed at high temperature around 150–200 K, when $\tau_{ie}(T) \sim \tau_e$. For the corresponding lenths $\ell_{ie}(T) = \sqrt{3D\tau_{ie}(T)}$, we get $\ell_e \sim 10–20$ Å. This value for ℓ_e is larger than in disordered metals for which $\ell_e \sim 3$ Å (see Tab. II). It corresponds to a fairly long elastic scattering time τ_e of a few 10^{-14} s compared to amorphous metals: $\tau_e \sim 10^{-15}$ s. This is consistent with τ_e value estimated from optical conductivity in the decagonal phase [40]. As a consequence, the average velocity between two scattering events is low ($v = \ell_e/\tau_e \sim 5\ 10^6\,\mathrm{cm/s}$) which is typically a few percent of the Fermi velocity in Al [19].

In these i-phases, we thus get a non classical regime compared to the picture of Table II: the conductivity is much lower than in an amorphous alloy, but seemingly it may have a significantly higher τ_e. The observation of QIE finally leads to the following conclusions about transport mechanisms: the electronic propagation is diffusive at a larger scale than the mean free path, the scattering mechanism at low temperature is predominantly elastic, since phase coherence is preserved on the trajectories.

3.4. Disordered insulator and Anderson localization

The weak localization effect must be distinguished from the Anderson localization which arises in the presence of very strong disorder potential [33, 34].
The former is a precursor of the latter. It deals with extended electronic wave
functions, whereas the latter concerns the spatial localization of the wave function itself. As depicted by Anderson, the envelope of the wave function decays
exponentially with distance, on a length ξ. The stronger the localization, the
smaller ξ.

For Anderson insulators, conduction at very low temperature proceeds via
electron hopping between exponentially localized states. In the mechanism of
variable range hopping (VRH) proposed by Mott [41], electrons may preferen-
tially hop to localized sites which are close in energy, but not necessarily close
spatially (Fig. 11). The balance between these two hopping processes yields
a hopping at variable range on typical distance $r_m > \xi$, where r_m scales as
$(\frac{\xi}{TN(E_F)})^{1/4}$: the lower the density of states, or inversely the less localized the
wave function, the higher the distance to find an appropriate state to hop to.
The derived conductivity is given by:

$$\sigma_{\text{VRH}}(T) = \sigma_0 \exp - \left[\left(\frac{T_0}{T}\right)^p\right] \text{ with } p = 1/4.$$

We find in this law the signature of localized states since σ_{VRH} is zero at
zero temperature. Including electron electron interaction may change the exponent to $p = 1/2$, which is only expected at very low temperature. The
activation temperature T_0 is a measure of the degree of localization: T_0 varies
as $1/\frac{1}{\xi^3 N(E_F)}$, where the localization length ξ diverges at the metal-insulator
transition, i.e. when the states become extended. Also, we expect localiza-
tion to be effective only if $N(E_F)$ is small. Indeed, the variable range hopping
conductivity was observed in systems containing insulators or semiconductors,
such as highly doped semiconductors, semiconductor-metal, insulator-metal alloys or even conducting polymers. It was then quite surprising to find that the
conductivity in some i-AlPdRe phases could follow this behaviour [20, 42, 43]
(see Sect. 6).

4. PERIODICITY AS AN APPROACH TO QUASIPERIODICITY

After this comparative overview between quasicrystalline phases and classical
systems, we now discuss the extent to which electronic properties can be related
to the long range order of quasicrystals. It will be shown that quasicrystals can
be well approached by periodic structures. Numerical calculations techniques

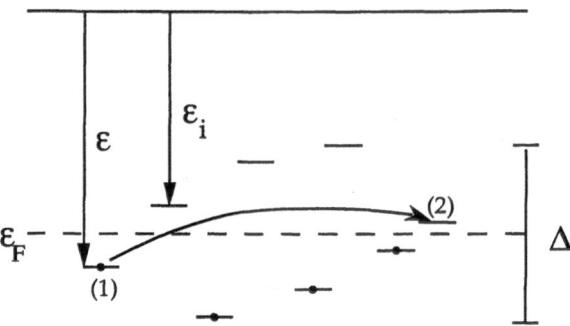

Fig. 11. — Variable range hopping for electrons in an Anderson insulator. The states are randomly distributed on sites at energies ϵ_i. One electron at site 1 may hop to a site 2 close in energy.

proves to be very usefull since in principle quasiperiodic structure suppresses the usual simplifications of periodic conditions.

4.1. Approximant phases

Linking quasicrystals and periodicity, we now turn to the periodic approximants of quasicrystals as defined in the first section. In the i-AlCuFe system for instance, in a tiny compositional range, four approximant phases were studied in the microcrystalline state [44]: cubic α-$Al_{57}Cu_{25.5}Fe_{12.5}Si_5$ (approximant order $\alpha_n = 1/1$, unit cell parameter $a = 12.33$ Å), rhombohedral R-$Al_{63.6}Cu_{24.5}Fe_{11.2}$ ($\alpha_n = 3/2$, $a = 32.14$ Å), orthorhombic o-$Al_{60.4}Cu_{29.9}Fe_{9.7}$ (along the 3 directions: $\alpha_n = 2/1 - a = 19.85$ Å, $\alpha_n = 11/7 - b = 116.34$ Å, $\alpha_n = 3/2 - c = 32.16$ Å), and pentagonal P1-$Al_{63.6}Cu_{24.5}Fe_{11.9}$ (only one periodic direction along a five fold axis $\alpha_n = 4/3$, $a = 52.31$ Å). The same local environments are present in the periodic and QC structures on typical lengths of $10-20$ Å allowing a real comparison between QC and periodic long range order, provided the composition doesn't affect too much the properties.

4.2. Experimental electrical conductivity in approximant phases

Figure 12 presents the temperature dependence of conductivity for the four AlCuFe approximant phases defined above. The *same* behaviour is found in these crystalline phases as for the i-AlCuFe phases, with low conductivity values and the same temperature dependence [45,46]. Similar results are found in the cubic α-AlMnSi phase ($\sigma_{4K} = 258$ $\Omega^{-1}\,cm^{-1}$) [4,5].

In AlGaMgZn [47], the lower order approximant ($\alpha_n = 1/1$, $a = 14.2$ Å) presents a metallic conductivity ($\sigma_{4K} \approx 17000$ $\Omega^{-1}\,cm^{-1}$), whereas the higher order one ($\alpha_n = 2/1$, $a = 23$ Å) has the same conductivity as the corresponding

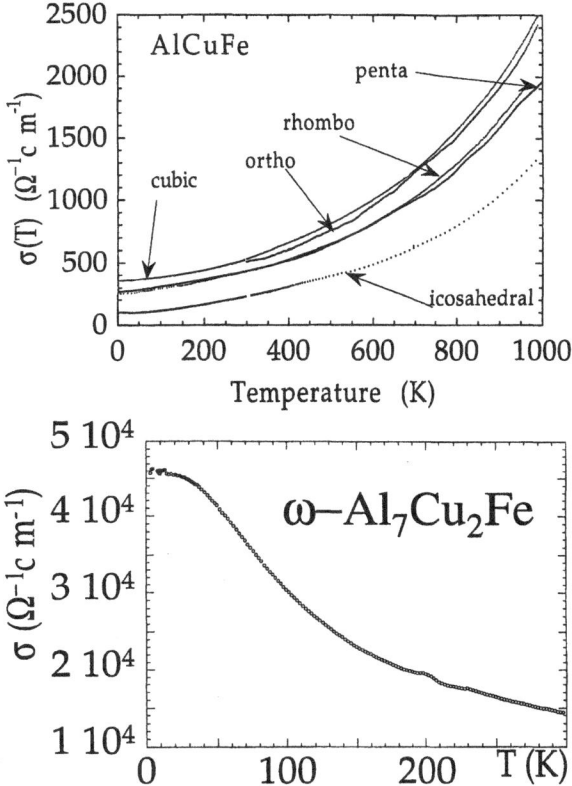

Fig. 12. — Temperature dependence of the electrical conductivity of (a) approximant phases in i-AlCuFe, defined in Section 4.1 (b) the non approximant tetragonal ω-Al$_7$Cu$_2$Fe phase.

i-phase ($\sigma_{4\,K} \approx 9000 \; \Omega^{-1}\,cm^{-1}$). As a remark it is worth noting the high conductivity values of this i-phase, for a system containing no transition metal. By analogy with the comparative conductivity values in these i- and approximants phases, it would be tempting to correlate the conductivity with the order of the approximant (or the unit cell size) in the AlCuFe approximants too. But we must keep in mind that the composition also has a strong influence on the transport properties even in this tiny compositional range [46].

From these data, we conclude to a similar conduction mechanism in the approximant and i-phases in AlCuFe. In contrast, a metallic behaviour is measured in the tetragonal ω-Al$_7$Cu$_2$Fe phase, of comparable unit-cell volume ($a = 6.34$ Å, $c = 14.87$ Å) and composition to the cubic approximant

α-AlSiCuFe phase, with $\sigma_{4\,K} = 4.6\ 10^4\ \Omega^{-1}\,\mathrm{cm}^{-1}$ and a negative temperature dependence. This ω-Al$_7$Cu$_2$Fe phase is not an approximant of i-AlCuFe. So the conductivity in approximant and i-phases is dominated by a mechanism at the scale of the common local environments. This yields a typical length of the order of the smaller approximant cell size, $\ell \sim 15{-}20$ Å. It is worth noting that this length is about the actual size of the packing units used in structural model for i-AlPdMn and i-AlCuFe [2], and is comparable to the estimated electronic mean free path in i-AlCuFe [19].

We outline that the electronic density of states at E_F found in good approximants also does not differ from the one of the i-phases of high structural quality, with low $N(E_F)$. Indeed, the R-AlCuFe phase, as well as approximants of simpler structure like the R-AlCuLi phase ($a = 13.9$ Å) and the cubic α-Al$_9$Mn$_2$Si$_2$ phase ($a = 12.68$ Å) present either the same γ value in specific heat measurements, or the same NMR or photoemission spectra as the corresponding i-phases [4, 24, 48].

4.3. Calculated electronic properties in approximants

The electronic structure was calculated by ab initio calculation (Linearized Muffin Tin Orbitals – LMTO) in model approximant structures [49–51] which are taken to be close to realistic quasicrystalline phases.

Figure 13 presents a typical result, for a model 1/1 approximant of i-AlCuFe, by comparison to the non approximant ω-Al$_7$Cu$_2$Fe phase. Two general features can be outlined in the 1/1 approximant: a deep pseudo-gap is observed around the Fermi level and the density of states is finely structured.

The presence of the pseudo-gap is in agreement with the experimental determination of a low density of states in well ordered i-phases and approximants (see Fig. 6 and Sect. 4.2). Calculations have also been performed on other non approximant crystals of similar composition. It is noted that the density of states at E_F is then significantly higher than in approximants [49].

The fine peaks however, of typical width a few 10 meV to $\sim$ 100 meV, are generic to all approximant phases and are not found in other crystals close in composition. This so called spiky structure may be of particular importance to account for some of the unusual properties of approximants and i-phases.

First of all, from stability arguments, the Fermi energy cannot be located on top of a sharp peak. Then, a small variation in atomic composition of the alloy may shift the the position of E_F from one valley to another. This can result in a strong deviation of the conductivity even if the density of states doesn't change significantly. The rapid variation of the density of states at an energy scale $k_B T$ could also induce anomalous temperature dependence like those observed in the conductivity, nuclear magnetic resonance, or magnetic susceptibility [25, 27]. Indeed in metals the density of states is taken to be a constant to analyse these properties. Finally it was shown that the spiky structure is associated to flat bands, *i.e.* a low energetic dispersion in reciprocal space (see Fig. 13). This yields a small electronic velocity in these structures, in agreement with the low conductivity found experimentally.

However, the measured density of states of quasicrystals doesn't show multiple gaps and peaks. Measurements at high resolution (a few meV) by tunneling spectroscopy experiments reveal a single narrow dip at the Fermi level of width about 50 meV down to ~15 meV [52-54]. But with photoemission studies, no other structures are observed (see for instance Fig. 7). The question of non homogeneity at the quasicrystal/oxide junction was addressed to understand a possible erasement of the fine structures [52]. Also Coulomb blockade effects or electron-electron interactions could be of importance. Indeed the latter are not well taken into account in LMTO calculations. But it is still a matter of debate whether those peculiar spiky features really are characteristics of quasicrystals.

For large model approximants, the intra band conductivity was also calculated, in the relaxation time approximation [49, 50]. Here $\sigma = (e^2 n/m^*)\tau$, where n is the electronic density, m^* the effective electron mass as determined by band calculation. The scattering time τ was chosen in the range of the experimental estimates: $\tau \sim 10^{-14}$ s (see QIE in quasicrystals, Sect. 3.3). The so calculated conductivity values are low, albeit significantly higher than found experimentally. However, no mechanism is provided for the physical origin of this scattering time. By introducing defects in the periodic approximant model, the density of states was shown to be smeared out [51], which can be related to the increase of conductivity in i-phases containing defects. Also the conductivity calculated for a model decagonal approximant is found anisotropic, and the ratio $\sigma_{//}/ \sigma_{\perp}$ is close to the experimental value for the decagonal phase [49].

In conclusion, some of the general features of the electronic properties in quasicrystals can be well approached using periodic approximants. This can be justified because it is usually admitted that the density of states essentially depends on the short and medium range order which are similar in both structures. Also, as we will now discuss, calculations and models for low dimensional quasiperiodic structures could show that the spectral properties converge rapidly as the unit cell size increases.

4.4. Low dimensional perfect quasiperiodic models

The specific properties of quasiperiodic lattices can be evaluated by exact calculations, or numerical analysis on low dimensional quasiperiodic models. Models and results are discussed in details elsewhere in this book, with the lectures by Kramer and Janssen. The article by Sire [55] gives a comprehensive overview.

Here we focus only on the general features of the density of states, nature of the states and conductivity. The electron spectrum in 1d is continuous singular. At higher dimensions, a set of pseudo-gaps survives, and the widest ones are shown to be related to intense diffraction peaks of the lattice [56]. These features are in accordance with the above ab initio calculations for large approximant structures. The density of states was shown to be quite similar in small size approximants compared to quasiperiodic lattices, with the main

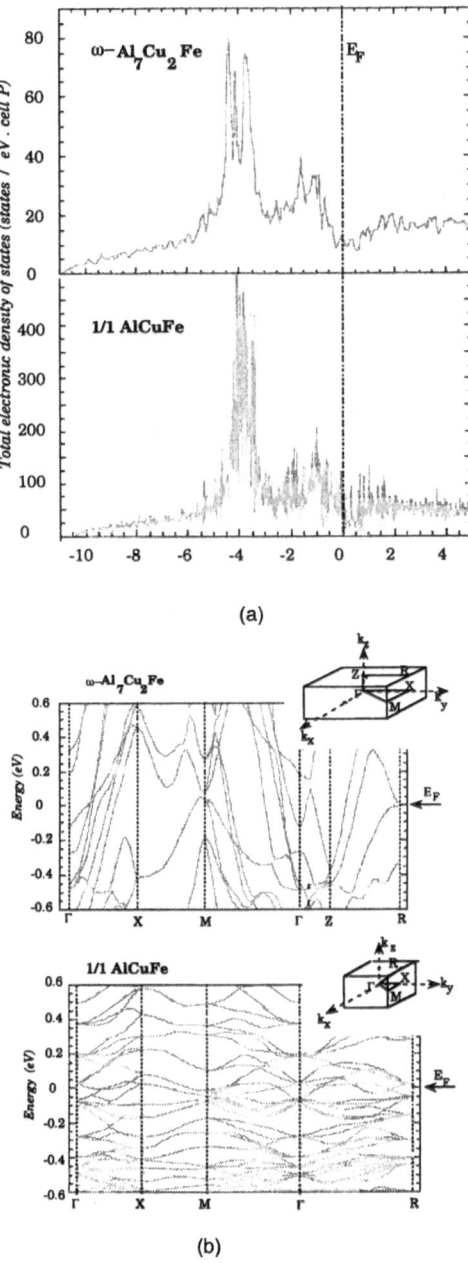

(a)

(b)

Fig. 13. — (a) Electronic density of states and (b) dispersion relation $E(k)$ calculated for a model 1/1 approximant of i-AlCuFe, and comparison with the non approximant phase ω-Al$_7$Cu$_2$Fe. From reference [49] and by courtesy of Trambly de Laissardière.

pseudo gaps being already present. The band width scales with the size L of the approximant unit cell like $\Delta E \sim 1/L^a$ with $a > 1$.

Specific states are found in 1d and 2d. These so-called critical states are neither exponentially localized like in the Anderson localization, nor extended, but they present a complicated structure which was descibed with multifractal analysis. However its envelope can be roughly described by a power law decay with distance. This can be understood following the simple topological argument taken from [55]. As seen in Section 1.1 (Conway theorem), for any part of size L in a quasicrystal an identical cluster can be found at a distance less than $2L$. So electronic tunneling could occur from one cluster to its not too far away copy, to the price of a damping factor ζ. If L is large enough we can expect the wave function to be similar on these identical clusters up to the factor ζ. There is thus a qualitative relation for the wave function

$$\psi_{2L} \sim \zeta \psi_L$$

leading to $\qquad \psi_L \sim L^\alpha$ with $\alpha \sim \ln |\zeta|/\ln 2$.

The same qualitative relation holds in a metal. There we have $|\zeta| = 1$, which corresponds to a Bloch state. Thus $\alpha = 0$ and $\psi_{2L} \sim \psi_L$: we recover that the states have the same order of magnitude everywhere, as it should be for extended states. For a disordered atomic potential, a cluster of size L is in average exponentially distant from a similar cluster. This prevents tunneling between them (at least in 1d), *i.e.* $|\zeta| \sim = 0$ and the states remain localized.

The scaling of bands has in consequence a low electronic velocity, a power law conductance and anomalous diffusion. In an approximant of unit cell L, the average velocity $v = \frac{1}{\hbar}\partial E(k)/\partial k$ is roughly $v \sim L\Delta E \sim L^{1-a}$. As L increases (to infinity for the limit of a quasiperiodic structure), the velocity tends to zero. Concerning the conductance, in 1D, it *decreases* with the approximant size L and varies as $g \sim L^{-\alpha}$. So the higher the order of the approximant the less conducting. The diffusion of a wave packet is anomalous too, with $\langle x^2 \rangle(t) \sim t^{2\nu}$ where ν is related to the spectral dimension of the spectrum. This result was generalized in 3D: the electronic propagation is neither ballistic ($\langle x^2 \rangle(t) \sim v^2 t^2$) as in the case of periodic systems nor diffusive ($\langle x^2 \rangle(t) \sim Dt$) as for disordered systems. The Boltzmann conductivity was calculated in 3D approximants. It turns out that the conductivity is not simply proportional to the density of states at E_F although local minima of the conductivity are located in local minima of the density of states, and that the conductivity was found to be lower by 2 or 3 orders of magnitude than in a crystalline metal of identical scattering time τ.

In conclusion, all the above mentionned properties of the perfect quasiperiodic lattices indicate low conductance and in general a similarity of behaviour between quasiperiodic lattices and periodic approximants even of quite modest size. However, it should be noted that here approximants do not behave as usual periodic metals. The wave function, and the properties derived from the scaling of the electronic bands call for specific features which are intermediate between those of normal periodic metals and those of randomly disordered solids.

5. QUASICRYSTALS AS ORDERED STRUCTURES OF HIGH SYMMETRY

5.1. Pseudo Brillouin zone

In the previous section, we started from the translational symmetry view point. This is clearly not sufficient, since as explained above periodic approximants look more similar to quasicrystalline phases than to other crystalline non approximant crystals. We now consider the effect of the orientational icosahedral symmetry, which is not found in non approximant periodic crystals.

First let us consider the diffraction effect in the case of the icosahedral symmetry. Following the discussion of Section 3.2, for quasiperiodic structures a pseudo-Brillouin zone can be designed by analogy to the crystal case. It is built by taking the planes in k space perpendicular and at half distance from the center to a set of intense diffraction peak vectors K. For i-phases, it has the icosahedral symmetry. Since the icosahedron is highly symmetrical, the diffraction condition $K/2 \sim k$ is fulfilled in many more directions than for a lower symmetrical Brillouin zone of periodic crystals. As for stability arguments we expect E_F in the pseudo gap, $N(E_F)$ in i-phases should be lower than in non-approximant crystals (for the same potential $|V_K|$). Note that as good approximants have roughly an icosahedral symmetry but slightly distorted to fit periodicity, the same mechanism applies. It is then understandable that the pseudo-gap at E_F is deeper in approximants than in ordinary crystals as calculated by *ab initio* methods (see Sect. 4.3). Consequently, the electronic stabilization term in i-phases can be strong, and i-phase can be viewed as an especially stable case of Hume-Rothery phases [57].

Another consequence of the diffraction effect deals with the small size of the Brillouin zone in large approximants. The latter is indeed defined by a set of K's in reciprocal space, which are inversely proportional to the unit cell size. Coming back to Figure 13, the proximity of the diffraction planes to one another partly contributes to the flatness of the $E(k)$ dispersion of the bands. However, the slope $v = \frac{1}{\hbar} \partial E(k)/\partial k$ in approximants is still much lower than in other crystals.

The i-phases were successfully described by the simple band Hume-Rothery picture first in the i-AlCuLi phase [57]. In that case, $2k_F$ was found to match with the K's of intense diffraction peaks. In the i-(Al,Ga)MgZn phase, singularities of the electronic properties were found [4] for each alloy composition for which $2k_F \simeq K$. In a Hume-Rothery picture, the Fermi surface is expected to be composed of electron and hole pockets (for the directions in which K is greater or smaller than $2k_F$ respectively, see Fig. 8). Electron and hole pockets compensate when the Fermi surface and the Brillouin zone exactly match. This could explain the low density of effective carriers n_{eff} deduced from the Hall coefficient R_H ($n_{eff} = e/R_H \sim 5 \; 10^{20} \, \mathrm{cm}^{-3}$ typically in i-AlCuFe), which is about 1/100 of a normal metal. Moreover, the electron-hole balance can easily

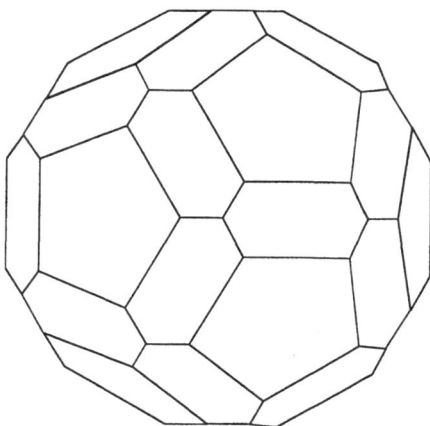

Fig. 14. — Schematic pseudo-Brillouin zone for an icosahedral symmetry, built from the two most intense diffraction peaks of the i-AlCuFe phase.

be changed by the displacement of E_F induced by composition variation, giving a sign reversal of R_H as observed in i-AlCu(Fe, Ru) and i-AlPdMn [4, 5].

This diffraction picture was extended in two directions for quasicrystals. The first one comes from the comparison between i-phases and amorphous phases of the same composition in the AlCuFe and AlPdRe systems [58]. It was indeed found a similar inverse Matthiessen rule for conductivity in both types of material. From this it was concluded that the main mechanism for electronic transport in both i- and amorphous phases comes from a diffraction effect similarly to the Faber-Ziman picture of Section 3.2. The position of the first halo of diffraction of the amorphous phase indeed fits with the most intense diffraction peaks of the i-phases, as expected from the close resemblance of the local environments in both structures. The wider diffraction peaks of the amorphous phase would lead to a less strong electronic diffraction (loss of orientational and translational order), explaining the higher conductivity compared to the i-phase.

The second extension is in the opposite direction. In quasicrystals, theoretically, there is a denumerable infinity of Bragg peaks in reciprocal space. Experimentally the resolution limit of the apparatus and together with the occurence of some disorder and of noise reduces the number of measured peaks. However, a large number of diffracting peaks should be considered for the construction of a pseudo-Brillouin zone instead of only the strongest ones. Physically, this dramatically increases the number of electron-hole pockets up to the limit of a fractal Fermi surface [59, 60]. Strong diamagnetic susceptibility and thermoelectrical power are predicted [60, 61].

More generally the above mentionned properties of the reciprocal space of a quasiperiodic structure also has interesting effects for all properties that are sensitive to K diffraction, since all values of $|K|$ are allowed. For instance, in the case of umklapp processes, two wave vectors k and k' respectively can merge into a single wave k''. The latter can be brought back to the Brillouin zone by adding a vector K of the reciprocal lattice (translational invariance). In the case of quasiperiodic lattices, there is no minimum K vector. Then a multitude of umklapp process are allowed [60,61]. Also, with no minimum $|K|$, gaps Δ could be in principle created at any energy, which is of interest for photonic band gap. Indeed one could imagine to realize artificial quasiperiodic structures with as small energy gaps Δ as required to forbid light propagation and get perfect mirrors for instance.

We finally remark that this picture doesn't take into account the exact nature (s, p or d) of the states. The correlation of transport properties with the transition metal content in i-AlCuFe, and the presence of d states close to the Fermi level indicate these should actually be considered [5]. In fact, sp-d hybridization effects associated to the Bragg diffraction effect were shown [62] to contribute also to the reduction of $N(E_F)$. It is also noted that the fine structure calculated in large approximants is significant only in the energy range where transition metal states are present. We now consider the nature of the atoms, in relation with symmetry considerations, by a study of the influence of the local atomic order.

5.2. Role of local atomic clusters

A simple transition metal (Mn) icosahedral cluster, modelling the clusters of i- and approximant phases, was imbedded in a metallic medium [63]. The interesting result presented in Figure 15 is that these clusters can lead to the formation of peaks in the electronic density of states like those found in ab initio calculations. More precisely these correspond to a resonance of the wave function due to scattering by the cluster. This effect of resonance is similar, at the scale of the cluster, to that of a transition metal impurity in Al (virtual bound states). The narrower the peaks, the longer the lifetime of the corresponding states. Other cluster geometries have also been studied, and in all cases, some confinement effects are present, but the strongest confinement are obtained in the case of the icosahedral symmetry. Finally the effect is enforced in a cluster of clusters, as seen in Figure 15b. A hierarchy of virtual bound states can thus be built, which is reminiscent of the idea of critical states on quasiperiodic tilings which are located on tiling units and show resurgences further apart on similar units.

This study leads to the conclusion that icosahedral clusters also are a key point to understand the electronic properties of the i-phases. Then it is conceivable that hopping could occur between these clusters apart from a typical distance L_0 [19]. In that case the diffusivity at time t is

$$D = L(t)^2/3t = L_0^2/3\tau$$

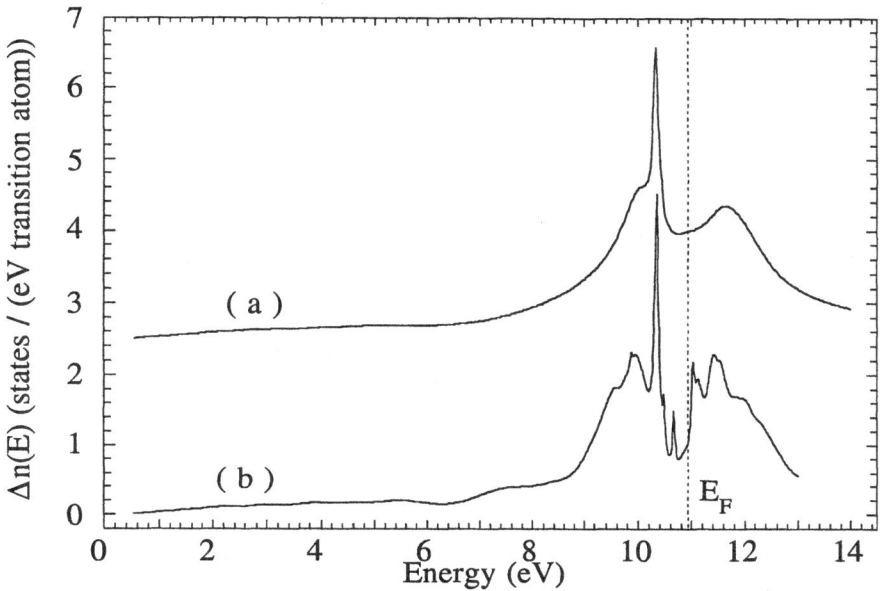

Fig. 15. — Electronic density of states of a model system consisting of a cluster imbedded in a jellium (a) Mn icosahedron cluster, (b) icosahedral cluster made of 12 Mn icosahedra. From reference [62].

in the limit $t \sim \tau$, with τ the scattering time. The electrical conductivity $\sigma = e^2 D N(E_F)$ then varies like $1/\tau$ (instead of $\sigma \sim \tau$ in a metal). By considering the classical relation

$$1/\tau = 1/\tau_e (\text{static impurities}) + 1/\tau_{ie} (\text{phonons}),$$

the experimental inverse Matthiessen rule follows [19]. This mechanism allows a diffusive propagation at long scale, which is required for the observation of quantum interference effects. It could apply as well to approximants or disordered phases containing similar packings.

6. TOWARDS A METAL-INSULATOR TRANSITION IN QUASICRYSTALS: COMPARISON WITH DISORDERED SYSTEMS

All along this paper, we put the emphasize on the low conductivity and low electronic density of states of quasicrystals and we tried to see how this behaviour could be approached by classical considerations related to periodicity (comparison with approximants) and icosahedral symmetry (point group symmetry and atomic environments). They all lead to reduced electronic density of states at E_F, and low conductivity. However these approaches do not take into account the role of disorder, which in alloys of metal-semiconductors is the key for the metal-insulator transition (Tab. II). We address the question of the possible way towards a metal-insulating (MI) transition because i-phases present the paradoxical behaviour of being ordered metallic compounds with overall conductivity features similar to disordered semiconductor-metal alloys. This point is much debated nowadays.

6.1.   Some experimental evidence for the approach to a metal-insulator transition in quasicrystals

We now present several experimental evidences for the approach to a metal-insulator transition in quasicrystals which are: a correlation between σ_{4K} and σ_{300K}, the role of transition metal and defects, and finally the low temperature conductivity and magnetoconductivity behaviour in i-AlPdRe which both indicate a plausible crossing of the MI transition. Here, we take the usual definition of an insulator, which is a zero conductivity at zero temperature. The MI transition is thus the point for a system state (as a function of disorder or composition for instance) where the conductivity extrapolated to zero temperature goes from a finite value to a zero value.

We noticed in Figure 3, that the conductivity of the i-phases of high structural quality follows the inverse Matthiessen rule: $\sigma(T) = \sigma_0 + \delta\sigma(T)$, with about the same $\delta\sigma(T)$ for all samples. Consequently, σ_{4K} can be written as $\sigma_{4K} = A (\sigma_{300K} - \sigma_M)$, where A is independent of temperature. The experimental σ_{300K} values (see Fig. 16) almost reach the $\sigma_M = 150 \ \Omega^{-1} \, cm^{-1}$ value for which the correlation between σ_{4K} and σ_{300K} predicts an insulator [18]. It is interesting to note that this correlation was first suggested on the basis of scaling theory for the conductance and was observed in disordered SiY close to the metal-insulator transition [64]. The analysis of the $\sigma(T, B)$ dependences indicate strong electron-electron interaction effects in the metallic regime of the quantum interference. Similar effects were also observed for disordered systems in the vicinity of the MI transition.

Which are the parameters leading to the MI transition in i-phases? First of all a high structural quality is required. Indeed, as already shown in Figure 3 for i-AlCuFe, the better the structural quality, the lower the conductivity. A second point is the composition of the i-phase, more precisely its transition metal content. As seen in Figure 16 and in Table III, the general trend is: the heavier the transition metal the lower the minimum measured conductivity [4, 5, 13, 14, 65]. Furthermore, other stable high structural quality i-phases, not containing transition metal atoms, have been studied. In i-AlMgZn, i-ZnMgRE (RE = Y, Gd) samples and even in single grains of the stable i-AlLiCu phase, much higher values of the conductivity have been measured.

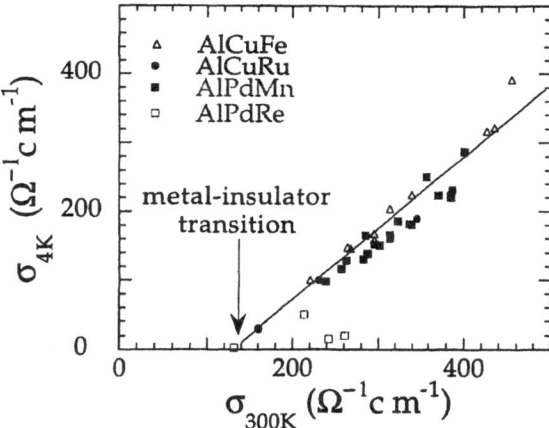

Fig. 16. — Correlation between $\sigma_{4\,K}$ and $\sigma_{300\,K}$ for the high structural quality i-AlCu(Fe,Ru) and i-AlPdMn phases.

Table III. — Lowest conductivity values measured at 4 K in stable icosahedral phases of the two systems AlCuMT (MT = Fe, Ru, Os) and AlPdMT (MT = Mn, Re).

AlCuMT		AlPdMT	
Fe	$100\ \Omega^{-1}\,\mathrm{cm}^{-1}$	Mn	$100\ \Omega^{-1}\,\mathrm{cm}^{-1}$
Ru	$30\ \Omega^{-1}\,\mathrm{cm}^{-1}$	/	
Os	$10\ \Omega^{-1}\,\mathrm{cm}^{-1}$	Re	$\leq 1\ \Omega^{-1}\,\mathrm{cm}^{-1}$

The minimum conductivity is $\sigma_{4\,K} = 1000\ \Omega^{-1}\,\mathrm{cm}^{-1}$, with small temperature coefficient $\sigma_{300\,K}/\sigma_{4\,K} = 1.1$ (from [66, 67]). However, there is a need for further structural investigations to completely rule out the influence of possible residual defects. This clearly outlines the role of the atomic potential for the tendency to localization.

6.2. Crossing of the metal-insulator transition in i-AlPdRe

In the i-AlPdRe samples, the transport properties are observed to be very much sample dependent. In the literature, the $\sigma_{4\,K}$ values are ranging between $3\ 10^{-2}$ and $3\ \Omega^{-1}\,\mathrm{cm}^{-1}$, and the temperature dependences summarized in the ratio $R = \sigma_{4\,K}/\sigma_{300\,K}$ are spread between 2 to ~ 200 [13–16], and even 300 [68]. In general, for a series of samples, the lower the values of $\sigma_{4\,K}$ the higher R.

Although the higher conducting i-AlPdRe samples are believed to be on the metallic side of the MI transition, we have several indications that the sample may cross the metal-insulator transition in the lowest conducting samples.

The low conductivity value ($\sigma_{4K} \leq 1 \; \Omega^{-1} \text{cm}^{-1}$) and the strong temperature dependence ($R \geq 100$) are both comparable to those observed in systems reaching the insulating regime. Also, the tunneling spectroscopy curves could be interpreted by the onset of a correlation gap in the density of states [16,54], as is expected at the MI transition. However the more convincing indication is the drop of the conductivity at very low temperature (below 0.5 K). As presented in Figure 4 for samples of low conductivity (high $R \sim 100$), the conductivity can be fitted with Mott's law $\sigma(T) = \sigma_0 \exp{-(T_0/T)^{1/4}}$ [20], which extrapolates to zero at $T = 0$. The very low T_0 value of about 1 mK indicates the onset of the hopping regime. So it is likely that these samples are very close to the MI transition, but on the insulating side.

We can also follow the approach to the transition in a series of i-AlPdRe samples of varying conductivity, by the breakdown of weak localization effects [69]. Indeed the magnetoconductivity $\delta\sigma(T, B)$ of samples of low R ratio (roughly $R \leq 15$) behaves very similarly to i-AlPdMn or i-AlCuFe. This is well interpreted [69] by quantum interference effects in the metallic regime, as presented in Section 3.3. For samples of higher R, the magnetoconductivity departs gradually from this behaviour. From QIE theories, it is indeed expected that the relative magnetoresistance $\Delta\rho/\rho = \frac{\rho(B,T_m) - \rho(B=0,T_m)}{\rho(B=0,T_m)}$ at a given temperature T_m and moderate field B, would increase as ρ^ν, with ν ranging between 1 to 1.5. In Figure 17, the $\Delta\rho/\rho$ data plotted as a function of R for $T_m \sim 1.8$ K and $B = 9$ Tesla clearly show this departure [15,70]. In Al-Al$_2$O$_3$ or MoGe disordered systems, the maximum of $|\Delta\rho/\rho|$ with R was correlated with the crossing of the MI transition [71,72].

However, the physical origin of an insulating type behaviour in a quasiperiodic compound is still a puzzling question. Indeed, exponentially localized wave functions, have not been found in theoretical calculations for perfect quasiperiodic lattices. The finding, in perfect quasiperiodic model structures, of features resembling those of disordered systems could be invoked to quantitatively understand the present data. For instance, somehow localized states were found – but with a power law decay – [55]. The energy level statistics can be well described by the random matrix theory usually applied to disordered systems [73]. Also the unconventional effect of disorder could be considered. For instance in 3d quasiperiodic lattices including disorder, the conductivity can decrease with disorder as it does in metals. However depending on the Hamiltonian parameters, the conductivity can also increase with disorder for energies lying in pseudo gaps [74]. Also the quantum diffusion was found to evolve with time from a non ballistic regime (truly quasiperiodic, see Sect. 4.4) to a diffusive regime at long time [74]. From a naive viewpoint, one could also imagine that the actual quasiperiodic potential is very efficient to reduce the electronic density of states. Then, what would be the effect in such a system of a residual

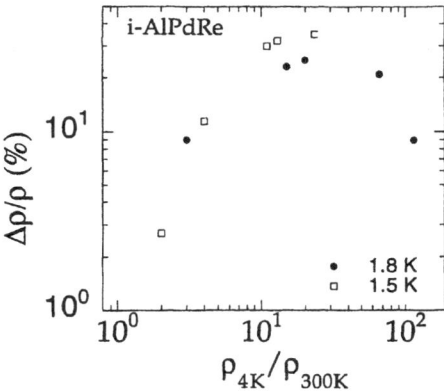

Fig. 17. — Relative magnetoresistivity $\Delta\rho/\rho = \frac{\rho(B,T_m)-\rho(B=0,T_m)}{\rho(B=0,T_m)}$ as a function of $R = \rho_{4\,K}/\rho_{300\,K}$ for various samples of i-AlPdRe. $B = 9$ Tesla and $T_m = 1.5$ K or $T_m = 1.8$ K for two sets of experiments. In MoGe and Al-Al$_2$O$_3$ systems, $|\Delta\rho/\rho|$ goes through a maximum with $\rho_{4\,K}/\rho_{300\,K}$, for samples at the metal-insulator transition. From references [15,68].

disorder of the order of 10^{16} to 10^{18} cm^{-3} $i.e.$ comparable to the concentration in doping centers in highly doped semiconductors?

Finally, the question of electron-electron interaction should also be addressed. Indeed, electron-electron correlations are in general expected to be important at the metal-insulator transition. It is predicted [75] that electronic correlations could enforce the localization effects in systems like quasicrystals where the electron velocities are low.

7. CONCLUSION

In this paper, we mainly addressed the question of the low electrical conductivity in the stable i-AlCu(Fe,Ru) and i-AlPd(Mn,Re) phases, which is one of the most outstanding property in this series of ordered alloys of metals. The conductivity decreases as structural perfection is improved and with heavier transition metal constituent. It finally likely undergoes a metal-insulator transition for the lower conducting i-AlPdRe samples. The electronic density of states at the Fermi level is low for a metallic alloy but is still of metallic type.

We have focussed on two aspects: the long range translational order (quasiperiodic in quasicrystals $versus$ periodic in approximants) and the icosahedral orientational order (icosahedral point group and icosahedral atomic clusters). For each of these complementary viewpoints low electronic conductivities and density of states are found. On the one hand, large crystalline approximants present a low conductivity in numerical simulations within the Bloch-Boltzman theory, and low dispersing bands are predicted. On the other hand in perfect quasiperiodic lattices, specific electronic wave functions have been found which tend to be localized. In relation with the icosahedral point group, we have discussed the effect of high multiplicity of Bragg peaks. It is expected to reduce the electronic density of states, and consequently the conductivity. We have also presented the role of local icosahedral atomic clusters which also could confine electrons.

Finally, we have emphasized the similarity of the transport properties such as quantum interference effects and Mott's type conductivity with that of disordered systems, on both sides of the metal-insulator transition. It seems paradoxical to observe such similarities between highly ordered alloys of metals and disordered highly doped semiconducting alloys. This is still not clearly understood. We have suggested several possibilities, which are the general proximity of the behaviour between quasiperiodic systems (even perfect ones) and random systems, or the non-conventional role of disorder in quasiperiodic structures. This calls for further investigations which we hope this paper will motivate.

Acknowledgments

the authors wish to acknowledge Didier Mayou for illuminating discussions. The LEPES-QC members are warmly thanked for their constant help and support. The authors are in debt to their collaborators abroad and in the framework of the CNRS-GdR CINQ.

REFERENCES

[1] Shechtman D., Blech I., Gratias D. and Cahn J.W., *Phys. Rev. Lett.* **53** (1984) 1951.

[2] Lectures on Quasicrystals, edited by Hippert F. and Gratias D. (Les Éditions de Physique, Les Ulis, 1994) 1-503.

[3] Janot C., Quasicrystals (A primer, Oxford Sci. Publication, Oxford, 1992) 1- 320.

[4] Poon S.J., *Adv. Phys.* 4 (1992) 303.

[5] Berger C., in [2] p. 463.

[6] Rapp Ö., Springer Series in *Solid State Physics* **126** (1998).

[7] *Beyond Quasicrystals*, edited by Axel F. and Gratias D. (North Holland, Amsterdam, 1995) 1-420.

[8] Cahn J.W., Shechtman D. and Gratias D., *J. Mater. Res.* **1** (1986) 13.

[9] Sokolov J.B., *Phys. Rev. Lett.* **57** (1986) 2223; Kitaev A.Yu., *JETP Lett.* **48** (1988) 299.

[10] Laridjani M, *J. Phys. I France* **6** (1996) 1347.

[11] de Boissieu M., Durand-Charre M., Bastie P., Carabelli A., Boudard M., Bessiére M., Lefebvre S., Janot C. and Audier M., *Philos. Mag. Lett.* **65** (1992) 147.

[12] Tsai A.-P., Inoue A. and Masumoto T., *Mat. Trans. JIM* **30** (1989) 463.

[13] Pierce F.S., Poon S.J. and Guo Q., *Science* **261** (1993) 737.

[14] Pierce F.S., Guo Q. and Poon S.J., *Phys. Rev. Lett.* **73** (1994) 2220.

[15] Gignoux C., Berger C., Fourcaudot G., Grieco J.C. and Rakoto H., *Europhys. Lett.* **39** (1997) 171.

[16] Lin C.R., Chou S.L., Lin S.T., *J. Phys. Colloq.* **8** (1996) L725.

[17] New Horizons in Quasicrystals, edited by A.I. Goldman, D.J. Sordelet, P. A. Thiel and J.M. Dubois (World Sci., Singapore, 1997); MRS Bulletin, Nov. 1997 48.

[18] Klein T., Berger C., Mayou D. and Cyrot-Lackmann F., *Phys. Rev. Lett.* **66** (1991) 2907.

[19] Mayou D., Berger C., Cyrot-Lackmann F., Klein T. and Lanco P., *Phys. Rev. Lett.* **70** (1993) 3915.

[20] Delahaye J., Brison J.P. and Berger C., *Phys. Rev. Lett.* **81** (1998) 4204.

[21] Z. Ovadyahu, *J. Phys. Colloq.* **19** (1986) 5187.

[22] Grenet T and Gignoux C., unpublished results.

[23] Martin S., Hebard A.F., Kortan A.R. and Thiel F.A., *Phys. Rev. Lett.* **67** (1991) 719.

[24] Hippert F., Kandel L., Calvayrac Y. and Dubost B., *Phys. Rev. Lett.* **69** (1992) 2086.

[25] Simonet V., Hippert F., Gignoux C., Berger C. and Calvayrac Y., Quasicrystals, edited by Takeuchi S. and Fujiwara T. (World Sci., Singapore, 1998) 696.

[26] Gavilano J.L., Ambrosini B., Vonlanthen P., Chernikov M. A. and Ott R.H., *Phys. Rev. Lett.* **79** (1997) 3058.

[27] Tang X.-P, Hill E.A., Wonnell S.K., Poon S.J. and Wu Y., *Phys. Rev. Lett.* **79** (1997) 1070.

[28] Belin E., Dankhazi Z., Sadoc A., Dubois J.M. and Calvayrac Y., *Europhys. Lett.* **26** (1994) 677.

[29] Schaub T., Delahaye J., Berger C., Guyot H., Belkhou R., Taleb-Ibrahim A. and Calvayrac Y., *Phys. Rev. Lett.* (submitted).

[30] Mizutani, *J. Phys. Cond. Matter* **21** (1998) 4609.

[31] Rivory J., Les Amorphes Métalliques (Les Éditions de Physique, Les Ulis, 1983) 497.

[32] Howson M.A. and Gallagher B.L., *Physics Reports* **170** (1988) 265.

[33] Anderson P.W., *Phys. Rev.* **109** (1958) 1942; Ladieu F. and Sanquer M., *Ann. Phys. Fr.* **21** (1996) 267.

[34] T.G. Castner, Hopping transport in Solids (Elsevier Sci., Amsterdam, 1991) 1.

[35] Hume-Rothery W., Coles B.R., *Adv. Phys.* **3** (1954) 149.

[36] Faber T.E., Ziman J.M., *Philos. Mag. 11* (1965) 153.

[37] D.M. North, J.E. Enderby and P.A. Egelstaff, *J. Phys. France I* (1968) 1075.

[38] Abrahams E., Anderson P.W., Licciardello D.C. and Ramakrishnan T.V., *Phys. Rev. Lett.* **42** (1979) 693.

[39] Altshuler B.L. and Aronov A.G., Electron electron interaction in Disordered systems (Elsevier, Amsterdam, 1985) 18.

[40] Basov D.N., Timusk T., Barakat F., Greedan J. and Grushko B., *Phys. Rev. Lett.* **72** (1994) 1937.

[41] Mott N.F., *J. Non-Cryst. Solids* **1** (1968) 1.

[42] Q. Guo and S.J. Poon, *Phys. Rev. B* **54** (1996) 12793.

[43] Wang C.R., Lin C.R., Lin S.T., Quasicrystals, edited by Takeuchi S. and Fujiwara T. (World Sci., Singapore, 1998) 583.

[44] Quiquandon M., Quivy A., Faudot F., Sâadi N., Calvayrac Y., Lefebvre S., Bessière, Quasicrystals, edited by Janot C. and Mosseri R. (World Sci., Singapore, 1995) 152.

[45] Berger C. and Mayou D., Cyrot-Lackmann, *ib.* p. 1423.

[46] Berger C., unpublished results.

[47] Edagawa K., Naito N. and Takeuchi S., *Phil. Mag. B* **65** (1993) 1011.

[48] Bruhwiller P.A., Wagner J.L., Biggs B.D., Shen Y., Wong K.M., Schatterly S.E. and Poon S.J., *Phys. Rev. B* **37** (1988) 6529.

[49] Trambly de Laissardière G. and Fujiwara T., *Phys. Rev. B* **50** (1994) 5999.

[50] Trambly de Laissardière G. and Fujiwara T., *Phys. Rev. B* **50** (1994) 9843.

[51] Krajci M., Hafner J., *Mat. Sci & Eng. A* **226-A228** (1997) 950.

[52] Davydov D.N., Mayou D., Berger C., Gignoux C., Neumann A., Jansen A.G.M. and Wyder P., *Phys. Rev. Lett.* **77** (1996) 3173.

[53] Escudero R., Lasjaunias J.C., Calvayrac Y. and Boudard M., submitted.

[54] Schaub T., Delahaye J., Gignoux C., Berger C., Fourcaudot G., Giroud F., Grenet T. and Jansen A.G.M., Proc. LAM10, to appear.

[55] Sire C., Lectures on Quasicrystals (Les Éditions de Physique, Les Ulis, 1994) 505.

[56] Lück J.M., *Phys. Rev. B* **39** (1989) 5834.

[57] Friedel J. and Dénoyer F., *C.r. Acad. Sci (Paris)* **305** (1987) 171.

[58] Haberkern R., Summer school on Quasicrystals, edited by Suck J.B., Chemnitz, Sept. 97, to appear.

[59] Burkov S.E., Varlamov A.A. and Livanov D.V., *Phys. Rev. B* **53** (1996) 11504.

[60] Kalugin P.A., *Phys. Rev. B*, to appear.

[61] Cyrot-Lackmann F., New Horizons in Quasicrystals, edited by A.I. Goldman, D.J. Sordelet, P. A. Thiel and J.M. Dubois (World Sci., Singapore, 1997) 216.

[62] Trambly de Laissardière G., Mayou D., Nguyen-Manh D., *Europhys. Lett.* **21** (1993) 25; Friedel J., *Philos. Mag. B* **65** (1992) 1125.

[63] Trambly de Laissardière G. and Mayou D., *Phys. Rev. B* **55** (1997) 2890.

[64] Sanquer M., Tourbot R. and Boucher B., *Europhys. Lett.* **7** (1988) 635.
[65] Honda Y., Adagawa K., Takeuchi S, Tsai A.-T. and Inoue A., *Jpn. J. Appl. Phys.* **34** (1995) 2415.
[66] Matsuda T., Ozaki T., Sato H. and Mizutani U, Quasicrystals, edited by Takeuchi S. and Fujiwara T. (World Sci., Singapore, 1998) 583.
[67] Kondo R. Hashimoto T., Edagawa K., Takeuchi S., Takeuchi T. and Mizutani U., *J. Phys. Jpn.* **66** (1997) 1097.
[68] Poon S.J., private communication.
[69] Ahlgren M., Rodmar M., Brangefält, Berger C., Gignoux C. and Rapp Ö., *Czecholovak J. Phys.* **46** (1996) 1989.
[70] Rodmar M., Oberschmidt D., Ahlgre M., Gignoux C., Berger C. and Rapp Ö., Proc. LAM10, to appear.
[71] Yoshizumi S., Geballe T.H., Kunchur M. and McLean W.L., *Phys. Rev. B* **37** (1988) 7094.
[72] Sin H.K., Lindenfeld P. and McLean W.L., *Phys. Rev. B* **30** (1984) 4067.
[73] Zhong J.X, Grimme U., Römer R.A. and Schreiber M., *Phys. Rev. Lett.* **80** (1998) 3996.
[74] Roche S. and Mayou D., *Phys. Rev. Lett.* **79** (1997) 2518.
[75] Mayou D., Quasicrystals, edited by Takeuchi S. and Fujiwara T. (World Sci., Singapore, 1998) 555.

Exact Electron States in $1D$ (Quasi-) Periodic Arrays of Delta-Potentials

Given by P. Kramer

Written by P. Kramer and T. Kramer

Institut für Theoretische Physik der Universität Tübingen, Tübingen, Germany

1. INTRODUCTION AND SCOPE

In the solid state physics of crystals, an important part is played by the band structure of the electronic states. This band structure arises from the representation of the periodicity in the electronic state space. Powerful computational methods were developed for the calculation of band structures, among them the linear muffin-tin (LMTO) method [17] in the atomic sphere approximation (ASA) [1]. In the physics of quasicrystals, it is believed that the electronic system plays an important part [19]. Here one is lacking the periodic symmetry. To still use the powerful methods of band computations, one must replace the quasicrystal by a periodic approximant. The question arises how such approximant computations approach a quasiperiodic limit. In a recent calculation [6] it is shown by a supercell analysis that an approximant computation may lead to artefacts in the electronic density of states (DOS). More detailed local properties of the electrons appear as derivatives of the electronic charge density in Mössbauer studies on quasicrystals [14].

In the present work we wish to analyse and compare electronic states in periodic and quasiperiodic potentials in a way free of approximations. In this way we can hope to address the similarities and differences of these systems from first principles.

To a first and very important approximation, the many-electron states in ordered crystalline or quasicrystalline solids are constructed from one-electron states, compare for example Ashcroft and Mermin [2], after appropriate antisymmetrization. The one-electron Hamiltonian then contains periodic and quasiperiodic potentials respectively. For a general account of band theory we refer to Blount [4]. Analytic properties and the approach due to Wannier [20] are discussed by Kohn [8]. For general electronic structures beyond periodicity, a local view was advocated by Heine [7] who proposed to throw out K-space.

In what follows we wish to elaborate and to compare electronic systems and their local structure in terms of the following concepts:

(i) **Local view of K-space in periodic potentials**:

A Hamiltonian with an infinite *periodic potential* has the discrete symmetry group Λ of its lattice. The electron states are characterized by the irreducible representations of Λ. These are labelled by the continuous set of Bloch vectors K from the Brillouin zone, that is, the first unit cell of the reciprocal lattice Λ^R. The eigenstates for fixed energy $E(K)$ are the Bloch states belonging to the bands.

A local formulation of K-space arises as follows: Wigner and Seitz [21] in 1934 introduced the *cellular method*: the one-electron Schrödinger equation with fixed energy E is to be solved exclusively on the finite unit cell of the lattice with the *local boundary conditions* that the solution, after propagation over a full primitive period say a_j, $j = 1, 2, 3$, picks up a pure phase factor. This phase factor when written in the form $f_j := \exp(iK_j a_j)$, $j = 1, 2, 3$ determines the Bloch labels. Two solutions with $K \to -K$ are degenerate. The energy $E(K)$ appears in bands, with Bloch labels ranging over the Brillouin zone, and in gaps where there are no states with Bloch type boundary conditions. This approach allows an extension to finite and to infinite systems. Within bands, the solutions can be matched and propagated as Bloch states over all the cells of the crystal.

(ii) **Local view of K-space in quasiperiodic potentials**:

For infinite almost periodic systems in $1D$ some general results with many references are discussed in [5]. For another detailed study of discrete systems with many references we quote Sütö [18]. For our purposes we consider an infinite *quasiperiodic potential* built from a few basic cells as in a tiling model for quasicrystals. For each cell one can apply the local analysis of Wigner and Seitz with local Bloch type boundary conditions. Quasiperiodicity now requires a matching of these local boundary conditions between the basic cells. It is not enough that, at fixed energy E, each cell separately admits Bloch type boundary conditions. We must also check if a patch of cells admits these boundary conditions. Again one can study the quasiperiodic extension of the system, following for example a geometric inflation of the tiling.

(iii) **Local view, positive and negative energy, motive clusters**:

Imagine the atomic nuclei of a unit cell in a periodic crystal to form a free *motive cluster*. This motive cluster is not unique since the unit cell of a crystal admits translations as well as transformation of shape, for example in going from the primitive to the Wigner-Seitz cell.

For a fixed choice of motive clusters we should not relax the positions of the nuclei to gain energy, as would be done in real molecules, since in the crystal the positions are controlled by *all* atomic neighbours.

Example: Take for example a linear crystal with two atoms A, B, in two spacings $a \neq b$ with the sequence AaBbAaBbAa... and period $a + b$. Choosing the unit cell as $AaBb$ or $BbAa$ respectively we define two different motive clusters AaB and BbA which generate the same crystal with identical band spectrum. Looking locally at the isolated clusters we get in general two different bound states, since the binding energy depends on the relative distances a and $b \neq a$ respectively. In a linear molecule we would relax the distance and end up in both cases at the same binding energy. This is not allowed in the crystal! In the linear crystal *the two different motive clusters generate identical periodic structures and bands!*

Consider then the one-electron bound states for the chosen motive cluster. It is true that, in order to treat it like a molecule, we should take along the nuclei and their Coulomb repulsion. However since we keep the nuclei at fixed positions we may neclect these interactions which of course influence the binding and the total energy. Independent of these interaction terms, the different boundary conditions of the electron states at positive or negative energy will enter into the computation of the many-electron states.

Since the Hamiltonian for fixed nuclear positions is the same as in the Wigner-Seitz method, the only difference is in the boundary conditions which for a bound state require the solutions to decay exponentially. Clearly now we must distinguish between the *positive and negative energy region*. The former admits scattering states, the latter bound states.

The infinite systems at positive energy should here be considered as an approximation to a finite macroscopic systems which still admits a scattering experiment. We could also consider half-infinite systems which allow at least for backscattering. For both finite and infinite systems, the distinction between positive and negative energy states is crucial. In considering finite patches of periodic or quasiperiodic systems one therefore should pay attention to the sign of the energy.

(iv) **Bound, Bloch and scattering states in periodic potentials**:

We shall term the negative energy region the *exact tight binding scenario* in what follows. As a partial differential equation of second order, the stationary Schrödinger equation has two fundamental solutions on a single cell. Suppose that the energy of a bound state lies within the range of energy for a band. Then it follows from the Schrödinger equation that *the bound state solution restricted to the unit cell must be an exact linear combination of the local Bloch states at the bound state energy.* In the positive energy region we can still obtain states with local Bloch boundary conditions. Now we can inquire how the band structure of these positive energy states is related to one-electron scattering states.

Example: The relation of bound and Bloch states in crystals is often considered in approximate rather than exact computations, as for example in the *tight binding approximation.* Here one starts exclusively from bound atomic l-orbitals of the isolated atoms, obtains the Bloch states from their tunneling to next neighbours, and labels the bands by the l-orbitals. If first-order degenerate perturbation theory applies, the energies in the bands result from the splitting of the single orbital whose degeneracy is proportional to the number of atoms. In this approximation it is clear that the energy of the local atomic orbital is within the corresponding band. In many cases the bound atomic orbitals of the atoms in the cell are separated in energy. Otherwise one can consider hybrid or motive cluster orbitals on the unit cell as the origin of band labels. The discrete version of this approximation leads to an eigenvalue problem and therefore looses the information on the sign of the energy.

The general arguments given above are valid *independent of any such approximation scheme.* They show that *in the exact tight binding scenario there must exists an exact local relation between bound and Bloch states.*

(v) **Bound, Bloch and scattering states in quasiperiodic potentials**:

For a quasiperiodic potential based on a tiling, we again wish to distinguish the negative and positive energy regions. One can consider at negative energy the bound states for finite patches which form part of a quasiperiodic structure and compare with states obeying local Bloch type boundary conditions on the same patch. We refer to such an energy range as a *band germ* and to the local Bloch states as *Bloch germs*. If the bound state energy admits Bloch type boundary conditions, it must be possible to express the bound states by pairs of local Bloch states (germs). The bound states may be related to local clusters.

In the scattering from quasiperiodic potentials at positive energy, the aim is again to compute the scattering matrix. For an exact computation one may inquire about the implication of inflation symmetry on the scattering matrix.

In what follows we wish to demonstrate the validity of these concepts in a way free of approximations. In dimension three we cannot yet implement exact examples for these concepts. The LMTO method may allow to examine some of our points. We choose here a comparative and specific study restricted to continuous periodic and quasiperiodic potentials in dimension one. On them we wish to demonstrate the concepts given above by exact computations. To do so we pay the price of choosing simple delta potentials even of the same strength. We explore the electron states in terms of local boundary conditions, in the spirit of the Wigner-Seitz approach, applied to finite strings (patches) which form part of periodic or quasiperiodic structures. We shall always start from finite patches and then look into their recursive extension. The solutions with Bloch boundary conditions on a finite string are systematically compared with bound state solutions on the same string.

The main tool of our analysis is the continuous transfer matrix which is well defined for piecewise constant potentials, including delta potentials as a limiting case. This matrix propagates by matrix multiplication a fundamental system of two solutions. We shall stress in what follows the polynomial dependence of the matrix elements on the strength parameter of the delta potential. This will allow us to order and analyze the matrix systems.

2.   FINITE PERIODIC STRINGS AT NEGATIVE ENERGY

2.1.   Preview: An energy gauge for crystals

As an introduction we present some material from [12]. We rephrase the well-known periodic case [15]. Consider first a finite string S with the transfer matrix M, described in more detail in Section 2.2. We define an *energy gauge* f as a function with value $f = 0$ on an energy interval such that $|\mathrm{tr}(M)| \leq 1$ and $f = 1$ otherwise, compare Figure 1. We call this a *band germ*. The condition on the trace assures that inside the band germ M has two complex conjugate eigenvalues of absolute value 1. Repeat the string n times to produce a new transfer matrix M^n. Since the existence of band germs is related to the eigenvalue problem, the range of energies for which $\frac{1}{2}|\mathrm{tr}(M^n)| < 1$ is *independent of n, the band germs stay the same on the new string*. The Bloch germs propagate through the string and pick up the same phase factors respectively after each transmission.

Consider next the bound states of the string M^n. In a very tight binding case we claim.

1 Prop: The bound states of the string M^n may be grouped into sets of n states, where the energies of each such set corresponds to a part $f = 0$ of the energy gauge f of the string M.

Proof: For very tight binding it suffices to use first-order degenerate perturbation theory, applied to the bound states of the single atoms: to this order, the bound states of the string are the eigenvalues of a matrix whose off-diagonal entries are the weak atom-atom cross terms. By standard matrix theory, the maximum level splitting will increase with n. But band theory tells us that in the limit $n \to \infty$ all energies stay inside the band, hence inside the initial band germ. Thus the energy of all these bound states for finite n must stay within the energy gauge of the (single) band germ.

We now have the following situation in the finite string M^n: in a band germ from the string M, there is for each energy value a pair of Bloch germs which can carry charge current. There are now n discrete bound states with the energy gauge as in the initial string.

A schematic view of the periodic scheme is given in Figure 1. Now we can extend the analysis to $n \to \infty$.

2 Prop: For an infinite periodic repetition of a fixed string, the energy gauge stays the same as for the initial string. The band germ generates a band. Within each band there is an *infinite set of (pairs of) Bloch states whose energies $E_K < 0$ fill up the original band germ.*

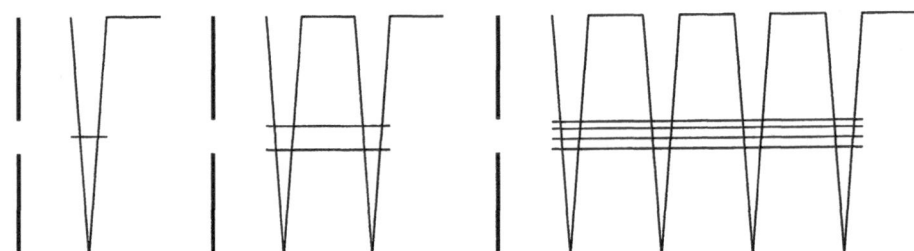

Fig. 1. — Periodic strings: the string S to the left has one attractive δ-well with a single bound state, followed by a tunnel. The vertical bar to its left shows the energy interval for the band germ. This interval is the energy gauge $f(S)$ comprising the bound state. The string S^2 in the middle has two attractive δ-potentials and two bound states. The energy gauge $f(S^2)$ is unchanged but comprises two bound states. The same energy gauge $f(S^4)$ for the string S^4 to the right comprises four bound states.

2.2. Bloch and bound states in a single band

We shall use the notation of [10]. Consider the $1D$ Schrödinger stationary equation on the line $x \in \mathbf{R}$,

$$\left[-\frac{\hbar^2}{2m}\frac{\mathrm{d}^2}{\mathrm{d}x^2} + V(x) \right] \phi(x) = E\phi(x) \tag{1}$$

with $V(x)$ a piecewise constant potential. Denote two fundamental solutions and their derivatives by $\phi_1(x), \phi_1'(x),\ \phi_2(x), \phi_2'(x)$. Choose the initial values $\phi_1(0) = 1,\ \phi_1'(0) = 0,\ \phi_2(0) = 0,\ \phi_2'(0) = 1$ which implies a Wronskian equal to 1. The 2×2 standard transfer matrix is defined as

$$M(x) = \left[\begin{array}{cc} \phi_1(x) & \phi_2(x) \\ \phi_1'(x) & \phi_2'(x) \end{array} \right], \tag{2}$$
$$M(0) = 1.$$

The transfer matrix obeys the first-order system of equations which is equivalent to the Schrödinger equation. A *second interpretation* of the transfer matrix, which we shall adopt in what follows, results because the matrix $M(x)$ can also be shown to *propagate the fundamental system*.

   The Schrödinger equation for piecewise constant potentials on a finite string can now be solved as follows [12]: we first determine the transfer matrix for simple building blocks of finite size at the fixed energy. Then we propagate the solutions over the full string by matrix multiplication. This matrix multiplication guarantees the continuity of the solutions and of their derivatives. Specific boundary conditions can finally be met by determining the appropriate linear combination of the two fundamental solutions. For real matrices with Wronski determinant 1 the transfer matrices belong to the matrix group $SL(2, R)$. More details on this group and the isomorphic group $SU(1, 1)$ can be found in [10]. One should be careful in applying the group concepts to the transfer matrices, as can be seen from the following transformation properties.

It will prove convenient to pass to a new fundamental systems of solutions by choosing different initial values. With respect to the matrix formed by the solutions, this transformation is achieved by *right multiplication with a constant matrix,*

$$M(x) \rightarrow M(x)C. \tag{3}$$

The *new transfer matrix which propagates the new system* $M(x)C$ is no longer equal to $M(x)C$, instead it *is given by* $C^{-1}M(x)C$. This propagating transfer matrix by itself does not admit an interpretation in terms of fundamental solutions and their derivatives!

We now pass to a new system of fundamental solutions with different initial conditions by right multiplication with the matrix

$$\tilde{R} := \sqrt{i}\, R = \sqrt{\frac{1}{2\kappa}} \begin{bmatrix} 1 & 1 \\ -\kappa & \kappa \end{bmatrix}. \tag{4}$$

The system corresponding to the matrix $M(x)\tilde{R}$ has the initial data $\tilde{R}$ at $x = 0$. At negative energy $E = -\frac{\hbar^2}{2m}\kappa^2$ we define a *tunnel* as a string with vanishing potential. In a tunnel the transfer matrix takes the form

$$M(x)\tilde{R} = \sqrt{\frac{1}{2\kappa}} \begin{bmatrix} \exp(-\kappa x) & \exp(\kappa x) \\ -\kappa \exp(-\kappa x) & \kappa \exp(\kappa x) \end{bmatrix}. \tag{5}$$

The transfer matrix which propagates this new basis system at negative energy is given by

$$\tilde{M}(x) \quad := \quad \tilde{R}^{-1} M(x)\tilde{R}. \tag{6}$$

For a bound state we require that an exponentially increasing function at negative energy on the left-hand tunnel, after passing an intermediate transfer matrix $\tilde{M}$, produces an exponentially decreasing function on the right-hand side. In the new basis, this property and the form equation (5) require that the intermediate transfer matrix obeys $\tilde{M}_{22} = 0$.

3 Prop: In the system $\tilde{M}$ of transfer matrices for finite strings, the condition for a bound state requires that the second diagonal element vanishes.

A tunnel of length b at negative energy $E = -\frac{\hbar^2}{2m}\kappa^2$, followed by an attractive delta-potential of strength u, we term the string S. We define $\delta := u/\kappa$ and $\lambda_1 := \exp(\beta)$, $\beta := \kappa b$ to obtain the transfer matrix of S as

$$\begin{aligned}
\tilde{M}_1 &= \begin{bmatrix} \left(1 + \frac{1}{2}\delta\right) & \frac{1}{2}\delta \\ -\frac{1}{2}\delta & \left(1 - \frac{1}{2}\delta\right) \end{bmatrix} \begin{bmatrix} \lambda_1^{-1} & 0 \\ 0 & \lambda_1 \end{bmatrix} \\
&= \begin{bmatrix} \lambda_1^{-1}\left(1 + \frac{1}{2}\delta\right) & \lambda_1 \frac{1}{2}\delta \\ -\lambda_1^{-1}\frac{1}{2}\delta & \lambda_1\left(1 - \frac{1}{2}\delta\right) \end{bmatrix}.
\end{aligned} \tag{7}$$

Here the first transfer matrix describes the delta-potential and the second one
the tunnel, compare [12]. This and all other transfer matrices we take as a
function of the dimensionless variable $\beta = \kappa b$, related to the energy and to
the length of the cell S. We also consider them as a function of the variable
$\gamma = ub$ related to the strength of the delta-potentials, and study the family of
systems with varying strength. Instead of γ we shall often use the ratio $\delta := \frac{\gamma}{\beta}$
because then the matrix elements of $\tilde{M}_1$ are (linear) *polynomials in the variable*
δ. Since we shall generate all other transfer matrices by matrix multiplication,
a general transfer matrix will be a *polynomial in the parameter* δ. This will be
the basis for a *polynomial method* for handling the analysis.

We introduce the following short-hand notation for the elements of a matrix
$\tilde{M}_i$

$$\tilde{M}_i \quad := \quad \begin{bmatrix} a_i & b_i \\ c_i & d_i \end{bmatrix} \tag{8}$$

$$x_i \quad := \quad \frac{1}{2}(a_i + d_i),$$

$$y_i \quad := \quad \frac{1}{2}(a_i - d_i).$$

The second row of equation (8) determines the *half-trace* $x_i = \frac{1}{2}\mathrm{tr}(\tilde{M})$. For the
matrix $\tilde{M}_1$ we find from equation (7)

$$x_1 \quad = \quad \frac{1}{2}(\lambda_1 + \lambda_1^{-1}) - \frac{1}{4}\delta(\lambda_1 - \lambda_1^{-1}), \tag{9}$$

$$y_1 \quad = \quad -\frac{1}{2}(\lambda_1 - \lambda_1^{-1}) + \frac{1}{4}\delta(\lambda_1 + \lambda_1^{-1}),$$

$$d_1 \quad = \quad \lambda_1 \left(1 - \frac{1}{2}\delta\right).$$

We shall assume that we are in the band germ and, moreover, that the upper
edge of the band occurs at negative energy. In Sections $4, 5$ we shall deal
with more general scenarios. For $|x_1| \leq 1$, the eigenvalues of $\tilde{M}_1$ are complex
conjugate of modulus 1 given by

$$\theta_1 = x_1 + i\sqrt{1 - x_1^2}, \ \theta_2 = x_1 - i\sqrt{1 - x_1^2}. \tag{10}$$

So x_i is the half-trace, and the local Bloch label K for fixed energy is obtained
from

$$\cos(Kb) = x_1(\beta, \gamma), \ 0 \leq K \leq \frac{\pi}{b}. \tag{11}$$

The values $\beta : x_1(\beta) = -1, 1$ determine the two band edges respectively.

The lower diagonal element by $d_1(\beta) = 0$ determines, compare [12] and what
was explained above, the *single bound state of the delta-potential* at $\delta = \frac{u}{\kappa_0} = 2$.
The variable λ_1, related to the tunnel length, drops out of this equation.

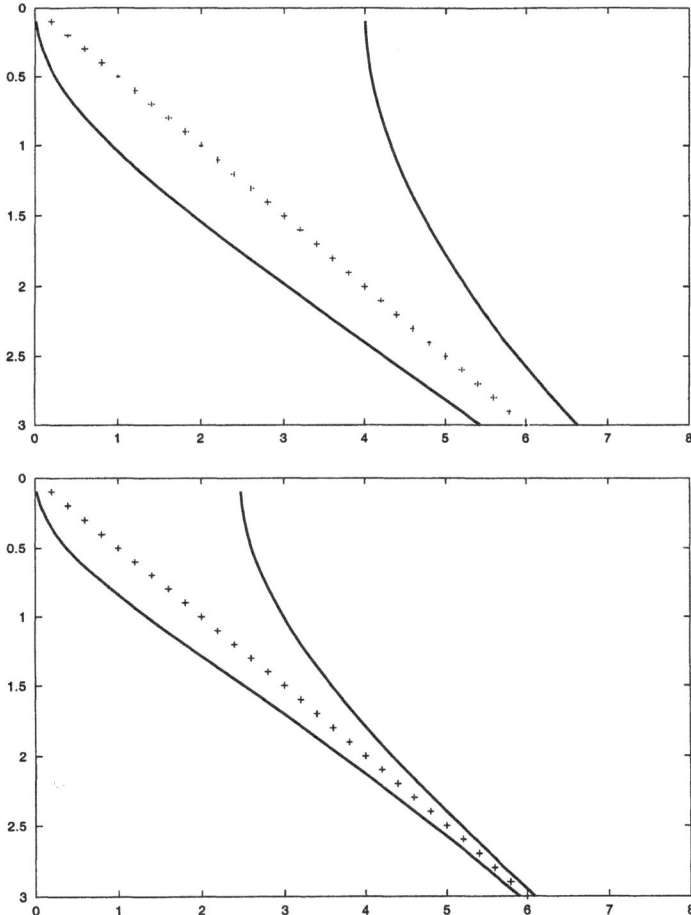

Figs. 2, 3. — The string of length $\kappa b = 1, \tau$ respectively produces two different band germs which encapsulate the same bound state. Here and in all the corresponding figures, the values of $\gamma = ub$ and $\beta = \kappa b$ determine the potential strength and the energy respectively, compare **Prop 3**. They are plotted as horizontal and vertical coordinates respectively. The upper and lower edges of the band germs are given as solid lines. The crosses between these lines determine the bound state energy in terms of β as a function of γ.

As $\tilde{M}$ is real, the eigenvectors can be chosen complex conjugate. We find

$$\tilde{M}_1 V = V\Lambda, \qquad (12)$$

$$V = \begin{bmatrix} p & \bar{p} \\ v & \bar{v} \end{bmatrix}$$

$$\Lambda = \begin{bmatrix} \theta_1 & 0 \\ 0 & \theta_2 \end{bmatrix}$$

with

$$\frac{v}{p} = \frac{\theta_1 - a_1}{b_1} = \frac{-y_1 + i\sqrt{1 - x_1^2}}{b_1} \tag{13}$$

$$= \delta^{-1}\left[(1 - \lambda_1^{-2}) - \frac{1}{2}\delta(1 + \lambda_1^{-2}) + 2i\lambda_1^{-1}\sqrt{1 - x_1^2}\right].$$

With the help of the matrix V we pass to the Bloch system of solutions defined by $M(x)\tilde{R}V$ since it obeys $M(0)\tilde{R}V = \tilde{R}V$, $M(b)\tilde{R}V = \tilde{R}V\Lambda$. In line with equation (12) we mark derivatives with respect to x with a prime and put

$$M_1(x)\tilde{R}V := \begin{bmatrix} \Phi_1 & \Phi_2 \\ \Phi_1' & \Phi_2' \end{bmatrix}, \tag{14}$$

$$\Phi_2 = \overline{\Phi}_1,$$

and obtain for $0 < x < b$ from equations (12–14) the Bloch state

$$\Phi_1(x) := \frac{p}{\sqrt{2\kappa}}\left[\exp(-\kappa x) + \delta^{-1}\left[(1 - \lambda_1^{-2}) - \frac{1}{2}\delta(1 + \lambda_1^{-2})\right.\right.$$

$$\left.\left. +2i\lambda_1^{-1}\sqrt{1 - x_1^2}\right]\exp(\kappa x)\right]. \tag{15}$$

To normalize the current density to the value $\frac{e\hbar}{2m}$ it suffices to choose $\det(V) = p\bar{v} - v\bar{p} = -i$ which together with equation (13) yields

$$p\bar{p} = \frac{\delta\lambda_1}{4\sqrt{1 - x_1^2}}, \quad p = i\frac{1}{2}\sqrt{\frac{\delta\lambda_1}{\sqrt{1 - x_1^2}}}, \tag{16}$$

where we have chosen a particular phase for p so that $\bar{p} = -p$. The two Bloch states obey the bilinear relation

$$(-i)(\overline{\Phi_l}\Phi_j' - \overline{\Phi_l'}\Phi_j) = (-1)^{l+1}\delta_{lj}. \tag{17}$$

Which can be used to define a *scalar product* for Bloch states.

4 Def: To any pair Φ_1, Φ_2 of complex solutions of the Schrödinger equation we can associate an indefinite hermitian scalar product $\langle \Phi_1, \Phi_2 \rangle$ by the left-hand side of equation (17).

The relation between the Bloch states and the exponential states $M_1(x)\tilde{R}$ is given from equations (12) and (14) by

$$\begin{bmatrix} \psi_1 & \psi_2 \\ \psi_1' & \psi_2' \end{bmatrix} := M_1(x)\tilde{R} = (M_1(x)\tilde{R}V)\, V^{-1} \tag{18}$$

$$= \begin{bmatrix} \Phi_1 & \Phi_2 \\ \Phi_1' & \Phi_2' \end{bmatrix} i \begin{bmatrix} \bar{v} & -\bar{p} \\ -v & p \end{bmatrix}.$$

A particular case arises if we choose the bound state energy $E = -\frac{\hbar^2}{2m}\kappa_0^2$ and construct the Bloch states equation (14) for $\kappa = \kappa_0, \delta = 2$. This bound state must increase exponentially in the unit cell towards the delta-potential and is given from equations (15, 16) and (18) for $0 < x < b$ by

$$\psi_2(x) = -\frac{1}{2}\sqrt{\frac{2\lambda_1}{\sqrt{1 - \lambda_1^{-2}}}}(\Phi_1(x) + \Phi_2(x)). \tag{19}$$

Which differs from the expression given in [12] equation (22) in the phase for the Bloch states and due to the new sequence of tunnel and potential. A natural real unbound companion of this bound state from equation (18) is

$$\psi_1(x) = i(\Phi_1(x)\bar{v} - \Phi_2(x)v), \tag{20}$$

because it can be easily be shown that the scalar product equation (17) for these two real states yields the Wronskian

$$\psi_1\psi_2' - \psi_1'\psi_2 = 1. \tag{21}$$

Here we wish to comment on the method of orthogonalized plane waves OPW in which Bloch states are required to be orthogonal to certain bound states [2] p. 206-208. We doubt that this requirement can be handled with the standard integral scalar product in Hilbert space, or with a scalar product that involves only integration over the unit cell: the first integration does not apply without qualification to unbound states, and the second one introduces terms at the boundary of the unit cell.

2.3. The string S^n

Now we pass from the string S by n-fold repetition to the string S^n with the transfer matrix $\tilde{M}_1^n$. This power of $\tilde{M}_1$ can be written as

$$\tilde{M}_1^n = V\Lambda^n V^{-1}. \tag{22}$$

Explicitly the four matrix elements from equations (12–14) are

$$a_n = a(\tilde{M}_1^n) = \cos(nKb) + y_1\frac{\sin(nKb)}{\sin(Kb)}, \tag{23}$$

$$d_n = d(\tilde{M}_1^n) = \cos(nKb) - y_1\frac{\sin(nKb)}{\sin(Kb)},$$

$$b_n = b(\tilde{M}_1^n) = \frac{1}{2}\lambda_1\delta\frac{\sin(nKb)}{\sin(Kb)},$$

$$c_n = c(\tilde{M}_1^n) = -\frac{1}{2}\lambda_1^{-1}\delta\frac{\sin(nKb)}{\sin(Kb)}$$

with y_1 taken from $\tilde{M}_1$ in equation (7).

2.4. Rational Bloch labels

Consider the transfer matrix equations (22, 23) for the unit cell. For fixed integer $n > 0$ at the points

$$nKb = \mu\pi, \ \mu = 0, \pm 1, \ldots, \pm(n-1), \pm n \tag{24}$$

it becomes $\tilde{M}_1^n = (-1)^n e$. For any such Bloch label it follows that the Bloch states have the period $2nb$ with symmetry group C_{2n}. We can relate these states to a chain of n atoms by imposing as boundary conditions the factor $(-1)^n = \pm 1$ for even and odd n respectively on the interval nb. Moreover the Bloch state for fixed μ on the unit cell transforms according to the representation

$$D^\mu = \exp\left(i\mu\frac{2\pi}{2n}\right) \tag{25}$$

for the generator of order $2n$ of the cyclic group C_{2n}.

There is *degeneracy* as we have a real hamiltonian: for opposite signs of Kb and μ we have pairs of Bloch states with equal energy, running to the left and right respectively. The set of rational states for fixed n yields a spectrum of (in part) conjugate eigenstates. The energy increases with $|\mu|$. The rational states are shown for $n = 2, 3$ in Figures 5, 6 below.

With respect to C_{2n}, each eigenstate is characterized by a real or by a pair of complex conjugate irreducible representations of C_{2n}. All these rational states occur in the band and separate it into n partial bands characterized by

$$K \geq 0: \ \frac{\mu\pi}{nb} \leq K \leq \frac{(\mu+1)\pi}{nb}, \ \mu = 0, 1, \ldots, (n-1), n \tag{26}$$

and similar expressions for $K \leq 0$.

2.5. Bound states and clusters of the string S^n

The bound states of the finite string S^n are characterized from equation (23) by

$$\begin{aligned} \mathrm{d}(\tilde{M}_1^n) &= 0, \\ \tan(nKb) &= \frac{\sin(Kb)}{y_1}. \end{aligned} \tag{27}$$

The free string forms a cluster with n atoms and hence at most n bound states. We now compare these bound states with the states of the (up to a sign) periodic string of n atoms: in the interval between two successive zeros of $\tan(nKB)$ with $nKb = \mu\pi, (\mu+1)\pi$, which correspond to the rational K labels associated with states of the periodic string, the value of $\tan(nKb)$ goes once to infinity. The right-hand side of equation (27) is a smooth function of β except at the point $y_1(\beta) = 0$. The left and right-hand side of equation (27) are displayed for the case $n = 10$ as functions of β in Figure 4.

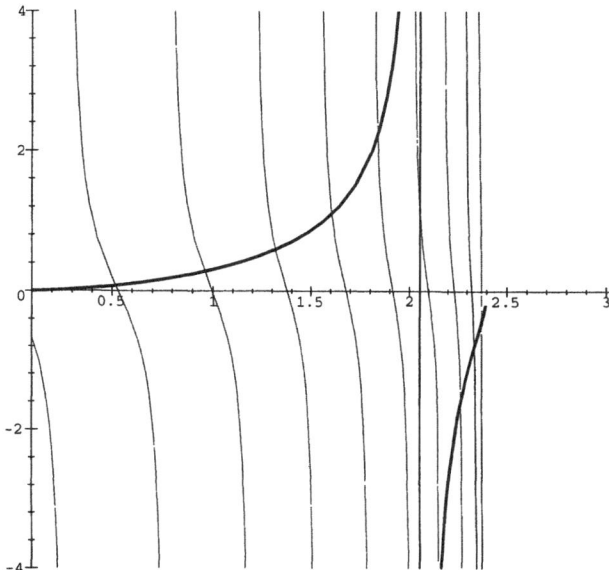

Fig. 4. — Left- and right-hand side of the binding equation (27) on a string of length 10 for $\gamma = 4$ as a function of β.

On any interval between consecutive zeros of $\tan(nKb)$ there is precisely one intersection point of the left and the right-hand expression of equation (27).

In terms of boundary conditions this implies that, in between the energies of two consecutive rational values of Kb with periodic boundary conditions, there is precisely one bound state of the n-atom cluster. This state, termed a *decaying state* in [12], is a single linear combination of two degenerate Bloch states on the interval of length nb and, when the n-atom string is cut out, admits exponential decay to the right and to the left.

5 Prop: Any set of rational Bloch labels equation (24) separates the band into n partial bands. Each partial band contains a single bound state of the free n-atom string.

The relation between rational Bloch labels and bound states is shown in the following Figures 5, 6.

Since the rational labels K form a dense set of energies, we must also have a dense set of bound or decaying state energies of free n-atom strings. Each bound state coincides on a cell of length nb with a linear superposition of two Bloch states at the same fixed negative energy similar to ψ_2 in equation (19). Each one has a companion state of a form similar to ψ_1 in equation (20), and these two states have a Wronskian as in equation (21).

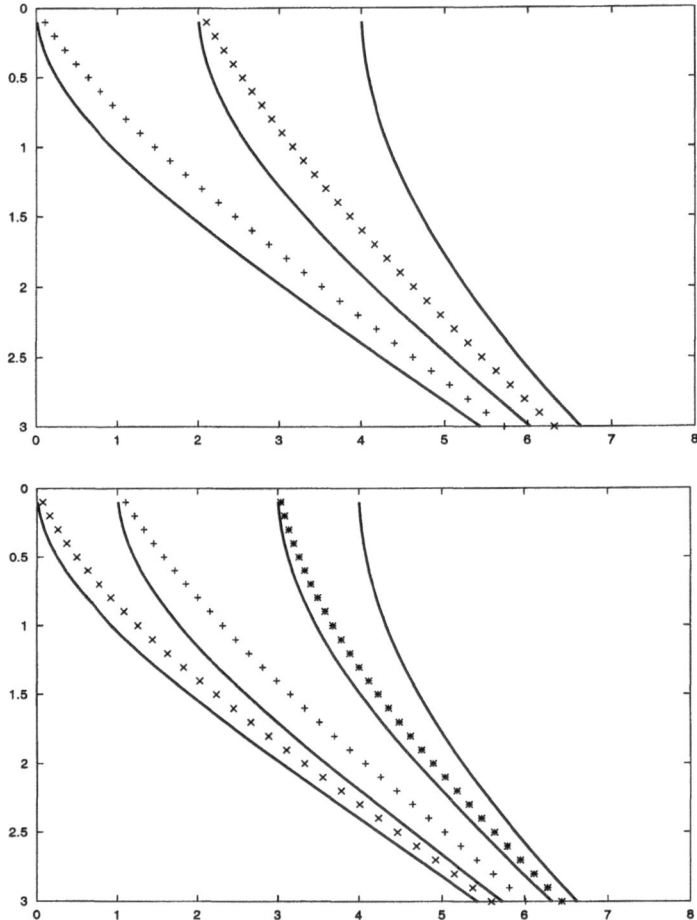

Figs. 5, 6. — The strings S^2 and S^3. In the supercell analysis, two or three band germs are glued together without gaps and overlaps. Each band includes a single bound or decaying state. The horizontal coordinate is the potential strength γ, the vertical coordinate the energy β. Notice the relation between the onset of additional bound states and the subband edges.

2.6. Participation number

From the relation between n-atom clusters and periodic states we can assign within a band a *participation number* n to the rational Bloch vectors of the form $K = \frac{\mu\pi}{nb}$, equation (24): this number determines the number n of atoms which participate in the Bloch states whose period up to a sign is nb, and at the same time assigns n bound or decaying states of the cluster. Small values of n indicate states which involve a few atoms and may be called localized with respect to the clusters.

2.7. Supercell interpretation

An alternative interpretation of the results given above can be given in terms of a *supercell scheme*: we consider the supercell of length nb formed from n centers. The corresponding Brillouin zones have the size $2\pi/(2bn)$. The positive Bloch labels for the n bands in this scheme are obtained from equation (26) as

$$0 \leq K(n,\mu) \;=\; K - \frac{\mu\pi}{nb} \leq \frac{\pi}{nb}, \tag{28}$$
$$\mu \;=\; 0, 1, \ldots, (n-1), n$$

and similar expressions for $K(n,\mu) \leq 0$. With the new Bloch label $K(n,\mu)$, these superbands are identical with the partial bands for the rational values of K.

6 Prop: The superbands for the supercell of length nb are glued together without gaps and overlaps and fill up the single band. Their Bloch states are identical to those of the partial bands of the rational scheme. Within each superband there is included a single bound state. This bound state coincides on the supercell with a bound state of the free n-atom cluster which forms the motive of the supercell. The alternating pattern of bound and periodic n-atom states illuminates the structure of the band.

2.8. Large n limit

Now we consider the limit $n \to \infty$. The rational K-labels form a dense set on the Brillouin zone. The length of the periodic chains and of the clusters increases with the participation number. As the rational numbers form a dense set on the Brillouin zone, we have pairs of Bloch states and bound states near any value of K in the band. By picking rational values of K characterized by n, we select a set of periodic states which pairwise encapsulate bound states of the finite string S^n.

From the equal spacing of the zeros of $\tan(nKb)$ with respect to the band label K, and from the inclusion property for bound states there follows.

7 Prop: The density of bound and of Bloch states in the limit $n \to \infty$ is a constant function of the Bloch label K. In this limit, the density of (bound) states per unit energy interval fulfills the well-known relation

$$\frac{\mathrm{d}N}{\mathrm{d}E} \sim \frac{\mathrm{d}K}{\mathrm{d}E}. \tag{29}$$

Although this is a smooth function in the limit, its rational approximants with large but finite n may look quite irregular as functions of β.

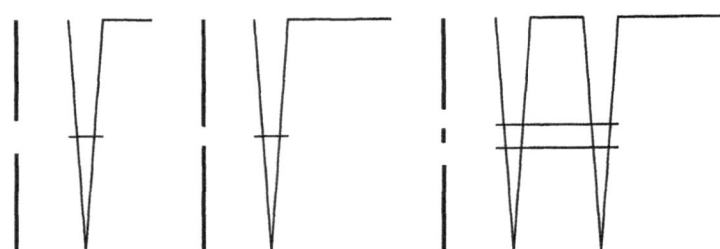

Fig. 7. — Fibonacci strings: the string S to the left and its energy gauge $f(S)$ are as in Figure 1. The string L in the middle has the same bound state within the gauge $f(L)$ of smaller width. The string SL to the right has two attractive δ-potentials and hence the same bound states as S^2 in Figure 1. In contrast to S^2, the string SL has two separate band germs. The corresponding two parts of the energy gauge $f(SL)$ comprise each one bound state but block the energy of the single-atom bound state.

3.   FINITE QUASIPERIODIC STRINGS AT NEGATIVE ENERGY

As an example of a quasiperiodic system we shall take the well-known Fibonacci system: we maintain delta-potentials of equal strength for the atoms but admit two different intervals S and L of length b and qb respectively. For the proper Fibonacci case we choose $q = \tau = \frac{1}{2}(1 + \sqrt{5})$, but for comparison with the periodic case we also consider the value $q = 1$.

The Fibonacci system we define algebraically by the recursive words W_m and initial data in the alphabet $\langle S, L \rangle$,

$$W_{m+1} := W_{m-1}W_m, \ W_1 = S, W_2 = L. \tag{30}$$

The word W_{m+1} contains f_{m+1} letters, f_{m-1} letters S, and f_m letters L. Here f_m are the integer Fibonacci numbers defined by $f_{m+1} = f_{m-1} + f_m$, $f_1 = f_2 = 1$.

3.1.   Preview: Energy gauge in Fibonacci strings

As an introduction we take from [12] the example of two strings S, L which represent the same attractive delta-potentials combined with tunnels of length b and τb respectively and then pass to the string SL. The energy gauges for S and L both contain the same bound state. The string SL has two bound states but may have, depending on the tunnel length, two or one band germ. In the first case there exists a region of negative energy where the single strings admit band germs but the combined string SL does not. This case is schematically represented in Figure 7.

3.2. Substitutional systems and their invariants

The Fibonacci system to be considered can be seen as a particular case of a substitution system. We shall restrict our attention to substitutional systems generated by automorphisms of the free group. For these algebraic concepts in general we refer to [13] and follow in detail [10]. Consider the free group F_2 with generators (alphabet) a_1, a_2. Its elements are all the words in the generators and their inverses. An automorphism of F_2 is defined as a map

$$\rho : a_1, a_2 \rightarrow f_1(a_1, a_2), f_2(a_1, a_2) \tag{31}$$

which is *algebraically invertible* so that a_1, a_2 can be expressed in terms of f_1, f_2.

Example: Consider the substitution $a_1, a_2 \rightarrow f_1 = a_2, f_2 = a_1 a_2$. Its inverse is given by $a_1 = f_2 f_1^{-1}$, $a_2 = f_1$. This substitution describes in abstract terms the Fibonacci systems to be discussed below.

For any automorphism of F_2 we have the general theorem.

8 Prop (Nielsen 1918): for ρ an automorphism of F_2, the commutator $\mathcal{K}(a_1, a_2) := a_1 a_2 a_1^{-1} a_2^{-1}$ obeys

$$\rho(\mathcal{K}) = w \mathcal{K}^{\pm 1} w^{-1} \tag{32}$$

with $w \in F_2$.

Example: For the Fibonacci automorphism one finds

$$\rho(\mathcal{K}) = \mathcal{K}^{-1}. \tag{33}$$

From this abstract algebraic set-up we can pass to matrix systems by first mapping the generators a_1, a_2 to two elements of a matrix group like $SL(2, R)$ for the transfer matrices and then performing the automorphism ρ induced on the matrix group. For certain classes of such induced automorphisms, one can obtain the (half-) traces of the images directly from the (half-) traces of the generators. This question is studied in the theory of *trace maps* for which we refer to Peyrière [16]. The Fibonacci automorphism and related systems belong to this class [10]. The Nielsen theorem in connection with the induced automorphisms now provides one or more *invariants of induced automorphisms*: consider the induced commutator $\mathcal{K}$ on the matrix group $SL(2, R)$ and take its trace. Under the Fibonacci automorphism the commutator as a matrix is transformed into its inverse. Since all matrices are unimodular, we get on the matrix group

$$\text{tr}(\rho(\mathcal{K})) = \text{tr}(\mathcal{K}^{-1}) = \text{tr}(\mathcal{K}). \tag{34}$$

9 Prop: The trace of the commutator is an invariant under the Fibonacci substitution.

The commutator $K(g_1, g_2)$ in a matrix group is a measure of the non-commutativity. In particular if the two matrices commute, the half-trace equation (34) must be 1. Conversely if $\mathcal{K}$ equals the unit matrix, the elements g_1, g_2 commute. We shall come back to this property in later sections.

3.3.   Recursive calculation of the transfer matrix

To represent the Fibonacci system by delta-potentials and the states by transfer matrices we define

$$\tilde{M}(S) := \tilde{M}_1, \quad \tilde{M}(L) := \tilde{M}_2. \tag{35}$$

We represent the Fibonacci string W_{m+1} of equation (30) by the transfer matrix

$$\tilde{M}_{m+1} = \tilde{M}_{m-1}\tilde{M}_m. \tag{36}$$

The transfer matrix $\tilde{M}_2$ has the same analytic form as $\tilde{M}_1$ given in equation (7) with the same variable δ and the same energy E but with the replacement

$$\lambda_1 = \exp(\beta) \rightarrow \lambda_2 := \exp(q\beta). \tag{37}$$

Similar systems have been studied extensively in the literature, compare references given by Kohmoto [9] and in [3]. Many of these studies used a discrete version or examined the recursion numerically. In what follows we shall use the recursive method to explore the *analytic structure* of the Fibonacci strings as functions of the variables β, δ. We recall that $\tilde{M}_1, \tilde{M}_2$ equations (7, 35) are linear polynomials with respect to δ. From this property and from equation (36) we get immediately.

10 Prop: The matrix elements of $\tilde{M}_m$ are polynomials of degree f_m with respect to the parameter δ, with coefficients which are functions of β *via* the expressions λ_1, λ_2.

For the finite Fibonacci string we wish to study the matrix elements of the transfer matrix $\tilde{M}_i$ and in particular their combinations x_i, y_i equation (8), $d_i := (x_i - y_i)$, since they yield information on the eigenvalues and bound states. Two methods are available for the computation:

(i) We can use the full matrix recursion equation (36), with the starting matrices $\tilde{M}_1, \tilde{M}_2$.

(ii) We can use recursion techniques for the half-traces and the other matrix elements, as was proposed in [10]. From this reference we deduce the following recursion equations for the combinations of matrix elements equation (8):

$$\begin{bmatrix} a_{m+1} & b_{m+1} \\ c_{m+1} & d_{m+1} \end{bmatrix} = \begin{bmatrix} a_m + d_m & 0 \\ 0 & a_m + d_m \end{bmatrix} \begin{bmatrix} a_{m-1} & b_{m-1} \\ c_{m-1} & d_{m-1} \end{bmatrix} \tag{38}$$

$$- \begin{bmatrix} d_{m-2} & -b_{m-2} \\ -c_{m-2} & a_{m-2} \end{bmatrix},$$

$$x_{m+1} = 2x_m x_{m-1} - x_{m-2},$$

$$y_{m+1} = 2x_m y_{m-1} + y_{m-2},$$

$$d_{m+1} = x_{m+1} - y_{m+1}.$$

To start these recursive relations we need the matrices $\tilde{M}_1, \tilde{M}_2$ given from equations (7, 35) and compute $\tilde{M}_3$ from

$$\tilde{M}_3 = \tilde{M}_1 \tilde{M}_2 : \tag{39}$$

$$a_3 = \lambda_1^{-1}\lambda_2^{-1}\left(1+\frac{1}{2}\delta\right)^2 - \frac{1}{4}\lambda_1\lambda_2^{-1}\delta^2,$$

$$b_3 = \frac{1}{2}\lambda_1^{-1}\lambda_2\delta\left(1+\frac{1}{2}\delta\right) + \frac{1}{2}\lambda_1\lambda_2^{-1}\delta\left(1-\frac{1}{2}\delta\right)$$

$$c_3 = -\frac{1}{2}\lambda_1^{-1}\lambda_2^{-1}\delta\left(1+\frac{1}{2}\delta\right) - \frac{1}{2}\lambda_1\lambda_2^{-1}\delta\left(1-\frac{1}{2}\delta\right)$$

$$d_3 = \lambda_1\lambda_2\left(1-\frac{1}{2}\delta\right)^2 - \frac{1}{4}\lambda_1^{-1}\lambda_2\delta^2.$$

These matrix elements are polynomials of degree 2 in the variable δ as expected. The half-traces x_m form a recursive system by themselves. This is the well-known *Fibonacci trace map*.

11 Prop: By use of the recursive relations equation (38) we can construct algebraic polynomial expressions of degree f_m for the half-trace $x_m(\beta,\delta)$ and for the matrix element $d_m(\beta,\delta)$, which we call. the *band polynomial* and the *bound polynomial* respectively. The edges of the band germs for the Fibonacci string W_m are given by the f_m *roots of the band polynomial equations*

$$x_m(\beta,\delta) = \pm 1, \tag{40}$$

while the bound states are given as the f_m *roots of the bound polynomial equations*

$$d_m(\beta,\delta) = 0. \tag{41}$$

The f_m roots of the polynomials equation (40) are real since they represent special half-traces of real matrices. The f_m roots of the polynomial equation (41) are real for sufficient strength of the delta-potential since then we must get f_m bound states.

For low values of m the roots of the polynomials can be obtained in closed algebraic form.

In Figures 8–11 we use the roots of equations (40, 41) to plot for the first 6 Fibonacci strings the edges of band germs and the bound states as functions of β, negative vertical coordinate, and γ, horizontal coordinate. For each value of m, we compare the periodic case $q = 1$ with the quasiperiodic case $q = \tau$.

For the value $q = 1$ we are back at periodic cases and would obtain glued systems of $2, 3, 5, 8$ superband germs respectively in a way similar to Figures 5, 6. These superband germs encapsulate the bound states of the free clusters. For the Fibonacci case $q = \tau$ there appears a system of f_m band germs, for any value of γ separated by gaps. The bound states for higher values of γ are encapsulated within these band germs. For large fixed values of γ the band germs become narrow and form a triple of subsets. The latter feature may be related to the occurrence of isolated 2-atom clusters at the short distance b in the Fibonacci strings. These may be considered as examples of motive clusters discussed in Section 1. The locally symmetric or antisymmetric hybrid states of these clusters can be related to the lowest and highest subset.

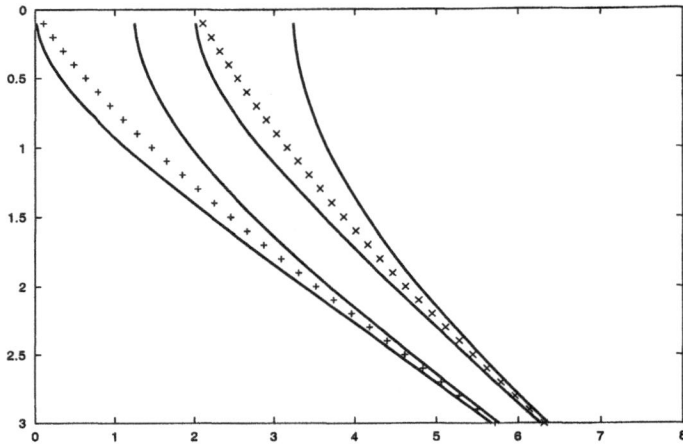

Figure 8

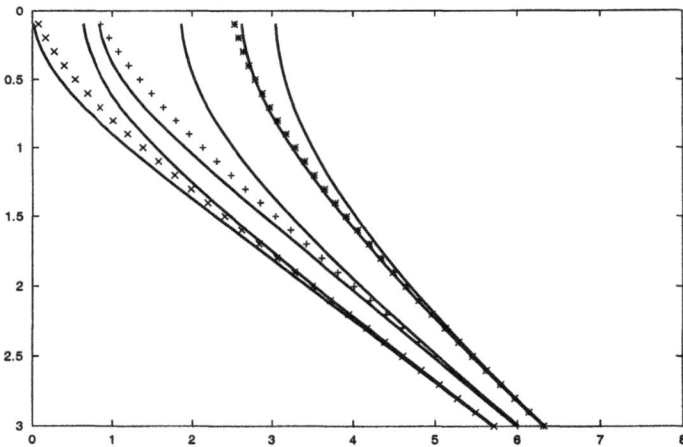

Figure 9

Figures 8–11 also demonstrate how the quasiperiodic system develops under the Fibonacci recursion or inflation. If one looks for fixed strength γ at the interchange of bands and gaps, one observes a permanent change of the density of states (DOS). In an approximant computation of this quasicrystal one would stop at a certain length and obtain the DOS from the band germ analysis of

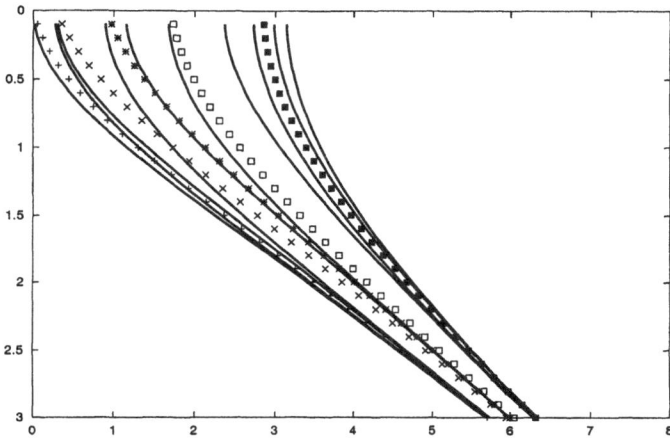

Figure 10

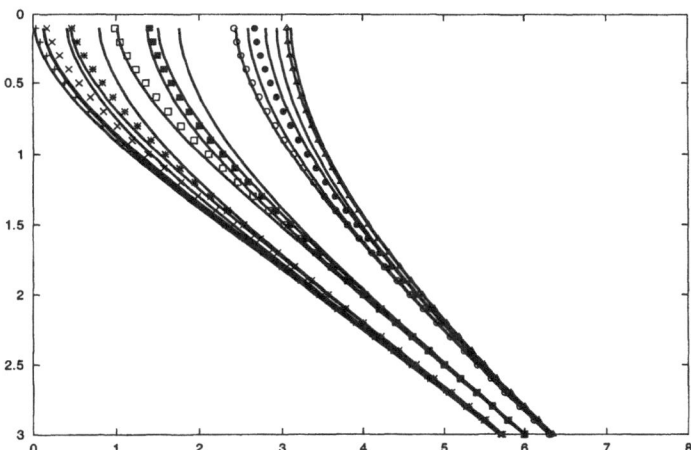

Figure 11

Figs. 8–11. — Band germs and bound or decaying states as functions of γ horizontal and β vertical for Fibonacci strings of 3-6 atoms. The bound states are not always inside the band germs. The band germs which were glued in the periodic string are now all separated.The number of band germs increases proportional to the number of atoms.

the corresponding string. In view of the present examples it is hard to believe
that this method describes the real DOS of the quasicrystal.

In a periodic potential we get a finite number of bands at least at negative
energy which can encompass a number of electron states proportional to the
number of atoms. In the Fibonacci strings notice that, in marked contrast to
the periodic case, when extending the Fibonacci string the number of band
germs which encapsulate bound states *increases proportional to the number
of delta-potentials, i.e. proportional to the number n of atoms*, each with one
electron, in the model. If this is so, then the DOS cannot be computed from
the gap and band germ structure of the string. As we have only n electrons, it
makes no sense to distribute them into a system of n bands. We conclude that
in approximant computations of the DOS one should keep track of the number
of bands in relation to the system size. Otherwise the conclusions on the DOS
in the proper quasiperiodic system are doubtful.

Now we turn to the significance of the conserved quantity related to the trace
of the commutator. This trace is related to the invariant I considered in [9]
and [3] by

$$\frac{1}{2}\text{tr}(\mathcal{K}) = 2I + 1. \tag{42}$$

If the commutator becomes the 2×2 unit matrix we get $I = 0$. In [12] we
computed the commutator for the present Fibonacci system with the result

$$\frac{1}{2}\text{tr}(\mathcal{K}) = 1 + \frac{1}{2}\left(\frac{u}{\kappa}\right)^2 (\sinh(\kappa(\tau - 1)b)^2. \tag{43}$$

This expression is always larger than 1. It implies that *in the negative energy
tight binding scenario the two transfer matrices never commute.* A different
result will appear for positive energy and scattering.

4. PERIODIC STRINGS AT POSITIVE ENERGY

We turn to electron states at positive energy. To this end we use the analytic
continuation from the variable κ to the variable k defined by

$$\kappa \to -ik. \tag{44}$$

For the energy we have now $E = \frac{\hbar^2}{2m}k^2 \geq 0$. For the variables β, δ we shall use

$$\beta \to -i\beta, \quad \delta \to i\delta. \tag{45}$$

So we maintain the symbol β as a real variable in what follows. We shall
plot the positive energy range $E = \frac{\hbar^2}{2mb^2}\beta^2$ at negative values of the new real
parameter β. Inserting the analytic continuation into the transfer matrix $\tilde{M}_1$
equation (7) we find

$$x_1 = \cos(\beta) - \frac{1}{2}\delta \sin(\beta), \tag{46}$$

$$y_1 = i\sin(\beta) + i\frac{1}{2}\delta\cos(\beta),$$

$$\lambda_1 = \exp(-i\beta) = \exp(-ikb).$$

Similar expressions apply for $\tilde{M}_2$.

4.1.   The S-matrix

The transfer matrix of a finite string can be rationally related to the scattering matrix S. For this purpose we shall use the transfer matrix in a form similar to equation (5) which corresponds to exponential functions. The analytic continuation equations (44, 45) is made both in $M(x)$ and with the matrix $R := \frac{1}{\sqrt{i}}\tilde{R}$ used instead of $\tilde{R}$. We redefine for positive energy

$$\tilde{M}(x) := R^{-1}M(x)R. \tag{47}$$

The negative energy tunnels with $\tilde{M}$ given by equation (6) become positive energy channels. The new fundamental system of solutions is

$$M(x)R = \sqrt{\frac{1}{2k}} \begin{bmatrix} \exp(ikx) & \exp(-ikx) \\ ik\exp(ikx) & -ik\exp(-ikx) \end{bmatrix} \tag{48}$$

and consists of free plane waves running to the left and right respectively. The transfer matrix in the channels becomes

$$\tilde{M}(x) := R^{-1}M(x)R \tag{49}$$

$$= \begin{bmatrix} \exp(ikx) & 0 \\ 0 & \exp(-ikx) \end{bmatrix}.$$

Consider now a finite string of length h with transfer matrix $\tilde{M}$ and with free channels on the left and right respectively. Denote the amplitudes on the left and right of the string by l_+, l_- and r_+, r_- respectively. These amplitudes are related by

$$\begin{bmatrix} r_+ \\ r_- \end{bmatrix} = \begin{bmatrix} a & b \\ c & d \end{bmatrix} \begin{bmatrix} l_+ \\ l_- \end{bmatrix}. \tag{50}$$

The elements of the scattering matrix are now determined by the ratio of amplitudes under certain boundary conditions and by phase factors which account for the length h of the string. For *scattering from the left* we obtain

$$r_- = cl_+ + dl_- = 0, \tag{51}$$

$$S_{++} = \frac{r_+}{l_+}\exp(-ikh)$$

$$= d^{-1}\exp(-ikh),$$

$$S_{-+} = \frac{l_-}{l_+}$$

$$= -d^{-1}c.$$

For *scattering from the right* we get

$$l_+ \;=\; dr_+ - br_- = 0, \tag{52}$$

$$S_{--} \;=\; \frac{l_-}{r_-}\exp(-ikh)$$

$$\;=\; -d^{-1}\exp(-ikh),$$

$$S_{+-} \;=\; \frac{r_+}{r_-}\exp(-2ikh)$$

$$\;=\; -d^{-1}b\exp(-2ikh).$$

Here S_{++} and S_{-+} are the amplitudes of *forward* and of *backward scattering* respectively. These expressions are given and discussed in [10] equations (50–54).

12 Prop: The scattering matrix S has elements which are rational expressions in the transfer matrix $\tilde{M}$. The full scattering matrix becomes

$$S \;:=\; \begin{bmatrix} S_{++} & S_{+-} \\ S_{-+} & S_{--} \end{bmatrix} \tag{53}$$

$$\;=\; \begin{bmatrix} d^{-1}\exp(-ikh) & d^{-1}b\exp(-2ikh) \\ -d^{-1}c & d^{-1}\exp(-ikh) \end{bmatrix}.$$

The exponential factors arise by requiring that for the free transfer matrix on the string of length h we must have $S = 1$. The expression for the S-matrix in terms of the transfer matrix is non-linear but rational. We must compute S from the transfer matrix of the full string, or otherwise evaluate all the backscattering effects from the components of the string. The unitarity of the S-matrix and its transformation under time reversal follow from properties of the transfer matrix [10].

Consider now the analytic continuation equation (44) in the inverse form

$$k \to i\kappa, \;\; \kappa \geq 0. \tag{54}$$

From inspection of equation (53) under this analytic continuation we find.

13 Prop: The poles of the scattering matrix equation (53) on the positive imaginary k-axis are given by $d(\kappa) = 0$. These poles determine the bound states of the system, fully in line with the analysis given in equation (41).

4.2. The S-matrix for the periodic string S^n

Again we consider a range of positive energy such that $|x_1| \leq 1$. This range may be called a band germ although it cannot correspond to bound states of the system. In this range we can find the eigenvalues and eigenstates of $\tilde{M}_1$ by analytic continuation of the expressions equations (8, 9). For the periodic string S^n with the transfer matrix $\tilde{M}_1^n$ we find the same expressions as in equation (23), but with y_1 and λ_1 now taken from $\tilde{M}_1$ in equation (46). We shall again introduce a band label K by

$$\cos(Kb) = x_1 = \cos(\beta) - \frac{1}{2}\delta\sin(\beta). \tag{55}$$

Consider first the backward scattering determined from equation (53) by c_n and d_n. The matrix element c_n oscillates rapidly with large n. We obtain *maximum backscattering* at the discrete values of the K-label

$$\begin{aligned}
Kb &= \mu\pi, \mu = 0, \pm 1, & (56) \\
c_n &= -i\frac{1}{2}\lambda_1^{-1}(-1)^{\mu(n-1)}n\delta, \\
d_n &= (-1)^{\mu n}(1 - ny_1(-1)^{\mu}), \\
\exp(-ikh) &= \exp(-inkb).
\end{aligned}$$

From

$$\cos(\mu\pi) = (-1)^{\mu}, \tag{57}$$

these maxima correspond to the *edges of the positive energy band germs* and become sharper with increasing n.

A *high-energy limit* is obtained for $\beta = kb \gg 1$. With $\delta = \gamma/\beta$, equation (55) enforces

$$\beta = kb = Kb + \nu 2\pi + \epsilon. \tag{58}$$

Together with equation (56), this equation yields *discrete values of k on the points of the reciprocal lattice*.

14 Prop: For large n and sufficiently high energy, the backscattering amplitude takes its non-vanishing values on the points of the reciprocal lattice.

To lowest order in ϵ we find

$$\begin{aligned}
x_1 &= (-1)^{\mu}, & (59) \\
y_1 &= i(-1)^{\mu}\frac{1}{2}\delta, \\
d_n &= (-1)^{\mu n}\left(1 - i\frac{1}{2}n\delta\right).
\end{aligned}$$

In the high-energy limit we then get from equation (53) at the band edges

$$\begin{aligned}
S_{++} &= \frac{1}{1 - i\frac{1}{2}n\delta}, & (60) \\
S_{-+} &= \frac{i\frac{1}{2}n\delta}{1 - i\frac{1}{2}n\delta}.
\end{aligned}$$

The limits $n \to \infty$ and of large energy should be distinguished from one another. In the limit $n \to \infty$ the absolute value of the backward scattering amplitude approaches its maximum $|S_{-+}| = 1$ allowed by unitarity, while the forward scattering amplitude goes to zero.

5. QUASIPERIODIC STRINGS AT POSITIVE ENERGY

The analytic continuation of the transfer matrix works for the Fibonacci strings. We simply must start the string with the analytic continuation of the starting matrices $\tilde{M}_1, \tilde{M}_2$ as given for $\tilde{M}_1$ in equation (46). Then we can apply the recursion technique of equation (38) to find an algebraic expression for the transfer matrix of the Fibonacci string W_m.

5.1. The Fibonacci-atlas

In principle we can construct the elements of the transfer matrix as polynomials of degree f_m with respect to δ for any order m. We obtain the corresponding $\mathcal{S}$-matrix from the relations equation (53). In general it will be hard to obtain in this way closed limiting expressions as was possible in the periodic case. Nevertheless there are particular regions where closed expressions can be derived. We illustrate these regions in Figure 12 by plotting the band germs of both transfer matrices $\tilde{M}_1, \tilde{M}_2$ for $q = \tau$ as functions of the variables β, γ in the range $\beta \geq 0, \gamma > 0, \gamma < 0$. This plot we call the *Fibonacci-atlas*.

Consider again the commutator $\mathcal{K}$ of the two transfer matrices. We must insert the analytic continuation to positive energy and obtain from equation (43)

$$\frac{1}{2}\mathrm{tr}(K) = 1 + \frac{1}{2}\left(\frac{u}{k}\right)^2 (\sin(k(\tau - 1)b)^2. \tag{61}$$

This expressions shows that there are periodic points where the half-trace becomes 1 and the invariant becomes $I = 0$. These points are given by

$$\beta(p) = \tau p\pi, \; p = 0, 1, \ldots \tag{62}$$

Moreover it was shown in [3] that the deviations $I - 1$ at these points starts quadratically with $(\beta - \beta(p))$. The values $\beta(p)$ are marked by dotted horizontal lines in Figure 12. If a line of this type intersects, depending on γ, with overlapping band germs for both transfer matrices, we obtain from $\lambda_2 = (-1)^p \lambda_1$ their proportionality,

$$\tilde{M}_2 = (-1)^p \tilde{M}_1. \tag{63}$$

15 Prop: At the values equation (61), the two transfer matrices $\tilde{M}_1, \tilde{M}_2$ commute and are even proportional to one another.

More precisely we require three conditions at these special points: the half-traces of both transfer matrices should be smaller than 1, and the commutator should become the unit matrix. These conditions are controlled by the Fibonacci atlas Figure 12: consider a dotted horizontal line corresponding to commuting transfer matrices. For not too large values of γ, it runs inside overlaps of the two band germs which imply that both $|x_1| \leq 1, |x_2| \leq 1$. Outside this region, that is outside the points marked by circles in Figure 12, we still have commuting transfer matrices, but their individual half-traces are larger

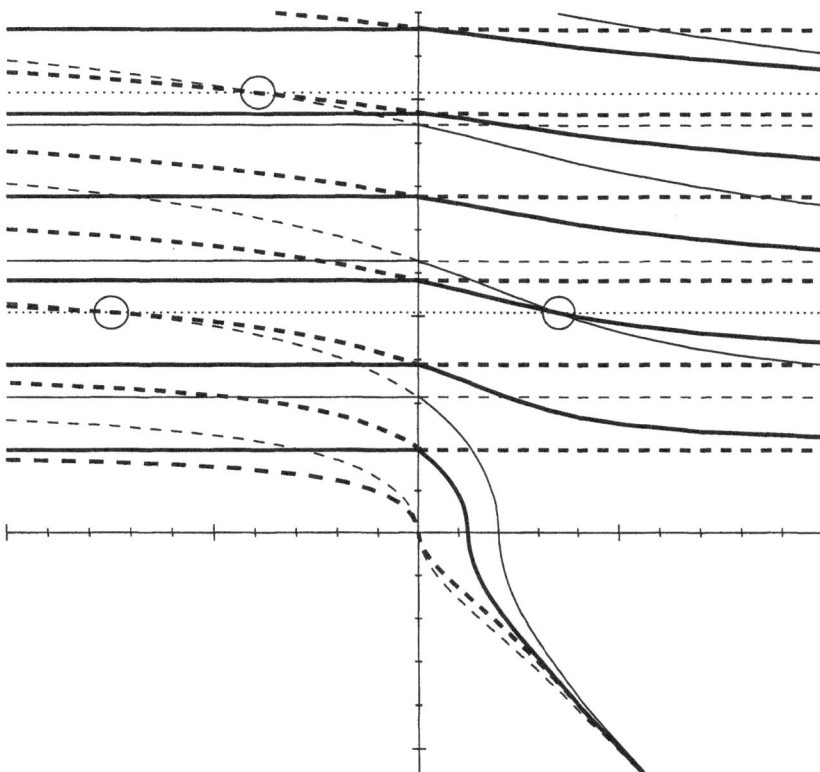

Fig. 12. — Fibonacci atlas: band germs of the transfer matrices $\tilde{M}_1, \tilde{M}_2$ for positive and negative energy as functions of the energy variable β and the strength variable γ. Upper band edges drawn as full lines, lower band edges as dashed lines, heavy lines for the matrix $\tilde{M}_2$. The two weak horizontal dotted lines mark values $\beta(p)$ where the two transfer matrices commute and give rise to extended states.

than 1. It follows from algebraic properties of the group $SU(1,1)$ given in [10] that a vanishing commutator implies group elements of the same class type.

We can now see the advantage of the present algebraic and polynomial approach compared to numerical studies of the trace systems: for example in [3] the points equation (62) were explored numerically and for fixed values of a strength $\gamma < 0$, corresponding to repulsive delta-potentials. The Fibonacci-atlas of Figure 12 now yields from closed algebraic expressions a full view on the regions of commutativity for all positive and negative values of γ and positive energy.

In a neighbourhood of such a commutative point we get, using the commutativity, for the full Fibonacci string W_m the transfer matrix

$$\tilde{M}_m = (-1)^{p f_{m-1}} \tilde{M}_1^{f_m}. \tag{64}$$

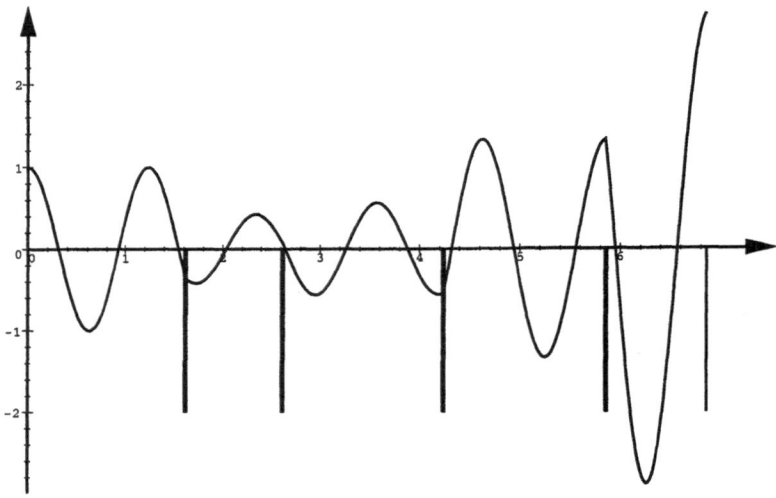

Fig. 13. — Electron state propagating through a Fibonacci string of attractive delta-potentials at a positive energy where the transfer matrices commute.

This transfer matrix agrees up to a factor with the transfer matrix of a periodic string of f_m delta-potentials with the spacing b. It follows that the S-matrix of the Fibonacci strings near these points has the form and limiting values discussed in Section 4.2.

16 Prop: In the neighbourhood of the discrete positive energies corresponding to equation (62), the transfer and S-matrix of the Fibonacci system are equivalent to those of a periodic string of length $f_m b$.

Some care is required because $\beta(p)$ in equation (62) is not rational whereas for the periodic case we used the discrete labels Kb and $\beta = kb$. Note however that the value $\beta(p)$ for $p = f_l$ and not too small values of l is well approximated by the integral multiple $\beta = f_{l+1}\pi$ of π.

Finally in Figure 13 we give the real wave function of an electron travelling at positive energy through a Fibonacci string of attractive delta potentials. The energy is tuned to a commutative value. At each passage through a delta potential, the derivative of the wave function jumps by a finite value.

6.   CONCLUSION

Our study of a very simple $1D$ electron system has demonstrated most of the points stated in general terms in the introduction, Section 1. The local boundary conditions were implemented and related. The negative and positive energy scenarios showed different behaviour. At negative energy, the bound states have a clear relation to the Bloch states. In going from periodic to quasiperiodic strings, the glued and gap-less systems of superbands open gaps and split into subbands which encapsulate bound states. Motive clusters appear in the bound state energy. With increasing length of the Fibonacci string, the subband structure changes in a non-trivial fashion, with a number of subbands proportional to the number of atoms, and so puts question marks to approximant calculations. At positive energy, the band edges in periodic strings are related to maxima of scattering amplitudes. For quasiperiodic Fibonacci strings we gave closed algebraic expressions. At positive energies related by a period wrt. k, the transfer matrices commute and give rise to maxima in the scattering amplitudes.

REFERENCES

[1] Andersen O.K., Jepsen O. and Sob M., Linearized Band Structure Methods, Springer Lecture Notes, edited by M. Yussouff (Springer, Berlin, 1987).

[2] Ashcroft N.W. and Mermin N.D., Solid State Physics (Saunders College, Philadelphia, 1976).

[3] Baake M., Joseph D. and Kramer P., *Phys. Lett. A* **168** (1992) 199-208.

[4] Blount E.I., Formalism of Band Theory, in: *Solid State Physics* **13**, edited by F. Seitz and D. Turnbull (Academic Press, New York, 1962) pp. 305-373.

[5] Cycon H.L., Froese R.G., Kirsch W. and Simon B., Schrödinger Operators (Springer, Berlin, 1987) pp. 197-216.

[6] Haerle R. and Kramer P., *Phys. Rev. B* **58** (1998) 716-720.

[7] Heine V., Electronic Structure from the Point of View of the Local Environment, in: *Solid State Physics* **35**, edited by F. Seitz and D. Turnbull (Academic Press, New York, 1980) pp. 1-127.

[8] Kohn W., *Phys. Rev.* **115** (1959) 332-344.

[9] Kohmoto M., *Int. J. Mod. Phys. B* **1** (1987) 31-49.

[10] Kramer P., *J. Phys. A* **26** (1993) 213-228.

[11] Kramer P., *J. Phys. A* **26** (1993) L245-L250.

[12] Kramer P., *J. Phys. A* **31** (1998) 743-756.

[13] Kramer P. and Garcia-Escudero J., Non-commutative Models for Quasicrystals, in: *Beyond Quasicrystals*, edited by F. Axel and D. Gratias (Springer and Les Éditions de Physique, Berlin and Les Ulis, 1995) pp. 55-73.

[14] Kramer P., Quandt A., Schlottmann M. and Schneider T., *Phys. Rev. B* **51** (1995) 8815-8829.

[15] Lieb E.H. and Mattis D.C., Mathematical Physics in One Dimension (Academic Press, New York, 1966).

[16] Peyrière J., Trace maps, in: *Beyond Quasicrystals*, edited by F. Axel and D. Gratias (Springer and Les Éditions de Physique, Berlin and Les Ulis, 1995) pp. 465-480.

[17] Skriver H.L., The LMTO Method: Muffin-Tin Orbitals and Electronic Structure (Springer Series in Solid State Physics, Berlin, 1984).

[18] Sütö A., Schrödinger difference equation with deterministic ergodic potentials, in: *Beyond Quasicrystals*, edited by F. Axel and D. Gratias (Springer and Les Éditions de Physique, Berlin and Les Ulis, 1995) pp. 483-549.

[19] Takeuchi S. and Fujiwara T., Proc. 6th Int. Conf. on Quasicrystals (World Scientific, Singapore, 1998).

[20] Wannier G.H., *Rev. Mod. Phys.* **34** (1962) 645-655.

[21] Wigner E.P. and Seitz F., *Phys. Rev.* **43** (1933) 804-810; **46** (1934) 509-524.

Course N° 5

Random Tiling Models for Quasicrystals

E. Cockayne*

Department of Applied Physics,
Yale University, P.O. Box 208284,
New Haven, Connecticut 06520-8284, U.S.A.

1. INTRODUCTION

Random tiling models for quasicrystals were introduced in 1985 [1] in the context of discussing the possible role of tiling entropy in stabilizing the then-recently-discovered quasicrystalline phase [2]. In fact, provided that certain conditions hold, entropic stabilization of a quasicrystalline tiling in three dimensions (3D) will produce a structure whose diffraction pattern contains Bragg peaks. An excellent and thorough overview of random tiling theory can be found in Henley's 1991 review article [3]. The aim of this lecture is to give a pedagogical review of the most important ideas and results in random tiling theory, including more recent developments.

This lecture is organized as follows: this Introduction presents the key concepts of random tiling models. Section 2 reviews the mathematics of quasicrystalline tilings and the random tiling hypotheses. Section 3 focuses on numerical and exact results for random tiling parameters. In Section 4, we present various random tiling models for real quasicrystal-forming materials. Finally, the physics of phase transformations involving random tilings are presented in Section 5 and the conclusions given in Section 6.

---

* *Current address*: Ceramics Division, Materials Science and Engineering Laboratory, National Institute of Standards and Technology, Gaithersburg, Maryland 20899-8520, U.S.A.

1.1. Basic definitions

Tiling: Grünbaum and Sheppard [4] give the following definition for a tiling: "a plane tiling T is a countable family of closed sets $T = \{T_1, T_2, \ldots\}$ which cover the plane without gaps or overlaps". In this lecture, we expand the definition of "tiling" to also include countable three-dimensional space-filling nonoverlapping closed sets.

Quasicrystal: We use the definition of Steinhardt and Ostlund [5]: "a quasicrystal is a quasiperiodic structure with a crystallographically disallowed orientational symmetry". This definition is not universally agreed upon; in particular, some authors view the inflation (rescaling or rescaling plus rotation) properties of quasicrystals to be of fundamental importance [6].

The two-dimensional quasicrystalline points groups include all groups with rotational axes of order n, $n \neq 1$, 2, 3, 4, or 6. In three dimensions, the quasicrystalline points groups are those axial point groups related to the two-dimensional quasicrystalline point groups, and, in addition, the icosahedral point groups. See the International Tables [7] for a complete list of the noncrystallographic point groups.

Random Tiling Ensemble: A random tiling ensemble is an ensemble of tilings formed by the same set of prototiles, (different orientations of each prototile are allowed). If the system of tiles is capable of forming a quasicrystal, then the random tiling ensemble is a "quasicrystalline random tiling ensemble". The term "random tiling" can be informally used to describe a "representative" member of a random tiling ensemble.

To make the concept of a random tiling ensemble more rigorous, we introduce the basic concepts of probability theory (see *e.g.* Ref. [8]). A "probability space" is a triple $(\Omega, \mathcal{F}, P)$, where Ω is a set, $\mathcal{F}$ is a σ field of subsets of Ω, and P is a probability measure on $\mathcal{F}$. A σ field is a collection of subsets of Ω that has $\emptyset$ as a member and is closed under complementation and countable unions. A probability measure assigns a value from the unit interval $[0, 1]$ to each member of σ such that $P(\emptyset) = 0$, $P(\Omega) = 1$, and $P(A \cup B) = P(A) + P(B)$ if $A \cap B = \emptyset$. For example, for the toss of a die, $\Omega = \{1, 2, 3, 4, 5, 6\}$, $\{2\}$ is a member of $\mathcal{F}$, and $P(1) = P(2) = \ldots = 1/6$ generates a legitimate probability measure.

A random tiling ensemble, then, is simply a probability space, where Ω a set of tilings that can be formed by a given set of prototiles, $\mathcal{F}$ is the field of all subsets of Ω, and P is a probability measure assigned to $\mathcal{F}$. One special type of random tiling ensemble is the "maximally random random tiling ensemble". In this case, P is such that all distinct individual tilings are equally probable. Note that definition of a random tiling ensemble says nothing about the nature of any process that may have been used to generate the member tilings. For a discussion of probability in *dynamical* systems, see Berthé [9].

(a)   (b)   (c)

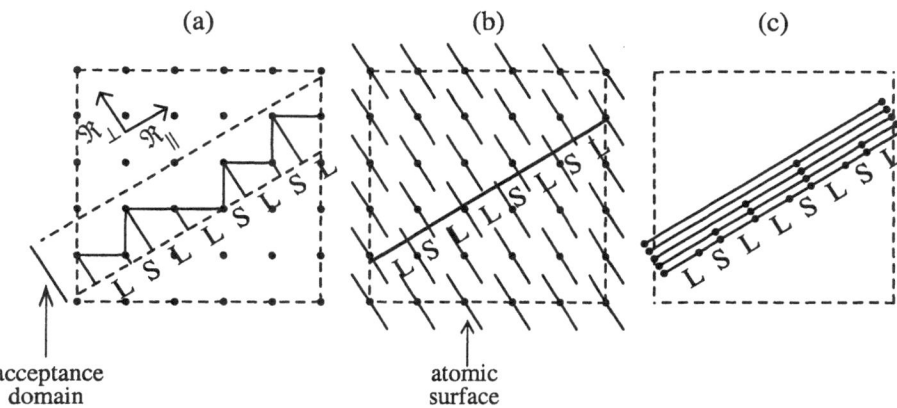

acceptance   atomic
domain   surface

Fig. 1. — Methods of generating quasiperiodic structures. (a) Projection method. The acceptance domain is a line segment. (b) Cut method. The atomic surfaces are line segments. The dots are not part of the atomic surfaces; they merely indicate the periodic arrangement of the atomic surfaces in hyperspace. (c) Inflation.

1.2. Generation of quasicrystalline tilings

There are various methods for generating quasicrystalline tilings. Examples include the projection method, the cut method, and inflation/deflation rules. We will illustrate each method for the example of the one-dimensional Fibonacci tiling and then give methods for generating quasicrystals in higher dimensions.

Figure 1a shows the projection method for generating the Fibonacci tiling. A strip of width $(\tau^2/\sqrt{1+\tau^2})a_0 \approx 1.3764\,a_0$ and irrational slope $1/\tau$ ($\tau \equiv (1+\sqrt{5})/2$) is cut through a two-dimensional square lattice of lattice parameter a_0; then all complete edges of the square lattice inside the strip are projected onto a line to form a chain of long (L) and short (S) line segments. The two-dimensional periodic space is separated into a "parallel" space $\mathcal{R}_\parallel$ in which the projected tiling lies, and a "perp" space $\mathcal{R}_\perp$ which is orthogonal to the parallel space. Since the slope of the cut is irrational, the projected tiling is aperiodic. The perp space cross section of the cutting strip is called the "acceptance domain". In this example, the acceptance domain is a line segment of length $(\tau^2/\sqrt{1+\tau^2})a_0 \approx 1.3764\,a_0$.

Similar to the projection method is the "cut" method shown in Figure 1b. Again a cut through a periodic structure in 2D yields a quasiperiodic tiling. By construction, each vertex of the tiling is formed when the parallel space intersects a periodic manifold in superspace. In cases such as this where the elementary domains of the vertex manifold are connected (or well approximated by connected domains), each connected domain is called an "atomic surface". The atomic surface in Figure 1b is in fact a line segment identical to the acceptance domain of Figure 1a[1].

---

[1] Technically, the atomic surface is the *inversion* of the acceptance domain. In

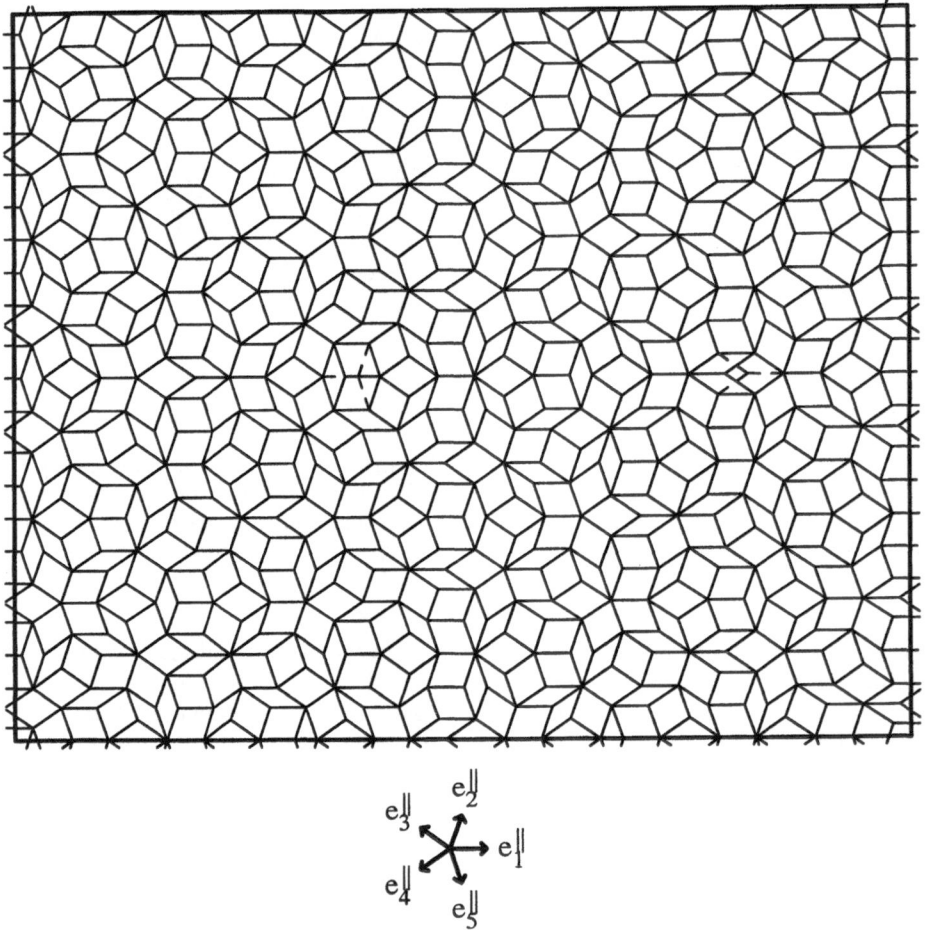

Fig. 2. — Unit cell of a periodic approximant to the Penrose tiling generated by the atomic surfaces in Figure 3. The projections of the lattice vectors of the 5D cubic hyperlattice into the 2D parallel space are shown at the bottom. Two possible phason flips are indicated by dashed lines.

   The cut method is more general than the projection method: (1) atomic surfaces need not be oriented strictly in perp space (as is the case for atomic surfaces that correspond to projection method structures); (2) the manifold that generates the vertices is not restricted to being a set of connected domains, or even well approximated by such a set. For more details on the mathematics of the cut method, see *e.g.* the article by Katz [10].

_____

the case of centrosymmetric acceptance domains, the two are identical.

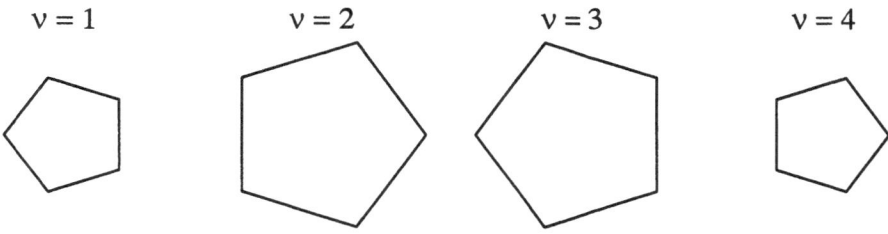

Fig. 3. — Atomic surfaces for the Penrose tiling.

Finally, we show how an inflation rule can generate the Fibonacci sequence. In Figure 1c, we have a series of lines broken into long and short tiles by the dots shown. The first line consists of a single long tile L, the partition of subsequent lines is governed by the following *inflation rule*:

$$
\begin{aligned}
L &\rightarrow LS \\
S &\rightarrow L.
\end{aligned}
\tag{1}
$$

In other words, each long tile is replaced by a long tile followed by a short tile in the next generation and each short tile is replaced by a long tile. The tiling obtained by repeated application of the inflation rules is identical to that shown in Figures 1a-b. The inflation technique is less general for creating quasicrystalline tilings than the other methods; however many important examples of quasicrystalline tilings do have inflation rules.

We now turn to similar techniques for generating quasicrystal tilings of higher dimensions. In a cut description with flat atomic surfaces, it is necessary and sufficient to specify (1) the hyperlattice (2) the orientation of the cut plane with respect to the hyperlattice, (3) the position(s) and (4) the shape(s) of the atomic surface(s).

We will illustrate this for the familiar Penrose tiling (Fig. 2). One choice of lattice is a 5D hypercubic lattice. The orientation of the cut plane is fixed by the 2D parallel and 3D perp components of the hyperlattice vectors $\mathbf{e}_i$:

$$
\mathbf{e}_i^{\parallel} = a_0 \big(\cos(2(i-1)\pi/5), \sin(2(i-1)\pi/5) \big);
\tag{2}
$$

$$
\mathbf{e}_i^{\perp} = a_0 \big(\cos(6(i-1)\pi/5), \sin(6(i-1)\pi/5) \big), 1/\sqrt{2} \big)
\tag{3}
$$

$1 \leq i \leq 5$. There is one atomic surface in the unit hypercell, centered on the node and shaped like a rhombic icosahedron [11].

There is already a subtlety. Note that the third components of the $\mathbf{e}_i^{\perp}$ vectors in (3) are all identical. Any node cut from an atomic surface centered on $\sum_i n_i \mathbf{e}_i$ has $\mathbf{r}_{\parallel} = \sum_i n_i \mathbf{e}_i^{\parallel}$. The position is indistinguishable from that of a

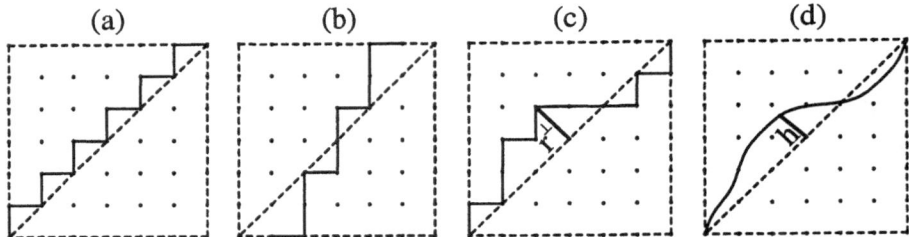

Fig. 4. — Fluctuating-surface picture of random tiling. (a) Zero phason strain. (b) Uniform global phason strain. (c) Representation of tiling with phason fluctuations. (d) Same as (c), after coarse graining. Phason coordinate $h(r)$ is differentiable, unlike $r_\perp(r)$.

node cut from the atomic surface at $\sum_i (n_i + 1)\mathbf{e}_i$. We define the *level* ν of a node in hyperspace or the projected structure to be $\nu \equiv (\sum_{i=1}^{5} n_i) \bmod 5$. Thus, without loss of generality, we can limit the hyperlattice to those nodes where $0 \leq \nu \leq 4$. The Penrose tiling (and any other decagonal tiling with the same set of edges and space group) can be described in terms of a set of five two-dimensional atomic surfaces, one for each level (see Fig. 3). The first two components of the perp space vectors (3) define the "phason space" and the third describes the "discrete perp space" [3].

1.3. Randomization of tilings

Elser showed in the case of the Penrose tiling, how a "flip" of a group of cells could create a new tiling [1]. Figure 2 shows the simplest flip moves for the 2D Penrose tiling (2DPT). By a series of flip moves, an ensemble of tilings can be created. If a flip move or set of flip moves exists that can generate all possible tilings of a given set of prototiles than the set of flip moves is "ergodic".

Using the projection representation, a random tiling member can be represented as a fluctuating surface in a higher dimensional space (Fig. 4). This led to the association of $\mathbf{r}_i^\perp$ with a phason variable $\mathbf{h}$ giving the deviation of the surface from the perfectly quasiperiodic line. By smoothing the surface appropriately, for example by convolution with a normalized Gaussian function $W(\mathbf{r}_\parallel)$,

$$\mathbf{r}^\perp(\mathbf{r}^\parallel) \rightarrow \mathbf{h}(\mathbf{r}^\parallel) = \int d\mathbf{r}'_\parallel \mathbf{r}^\perp(\mathbf{r}'_\parallel) W(\mathbf{r} - \mathbf{r}'_\parallel), \qquad (4)$$

one obtains a differentiable approximation to the true surface, indicated in Figure 4d[2].

The phason strain tensor $\mathbf{E}$, defined as $\nabla_{\mathbf{r}^{\parallel}}\mathbf{h}(\mathbf{r})$, is a quantity of key importance in random tiling theory. Phason strain is zero by definition for sufficiently uniform tilings in which the different possible orientations of each prototile occur equally often, as in the tiling represented by Figure 4a. As an example of nonzero phason strain, consider the tiling represented by Figure 4b. For each unit distance along $\mathbf{r}^{\parallel}$ in the tiling, the coordinate $\mathbf{h}$ of the associated fluctuating surface increases by $1/3$. The phason strain tensor for this tiling thus has a single component with value $1/3$.

In Section 2.1, we will look at the harmonic expansion of entropy density in coordinates of phason strain. Now, we wish to describe in more detail various tilings systems which are important both for modeling quasicrystals and in illustrating the geometrical differences between different tiling systems.

1.4.   A zoo of tiling models

For ten-fold, twelve-fold, eightfold, and icosahedral symmetry, simple polygonal (polyhedral) acceptance domains exist which generate tilings consisting of rhombi or rhombohedra. For each set of rhombi (rhombohedra), a variety of random tiling ensembles can be generated by imposing various geometric constraints. For example, as described in Section 1.2, the vertices of the decagonal quasicrystal random tiling system formed by fat and skinny Penrose tiles can each be labeled by their level ν. For the maximally random tiling system, all values of ν are possible.

Two important subensembles consist of those where only four values of level are allowed and those where only three values are allowed. For the four-level tilings, those vertices corresponding to levels 1 and 4 can be deleted leading to the "two-level" random tiling system (Fig. 5). The three-level tilings are also called binary tilings because large atoms can decorate level 0 vertices and small atoms can decorate the vertices at level $+1$ and -1 (Fig. 6)[3].

For rhombus tilings, an additional subensemble is that where all of the thin rhombi are grouped in pairs, as in Figure 7. Such tilings have been found for 8-fold [13], 10-fold [14, 15] and 12-fold [16] symmetries. In each case, when the middle shared vertex of each rhombus pair is deleted, a "compact tiling" is formed. We define a compact tiling here as a tiling of a set of prototiles for which it is impossible to obtain another tiling of the same set of prototiles by moving a single vertex. Figure 8 shows an example of an 8-fold compact

---

[2] In the case of a random Penrose tiling, a problem arises in the coarse graining procedure because the discrete perp space coordinate ν is only defined modulo 5. A solution to this problem that applies equally well to other decagonal tilings is to replace ν with the complex variable $e^{2\pi i\nu/5}$ before performing the coarse-graining procedure.

[3] We use $\nu = -1, 0, 1$ here rather than $\nu = 0, 1, 4$ to make the symmetry more obvious.

E. Cockayne

Fig. 5. — Unit cell of a periodic approximant to the two-level tiling, related to the Penrose tiling via the removal of certain vertices. A possible "bowtie" phason flip is indicated.

tiling. Note that the decagonal tiling in Figure 7 (after removal of the shared vertices of rhombus pairs) is both compact and binary. The compact tilings have various special properties:

(1) They are solutions of quasiperiodic sphere packing problems [13, 14, 17].

(2) All known deterministic tilings lead to fractal [13, 18–20] acceptance domains (see Fig. 9).

(3) The "flip" for going from one tiling to another consists of a closed chain necessarily involving the movement of more than one vertex. This move was named a "zipper" by Oxborrow and Henley [21]. See Figure 8 for an example of a zipper in the compact octagonal tiling.

Among icosahedral tilings, the only known compact tilings are the canonical cell tilings [22]. Canonical cell tilings have property (3) above. It is strongly conjectured that they also have properties (1) and (2). Unlike the other random

Fig. 6. — Unit cell of a periodic approximant to the binary decagonal tiling. A possible phason flip is indicated in the center. Dark and light atoms indicate the large and small atoms, respectively, of a toy model for quasicrystal stability [12].

tiling systems considered in this work, the density of vertices in a canonical cell tiling is not fixed. The mathematics and geometry of the canonical cell tiling system are thus even more complex than for the 2D compact tilings.

2. MATHEMATICS OF RANDOM TILINGS

2.1. Entropy density and phason elastic constants

The entropy density σ of a random tiling ensemble is defined here as the log number of distinct tilings per node in the random tiling ensemble, in the thermodynamic limit:

$$\sigma = \lim_{N_n \to \infty} (\ln N_t)/N_n, \tag{5}$$

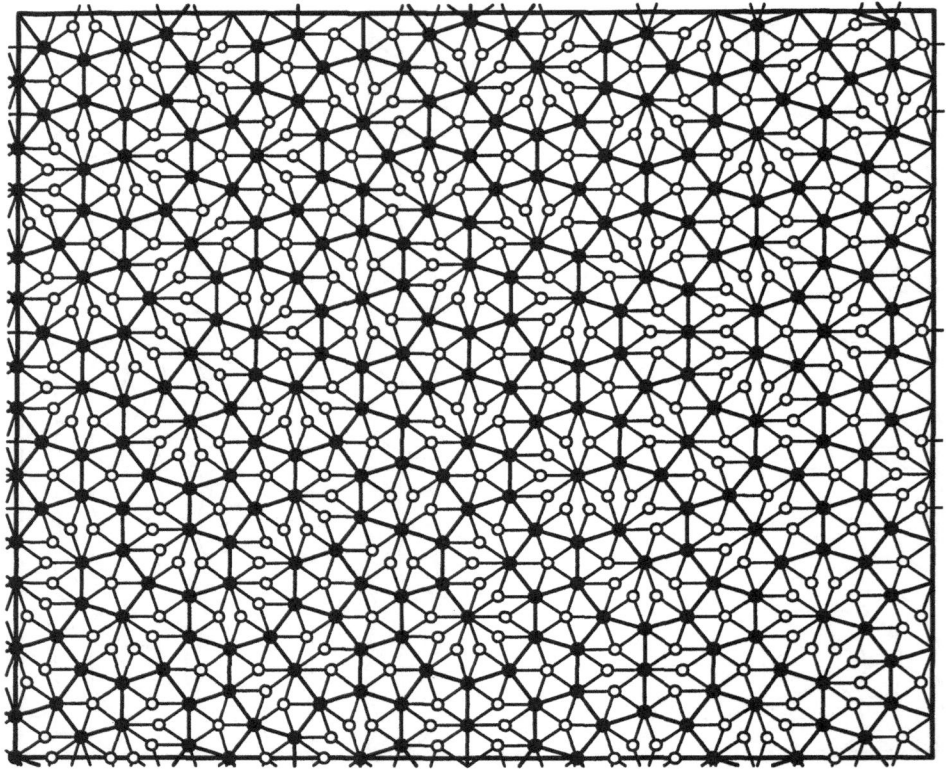

Fig. 7. — A member of the rectangle-triangle random tiling ensemble. Compare with Figure 6. The *zero-level network* is highlighted.

where N_t is the number of distinct tilings and N_n is the number of nodes. This definition is closely related to the definition of the topological entropy of a sequence [23] discussed by Berthé [9].

Several authors developed the following set of random tiling hypotheses which lead to the conclusion that tiling entropy can stabilize the quasicrystalline state [3]:

- A quasicrystal-forming system can be described as a tiling with two or more distinct tiles which each occur in different orientations.
- All such tilings nearly degenerate in energy.
- Entropy is maximized by a structure which has the highest symmetry allowed by the local ordering.
- Phason fluctuations are governed by gradient squared "elasticity".

The key hypotheses follow if (1) the expansion of the entropy density around phason strain $\mathbf{E} = 0$ is analytic, (2) the matrix of second-order coefficients

(a)   (b)

Fig. 8. — Octagonal hexagonal-square random tiling system. (a) Example of periodic approximant in this system. (b) Same as (a) after a single zipper update move.

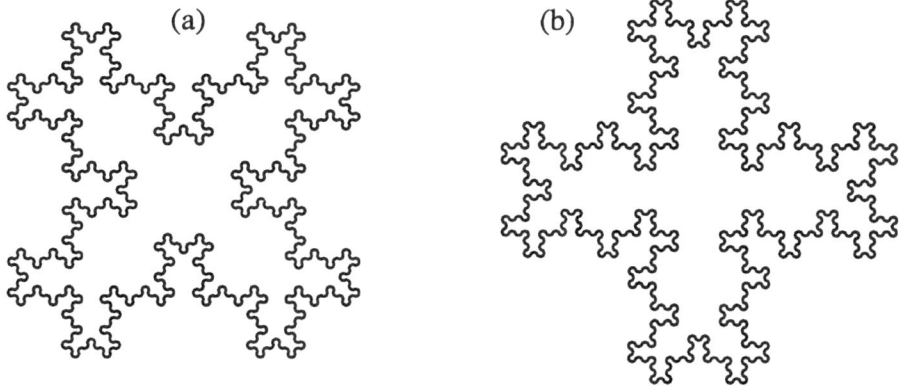

Fig. 9. — Fractal atomic surfaces for a deterministic octagonal square-hexagon tiling [13]. (a) Atom surface for even nodes. (b) Atomic surface for odd nodes.

is negative definite. Generally, the linear terms in the expansion vanish by symmetry; the harmonic form of the phason elasticity then simply reflects the lowest order terms in the analytic expansion of the entropy density. In analogy with ordinary elasticity theory, we write the general expression

$$\sigma = \sigma_0 - \frac{1}{2} \sum_{ijkl} K_{ijkl} E_{ij} E_{kl}. \qquad (6)$$

Note the following differences with ordinary phonon elastic theory:

(1) $K_{ijkl} \neq K_{jikl}$. Parallel and perp space coordinates can not be exchanged.
(2) There is no special significance to E_{ii}; some arbitrary choice determines the relative orientation of parallel and perp space.
(3) For K_{ijkl} that form a positive-definite matrix, the entropy density is a maximum, not a minimum at zero strain.

There are in principle 81 elastic constants for a quasicrystal with 2D parallel and perp spaces. Symmetry greatly reduces the number of independent phason elastic constants; there are typically only 2 or 3. Geometric restrictions on a tiling ensemble can further restrict the number of independent phason elastic constants. For example, the decagonal tiling systems such as the two-level and binary tiling systems, where the number of levels is limited, have no discrete perp space gradient, thus eliminating the corresponding phason elastic constant. In the case of 8-fold and 12-fold compact tiling systems, the phason field is irrotational [21], again reducing the number of phason elastic constants by one. Intriguingly, this is not the case for the 10-fold compact tiles, where there is no obvious constraint on the phason field. For the icosahedral symmetry groups, there is also a harmonic elastic term coupling phason strain and ordinary strain [24]. The phason elastic constants and the entropy at zero phason strain are system-specific parameters. However, many key features can be worked out without regard to the parameters of specific systems. This will be the subject of the following section.

2.2. Long-wavelength behavior and stability

Many results follow for a random tiling system with a maximum entropy density at zero phason stain and a harmonic form for phason elasticity. In this section, we calculate the form of the long-wavelength phason fluctuations and show that equilibrium phason fluctuations for the 2D quasicrystals are sufficient to destroy long-range order, while in 3D, phason fluctuations are bounded.

The arguments for the presence or lack of long range order in a quasicrystal as a function of dimensionality are similar to those in the Mermin-Wagner theorem for long-range order in the classical harmonic solid. Let us begin by taking the Fourier transform of the phason field: $\mathbf{h}(\mathbf{r}) \rightarrow \tilde{\mathbf{h}}(\mathbf{q})$. It can be shown that the relative weight of a given mode amplitude $\tilde{h}(\mathbf{q})$ in the partition function is $\exp(-\frac{1}{2}K|\tilde{h}(\mathbf{q})|^2 q^2)$, where K is the corresponding phason elastic constant [3, 21]. Thus, the partition function becomes

$$Z = Z_0 \prod_{\mathbf{q}} \left[\int d^2 \tilde{\mathbf{h}}(\mathbf{q}) \exp\left(-\frac{1}{2}K|\tilde{h}(\mathbf{q})|^2 q^2\right) \right]. \tag{7}$$

The mean square phason amplitude for a given wavevector $\mathbf{q}$, $\langle \tilde{h}^2(\mathbf{q}) \rangle$, can be determined by finding its expectation value in (7). One obtains

$$\langle \tilde{h}^2(\mathbf{q}) \rangle \sim \frac{1}{K^2 q^4}. \tag{8}$$

The root mean square (*rms*) fluctuations thus go as $1/q^2$, exactly as in the case of phonons in a classical harmonic crystal at finite temperature. The total *rms* fluctuation between points separated by large $\mathbf{r}$ varies as

$$\int d^d q \ (1/q^2) \sim \int d|q| \ q^{d-3}. \tag{9}$$

The expression (9) diverges at $\mathbf{q} = 0$ for $d < 3$; thus there is no long range order for a random tiling quasicrystal in two dimensions. To show that there is long range order in 3D, note that there an effective large $\mathbf{q}$ cutoff to the integral in (9) that is approximately the reciprocal of the typical tile edge length scale; thus the integral converges and the long- range phason fluctuations are bounded.

2.3.   Diffraction

The diffraction pattern intensities of a quasicrystal are given by the Fourier transform of the hyperspace Patterson function [25]. The Bragg peaks are labeled by hyperspace reciprocal lattice vector $\mathbf{G}$ and are located in the physical diffraction pattern at position $\mathbf{G}_\parallel$, where $\mathbf{G}_\parallel$ is the projection of $\mathbf{G}$ onto parallel space. Since the set of diffraction peaks of a quasicrystal is dense [26], diffuse scattering can not be distinguished from the Bragg peaks based solely by location in reciprocal space, in contrast to ordinary crystallography, where the distinction between the Bragg peaks and the diffuse scattering is clear. For real quasicrystals, however, there is a clear distinction between Bragg peaks and diffuse scattering: the Fourier transform of the Patterson function of the *average* structure gives the Bragg peak intensities and the Fourier transform of the singular part of the Patterson function due to short-range correlations gives the diffuse scattering intensity distribution. Note that the same definition holds for ordinary crystals.

Assuming a harmonic form of phason elasticity leads to specific predictions for the form of diffraction in a quasicrystal [3]. Using arguments similar to those in Section 2.2, Henley showed that, in three dimensions, the intensity near a quasicrystal Bragg peak at $\mathbf{G}_\parallel$ is given by

$$I(\mathbf{G}_\parallel + \mathbf{q}) = I_0 e^{-w G_\perp^2} \left[f(\mathbf{G})\delta^3(\mathbf{q}) + \frac{G_\parallel^2}{2K q^2} \right]. \tag{10}$$

This expression is valid for small $\mathbf{G}_\perp$, where there is a "phason Debye Waller factor" w that, roughly speaking, reduces the intensity of the Bragg peaks relative to the intensities of the same peaks in the corresponding quasicrystal

having no phason fluctuations. Furthermore, around each Bragg peak, there
are $1/q^2$ diffuse scattering wings.

In two dimensions, the intensity near a diffraction peak $\mathbf{G}_\parallel$ is given by

$$I(\mathbf{G}_\parallel + \mathbf{q}) \sim q^{-\eta_\mathbf{G}}; \eta_\mathbf{G} = \frac{|G_\perp|^2}{2\pi K}. \tag{11}$$

Consistent with the argument that there is no long-range order in a truly random 2D quasicrystalline tiling, there are no Bragg peaks in the corresponding diffraction pattern. Indeed, the peak intensity at $\mathbf{G}_\parallel$ varies as a power law $L^{2-\eta_\mathbf{G}}$ in the system size L [3].

Given (10), it should be straightforward in principle to recognize a random tiling icosahedral quasicrystal by performing the appropriate diffuse scattering experiments. Boudard *et al.* measured the shape anisotropy of the diffuse scattering near Bragg peaks in icosahedral Al-Pd-Mn [27], where they determine a ratio for two phason elastic constants. However, this interpretation is not certain [28, 29]. Other sources of entropy and possibly temperature-dependent structural phase transformations can also contribute to the diffuse scattering. Very careful experiments will be necessary to definitively determine if any existing quasicrystal has the diffraction properties expected of a quasicrystal stabilized by random tiling entropy.

3.   RANDOM TILING RESULTS

A variety of techniques have been applied to obtain results for the random tiling entropy and phason elastic constants of various random tiling systems. This section will review simulation methods, combinatorial methods, transfer-matrix methods, and finally the Bethe Ansatz method, which has led to exact solutions of several compact tiling systems.

3.1.   Monte Carlo simulation

To solve a random tiling system by simulation, one finds a flip or zipper move which is sufficient for ergodicity and then repeatedly applies this in a Monte Carlo simulation. The advantage of the Monte Carlo method is that it can be very easy to set up the code for the flips and the evaluation routines. The disadvantage is that long simulation times are required to get good statistics. Phason elasticity can be found by measuring phason correlations either directly in the simulation or indirectly through the simulated diffraction pattern [3]. It is more difficult to measure the entropy; that requires imposing an artificial Hamiltonian and an artificial temperature in the simulation and then integrating the simulated specific temperature-dependent specific heat $C(T)$ [3]:

$$S(\infty) - S(0) = \int_0^\infty \frac{\mathrm{d}T}{T} C(T). \tag{12}$$

The difficulty lies in determining the "ground state" entropy S(0) and in the sensitivity of (12) to statistical error in the low-temperature regime. Monte Carlo methods have been applied to a variety of tilings where generally two-digit precision is obtained for the entropy per node and the phason elastic constants [3, 21].

3.2. Combinatorics

Combinatorics can be used to obtain an exact solution for the 1D (nonquasicrystalline) random tiling system. Consider a segment with k_1 A tiles and k_2 B tiles. The number of different tilings N_t is given by the binomial coefficient

$$N_t = \left(\begin{array}{c} k_1 + k_2 \\ k_1 \end{array} \right) = \frac{(k_1 + k_2)!}{k_1! k_2!}.$$

Using Stirling's formula $k! \sim \sqrt{2\pi k} e^{-k} k^k$, and taking the thermodynamic limit, the entropy density σ becomes $-p_1 \ln p_1 - p_2 \ln p_2$, where p_i is the relative frequency of tile i: $p_i = k_i / \sum_i (k_i)$. The entropy density is a maximum $\sigma_{\max} = \ln 2$ at $p_1 = p_2 = \frac{1}{2}$, where the symmetry is highest (see dashed lines in Fig. 4). Defining the phason strain $E \equiv p_2 - p_1$, the entropy density to quartic order takes on the form

$$\sigma = \ln 2 - \frac{1}{2} E^2 - \frac{1}{12} E^4 + \cdots \tag{13}$$

This system satisfies the random tiling hypotheses; furthermore we have an exact value for the phason elastic constant $K = 1$.

Because the 1D (nonquasicrystalline) random tiling system can be solved exactly by combinatorial formulas, work has been underway to find exact combinatoric expressions for quasicrystal tiling systems. An important step along this path is MacMahon's classic solution of the (nonquasicrystalline) 2D 6-fold 60° rhombus random tiling system with fixed boundary conditions [30] (see Fig. 10):

$$\sigma(p_1, p_2, p_3) = \frac{1}{2P} \left(\sum_{i=1}^{3} p_i^2 \ln p_i - (1 - p_i)^2 \ln(1 - p_i) \right), \tag{14}$$

where $P \equiv p_1 p_2 + p_2 p_3 + p_3 p_1$. This expression has a maximum value of $2 \ln(3/2) - (\ln 3)/2 \approx 0.2616$ at $p_1 = p_2 = p_3 = 1/3$. Once again, the maximum entropy is associated with the highest symmetry.

Mosseri *et al.* [31] have worked on the equivalent 3D problem of packing identical rhombohedra that can take on four different orientations. This problem is already much harder than the ones already discussed. Various inequalities have been established [32]. It is conjectured that S/N is a maximum at $p_i = 1/4$ and that this maximum value is $(44 \ln 2 - 27 \ln 3)/6 \approx 0.1393$. Destainville *et al.* [32] have obtained results for the entropy density of various patches with up to about 300 tiles; the entropy density tends to converge slowly in the direction of the conjectured value.

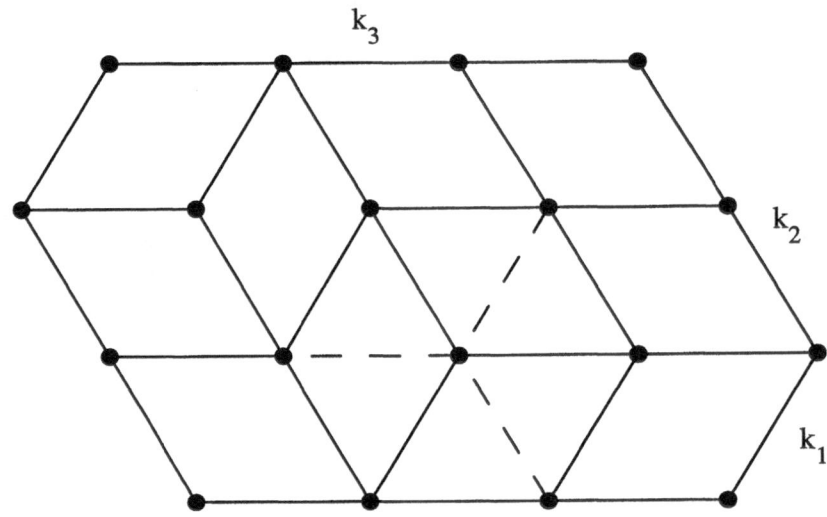

Fig. 10. — 60 degree rhombus random tiling system. A patch is shown with $k_1 = 1$; $k_2 = 2$; $k_3 = 3$. The dashed lines show the flip move for this system.

True quasicrystal combinatorics are even more difficult, but some progress has been made [33]. Note that the combinatorial method is distinct among the methods described in this section in relying on fixed boundary conditions rather than periodic boundary conditions. Joseph and Baake [34] found that the entropy density of a random tiling ensemble with fixed boundary conditions is less than that for the corresponding system with periodic boundary conditions[4].

3.3. Transfer matrix method

An approach that has proven more fruitful in getting very precise results for the random tiling properties is the transfer matrix method. In the transfer matrix method, a tiling is built from a patch that grows row by row. For simplicity, it is useful to have the patch terminate at a "dead surface" after each row is added. A "dead surface" is a surface for which no additional tile positions are forced. Each possible row is enumerated, as well as the number of ways to go from one type of row to another. In order to keep the number of rows finite, the cross section of the patch must effectively be finite, *e.g.* by having periodic boundary conditions.

The transfer matrix is then written, whose elements T_{ij} are the number of ways to go from row type i to row type j upon the addition of one row. Assuming that it is possible to eventually reach every row type from any other

---

[4] Except in 1D, where the two results are equal.

Fig. 11. — Dead surfaces for the $3 \times \infty$ domino random tiling (periodic boundary conditions in x direction). The growth direction is toward the top.

row type, the log of the dominant eigenvalue of the transfer matrix is equal to the entropy per dead surface. The limit as the size of the cross section approaches infinity is the entropy of the random tiling system per dead surface. Chemical potentials can be applied to tiles of different orientation [35] to obtain the change in entropy with phason strain, and thus the phason elastic constants.

Example 1: Random AB series. This example is trivial. Either A or B can follow either A or B; thus the transfer matrix is

$$\begin{pmatrix} 1 & 1 \\ 1 & 1 \end{pmatrix}.$$

The dominant eigenvalue is $\lambda_+ = 2$; thus the entropy per letter is $\sigma = \ln \lambda_+ = \ln 2$. This system was solved in Section 3.2 with fixed endpoint boundary conditions, with the same result for σ.

Example 2: Random domino packing in $3 \times \infty$ cell with periodic side boundary conditions. The random domino random tiling system is not a *quasicrystalline* random tiling system as per the definitions in Section 1.1 because it is not possible to generate noncrystallographic symmetry from dominoes packed edge to edge. Nonetheless, this simple yet nontrivial example illustrates the flavor of the transfer matrix method as applied to quasicrystalline random tilings. Four dead surfaces (*i.e.* distinct ways to terminate a layer of dominoes such that no additional domino positions are forced) for the $3 \times \infty$ domino tiling are shown in Figure 11[5].

The transfer matrix for this problem is

$$\begin{pmatrix} 1 & 1 & 1 & 1 \\ 1 & 1 & 0 & 0 \\ 1 & 0 & 1 & 0 \\ 1 & 0 & 0 & 1 \end{pmatrix}.$$

---

[5] There are three additional dead surfaces, where one vertical domino is added in one of three possible positions to dead surface 1. An alternate solution to the problem is to diagonalize the 7×7 transfer matrix representing *all* of the dead surfaces; it leads to the same result for entropy per node as the simpler solution presented here.

The dominant eigenvalue is $\lambda_+ = 1 + \sqrt{3}$; thus the entropy per dead surface is $\ln \lambda_+ = \ln(1 + \sqrt{3})$. The number of additional dominoes per dead surface ranges from 1 (dead surface 2, 3, or 4 to dead surface 1) to 3 (dead surface 1, 2, 3, or 4 to itself). The mean number of additional dominoes per dead surface, weighting layers according to the frequency of the parent dead surface in the dominant eigenvector, is $(6 - \sqrt{3})/2$ and there are 2 nodes per domino; thus the entropy per *node* is $\sigma = (\ln \lambda_+)/(6 - \sqrt{3}) \approx 0.2355$.

Transfer matrix methods have been applied to various models [35, 36], where generally three figure precision is obtained, an order of magnitude better than is generally obtained from simulation. One practical difficulty is that the enumeration of all possible rows grows exponentially in the system cross section, thus limiting the system size that can be studied. For more information on the mathematics of iterative matrix multiplication, see the article by Peyrière [37] and references therein.

3.4. Bethe Ansatz method

The Bethe Ansatz method can be applied in some cases to convert a tiling problem into an equivalent problem of the quantum mechanics of interacting particles. The world lines of the particles and their interaction statistics correspond to geometric features of the tilings and are different in different models. The many-body properties of the quantum states of the Bethe Ansatz model are then related to the tiling parameters of the random tiling model.

Widom first used this method, obtaining extremely precise results for the dodecagonal rectangle-triangle system [38]. Later, Kalugin solved this system analytically, obtaining exact expressions for the tiling entropy and phason elastic constants [39]. De Gier and Nienhuis then found a Bethe Ansatz solution for the octagonal rectangle triangle tiling entropy density and phason elastic constants [40, 41] and later a Bethe Ansatz solution for the entropy density of the decagonal rectangle-triangle system [42]. It is interesting that the Bethe Ansatz method has to date led to exact solutions only for compact tilings.

Results for entropy density σ of the exactly-solved quasicrystalline tiling models are shown in Table I. It is amusing that the argument of the final logarithm in each case is the inflation scale for the given symmetry. The entropy per node of the compact tilings is 0.26–0.33 times that of the corresponding unrestricted rhombus tilings [3], showing the magnitude of the restriction that rhombus pairing imposes.

4. ATOMIC MODELS FOR QUASICRYSTALS

This section concerns itself with recent models for real quasicrystal forming systems. In general, the models are broadly consistent with experimental diffraction studies and contain few if any unphysical interatomic distances. Most of the models allow a description in terms of symmetric clusters of atoms

Table I. — Exact solutions for entropy density of compact quasicrystal tiling models.

Symmetry	σ	Approx. value
8	$(2\sqrt{2} - 2)(\ln 4 - \sqrt{2}\ln(1 + \sqrt{2}))$	0.1159
10	$\frac{1}{2}(\ln(5^5/4^4) - 2\sqrt{5}\ln\tau)$	0.1750
12	$(\ln 108 - 2\sqrt{3}\ln(2 + \sqrt{3})$	0.1201

occupying the vertices of a tiling. See Figures 12–13 for examples of the clusters that will be described in this section.

Before considering a real quasicrystal-forming system, it is interesting to note that Widom *et al.* [12], as well as Lançon and Billard [50] showed that for a special set of Lennard-Jones potentials, a binary mixture of large (L) and small (S) atoms appears to have a decagonal binary random tiling ground state (*i.e.* the random tiling ensemble members form a set of degenerate ground states). The minima of the Lennard-Jones potentials that produce this ground state are at $r_{LS} = 1.0$, $r_{LL} = 2\sin(\pi/5) \approx 1.176$, and $r_{SS} = 2\sin(\pi/10) \approx 0.618$, and the potential well depths are in the proportion $E_{LS} : E_{LL} : E_{SS} = 1 : 1/2 : 1/2$. The potential radii are nonadditive, however it should be noted that real atoms do not have strictly additive radii either.

The decagonal models to follow are planar quasiperiodic and periodic in the third direction. A problem thus arises for entropic stabilization scenarios: namely the fact that entropy varies as L^2, while energy varies as L^3, where L is the system dimension [51]; thus if the ground state is crystalline, than the crystalline phase will be stable in the thermodynamic limit at any temperature. One scenario for preserving equilibrium entropic stabilizations is to also have phason fluctuations along the periodic direction; that is to break strict translational symmetry. More direct experimental evidence is needed to test this scenario.

4.1. HREM/diffraction-based

One method for devising random tiling models is to interpret high resolution electron microscope (HREM) images as tilings, then to decorate each prototile in the tiling based on what is known from diffraction experiments or related crystalline structures that have been solved. Numerous images of decagonal tilings support interpretation as hexagon-boat-star (Fig. 5) or hexagon-boat-star- decagon tilings, all with approximately the same edge length ≈ 6.5 Å or some power of τ^2 times this unit. These tilings are summarized in [52] and various decoration models for these tilings are given in [53]. The models show that within the same tiles, many different arrangements of atoms occur in different layers. By having different stacking sequences within the tiles, related decagonal phases can form structures with different repeat distances in the

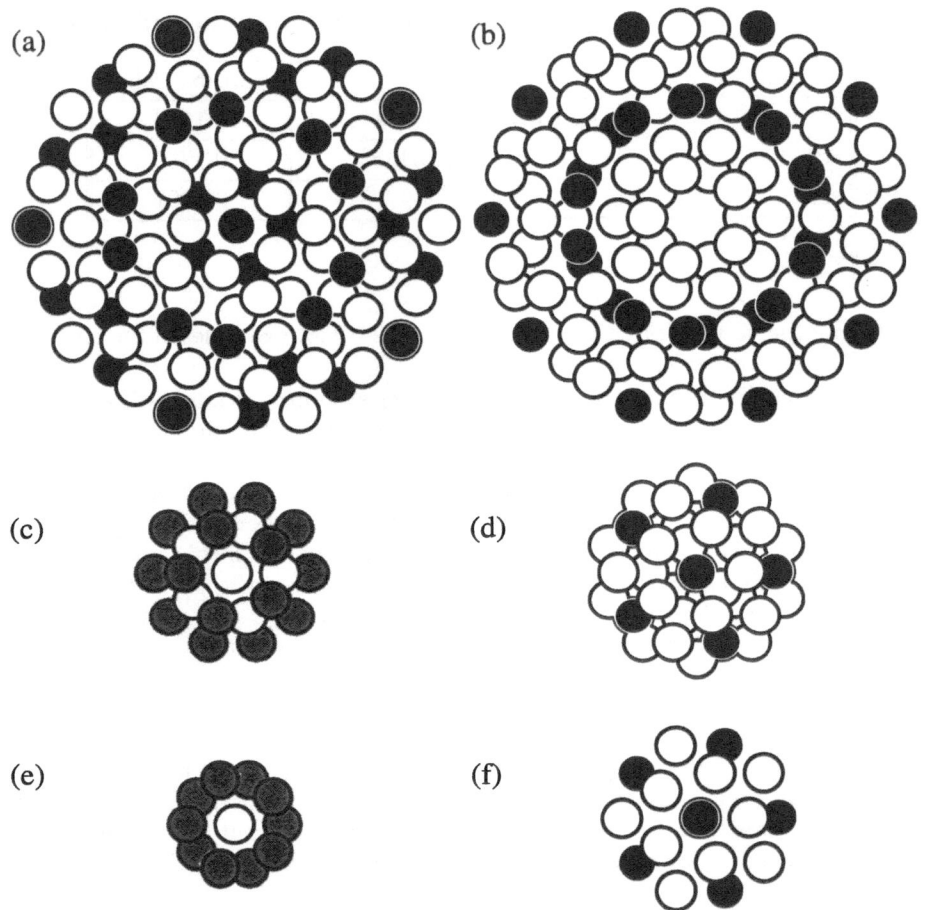

Fig. 12. — Examples of clusters used in modeling quasicrystals. (a) Krajčí *et al.* (d-AlPdMn) [43], (b) Steurer *et al.* (d-AlNiCo) [44], (c) Bergman cluster (i-AlCuFe) [45, 46], (d) Mackay icosahedron (i-AlMnSi) [47], (e) Columnar cluster (d-AlCuCo) [48], (f) Pentagonal bipyramid (d-AlCo) [49]. White atoms are Al, black transition metal and gray either Al or TM, depending on environment. See original references for full details.

periodic direction. For example, the AlNiCo model of Steurer *et al.* [44] has 4 Å periodicity and 2 layers per repeat distance; a set of three layers is shown in Figures 12–13b. Three layers are also shown for the Krajčí *et al.* model of AlPdMn [43], which has 6 layers per repeat distance (Figs. 12–13a). Note that the AlPdMn model contains layers almost identical to those in the AlNiCo model, interspersed with layers of a different kind, leading to a longer period in the AlPdMn model.

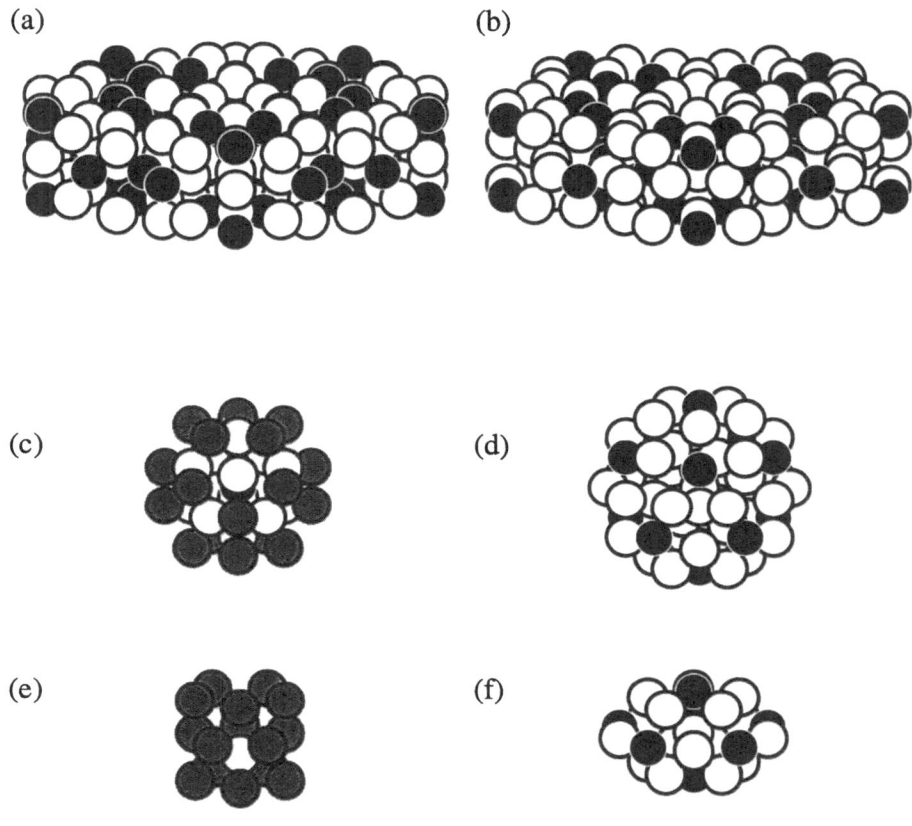

(a)   (b)

(c)   (d)

(e)   (f)

Fig. 13. — Side views of the clusters shown in Figure 12.

Consider now tiling models for d-Al-Ni-Co. HREM images of different samples exhibit a variety of tilings of edge length 20 Å, with visual indications of some degree of randomness [54]. The diffraction studies of Steurer et al. [44,55] are consistent with a symmetric cluster of diameter $\approx$ 20 Å located at positions corresponding to the vertices of the HREM tilings. Thus a model for Al-Ni-Co consists of placing the Steurer et al. clusters (Figs. 12–13b) on the vertices of a binary tiling (black dots in Fig. 6)[6].

---

[6] Additional atoms are required between the clusters to complete this model (and other models discussed here). These positions are generally well defined. See references [43–49,55,56] for full details.

In the case of AlPdMn, a further subtlety occurs. Again, a description at the level of 20 Å clusters is possible, but in this case, the tiling appears to be the zero-level network of a compact binary tiling (thick lines in Fig. 7). The author has investigated published HREM images of d-AlPdMn [57,58]. These images can be mapped onto binary tilings of edge length $\approx$ 17 Å, where the large atoms represent large white spots in the images and the small atoms represent small white spots. All small atoms occur either singly or as a part of sinuous chains connecting atoms $\approx$ 10 Å apart. The lengths of all small atom chains, in units of $\approx$ 10 Å segments, were counted. The results are shown in Figure 14. Remarkably, there are very few chains of odd length. There is thus a strong experimental signature for thin rhombus pairing, *i.e.*, a compact tiling description for d-AlPdMn. For comparison, the same enumeration for the (unrestricted) binary tiling simulation of Widom *et al.* [12] is shown in Figure 14c. In this case, there is no distinction between the frequencies of even length and odd length chains[7]. Clusters for an AlPdMn model will be discussed in Section 4.2.

4.2.   Models based on realistic interatomic forces

Next, we consider models based on electronic structure considerations, either at the "muffin tin" level or at the level of effective pair potentials. These kinds of models have the advantage that local structures which are energetically unfavorable, such as atoms too close together, are automatically excluded. Models based on the refinement of diffraction data, without additional constraints imposed by hand, often contain split atoms [55].

In 2D, Cockayne *et al.* have studied AlCo with 8 Å periodicity and Al-Cu-Co with 4 Å periodicity. The 8 Å model [49] is based on the pentagonal bipyramid (PB) cluster shown in Figures 12–13f. PB clusters decorate the vertices and the projected cluster axes form a hexagon-boat- star (HBS) tiling in the approximants studied (Fig. 5). Tiling flips are observed to occur in the simulations at a simulation temperature of 1000 K. Additional sources of entropy are found besides the random tiling entropy. The largest is Al disorder (although this could perhaps be related to tiling entropy for a tiling based on a smaller length scale). There is also disorder in the centering of the clusters along the z direction, leading to an additional, nonphason, random variable for this system.

The 4 Å model was first done as a binary simulation [59] and later mock ternary potentials were used to obtain a fully ternary model [48]. Like the 8 Å model, it is based on a HBS tiling; however the cluster is now a pentagonal antiprism column (Figs. 12–13e). The assignment of species to the clusters depends on local tiling geometry; however along each projected tile edge lies both Cu and Co. In fact the Cu-Co ordering leads to a description that is very

---

[7] Chains of length 10 are a special case because they frequently occur as closed rings, unlike the other chains observed.

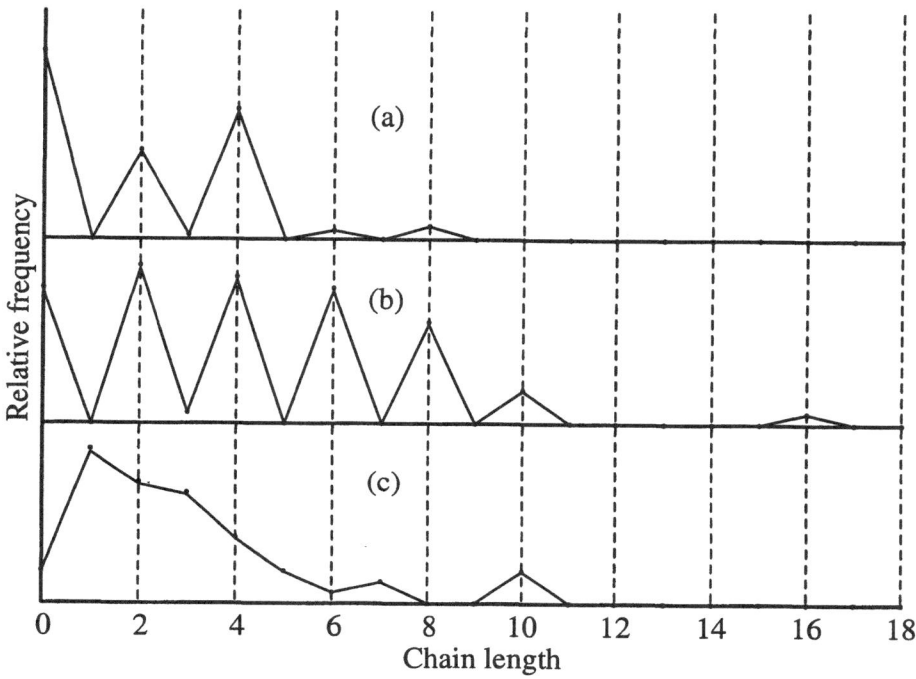

Fig. 14. — (a,b) Chain length statistics in Al-Pd-Mn. (a) From HREM image of Hiraga [57]. (b) From HREM images of Duneau and Audier [58]. (c) Chain length statistics in unrestricted binary simulation of Widom *et al.* [12].

much like the double arrows in the matching rules [60] for the original Penrose tiling (PT) (see Fig. 15). In the PT matching rules, arrows decorate each tile. If the arrow orientations on two tiles are required to match when the tiles join, then the PT is the only space-filling tiling that can be made. Most, but not all, of the arrow orientations on individual tiles in the simulations agree with those required for the PT matching rules.

Mihalkovič *et al.* did a pair potential simulation on a canonical-cells model of metastable AlMn [56]. It is based on a canonical cell tiling of edge length ≈ 12 Å. Mackay icosahedra (see Figs. 12–13d) decorate each vertex of the tiling. Various assignments of species to sites in the cells were tested to find the most favorable decorations. Relaxation preserved the tile decoration description. For the best model, there was little variation in energy as tiles rearrange; however numerical estimates of entropy differences *vs.* energy differences are not in favor of entropic stabilization of a quasicrystalline tiling. Still, it was noted that this is consistent with the observed metastability of the AlMn quasicrystal.

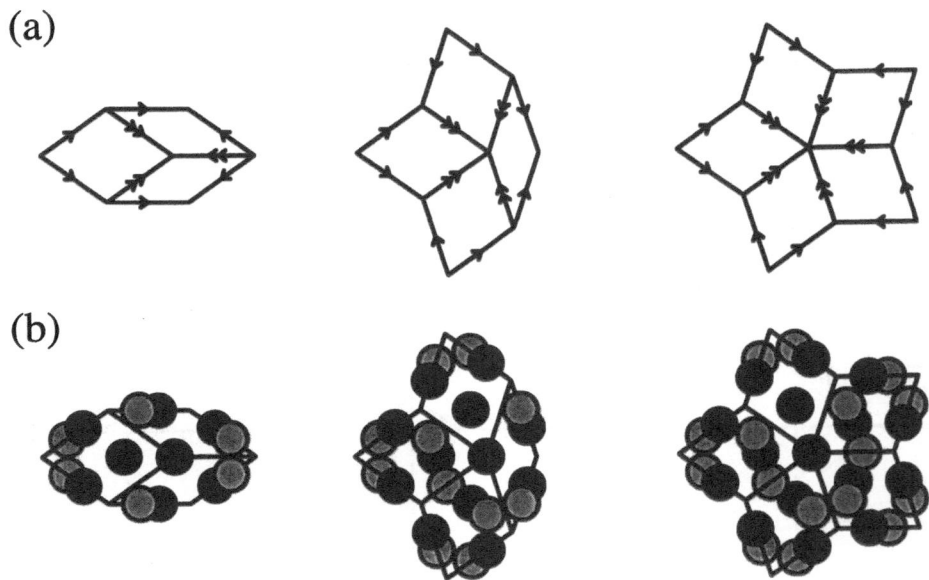

Fig. 15. — Relationship between matching rule arrows in Penrose tiling and Cu-Co
ordering in the Cockayne-Widom model for Al-Cu-Co [48].

Finally, we consider the AlPdMn decagonal phase, which Krajčí *et al.* studied
via muffin-tin electronic structure calculations. They found decorations of the
cell positions which are based on various clusters of pentagonal and decagonal
symmetry (see Figs. 12–13a). In contrast to the AlNiCo model, the tiling is
the rectangle-triangle tiling (Fig. 7).

4.3. Other models

In the icosahedral phase, interpreting HREM images as tilings is more difficult
because there is no periodic direction to exploit as a projection axis. Thus most
detailed icosahedral quasicrystal models are based on refinement of diffraction
results alone, which, at present, do not resolve unambiguously the appropriate
tiling model(s). Elser [46] exploited this uncertainty to produce a random
tiling model for AlCuFe based on a different tiling than previous icosahedral
models, namely a random three dimensional Penrose tiling (3DPT) of edge
length $\approx$ 7 Å. In contrast with the Mihalkovič *et al.* model [56], the nodes are
decorated both with Mackay icosahedra and Bergman clusters (Figs. 12–13c–
d). The 3DPT is bipartite; alternate vertices can be labeled "even" and "odd".
In the Elser model, there are Mackay icosahedra on the even vertices and
Bergman cluster on the odd vertices. Atomic packing considerations require
some rules for the atomic decoration in a few problem regions. In some regions,

there are more than one possible atomic decoration. The atomic displacements corresponding to the 3DPT flip move were worked out in detail.

5. QUASICRYSTAL PHASE TRANSFORMATIONS

This section deals with quasicrystal phase transformations. First we will consider models in which there are matching rules [10] favoring a quasicrystalline ground state and the phason unlocking transformation to a random tiling system that they undergo as the temperature is raised. Then we will consider systems in which the ground state is crystalline and the high temperature state quasicrystalline.

5.1. Phason unlocking

Consider the 2D and 3D Penrose tilings. In each case, there is a set of matching rules which force a quasicrystalline state. For a typical random tiling of the same tiles, there will be places where the matching rules are violated. Assigning an energy of 0.5 to each matching rule violation, the minimum excitation energy is 1 (a single flip from the ground state creates two mismatches). It was realized early on [61], that these model Hamiltonians would lead to a phason unlocking transition to a random tiling (stabilized by entropy) at a temperature T_c, with $T_c = 0$ for the 2DPT and $T_c > 0$ for the 3DPT.

By investigating the peak in an effective susceptibility for the 3DPT via Monte Carlo simulation, Dotera and Steinhardt [62] found that $T_c \approx 1.52$ for the 3DPT. Gähler [63] obtained a similar value for T_c, investigating the specific heat peak and changes in the autodiffusion rates of the 3DPT model as a function of temperature. The 2DPT was found in this simulation to undergo a phason-unlocking transition at $T_c = 0$, as expected.

5.2. Quasicrystal ↔ (micro)crystal

A different type of transition will occur if the ground state is crystalline. With certain dynamical constraints, the system will be expected to undergo a transition from a decagonal to a microcrystalline state consisting of various crystalline domains with different orientations related by symmetry. This state has been observed in AlCuCo [64]; a similar state has been observed in AlNiCo [65]. In the case of AlCuCo, the transition is reversible and possibly first order [66].

The Landau theory for the local crystal ↔ quasicrystal transition is rather simple. Let phason strain E be the order parameter. For a crystalline ground state, the free energy as a function of phason strain is given by

$$F \sim A - BE^2 + CE^4 - TV(\sigma_0 - KE^2). \tag{15}$$

The final term is the random tiling entropy term. With B and C both positive, the free energy has a double-well minimum and the quasicrystal ($E = 0$) is unstable. As temperature increases, the sign of the quadratic part eventually becomes positive, at which point the quasicrystal becomes lowest in free energy. For more general expansions of the free energy, one finds that the transition can be either first order or second order, depending on the sign of the fourth order coefficient in the expansion.

The author [67] has used a random tiling model to simulate a microcrystalline-quasicrystalline phase transition. In this model, the energy for a given tiling is set to $-N$, where N is the number of vertices that belong to the microcrystalline domains. In other words, energy in the model favors a state that is locally crystalline. The system was studied *via* Monte Carlo, with simple random tiling flips as the update move. At high simulation temperature, the simulated system is quasicrystalline. At low temperature, more than 65% of the vertices of the system belong to microcrystalline domains. The peak positions in the corresponding diffraction pattern agree very well with those shown by Fettweis *et al.* [66] for AlCuCo in its microcrystalline state.

6. CONCLUSIONS

This Lecture gave an overview of random tiling models for quasicrystals. It was shown that tilings in tiling systems that can form quasicrystals are related to fluctuating surfaces embedded in a higher-dimensional space, and that associated with these fluctuations is a phason variable. Simple hypotheses about the form of the entropy density as a function of phason strain lead to several conclusions; the most remarkable being that a three-dimension icosahedral random tiling system with all tilings equiprobable will form a quasicrystal with Bragg peaks in its diffraction pattern. Several random tiling models capable of forming a quasicrystal have been solved exactly; the entropy is indeed a maximum for the quasicrystal in all these cases. Simulations on other known quasicrystalline tiling models suggest that the random tiling hypotheses hold in these cases too. Models for real quasicrystals were presented, consisting of atomic clusters decorating the vertices of a tiling. High resolution electron micrograph images provide the most direct evidence for the existence of tiling fluctuations in real quasicrystal-forming systems. Finally, several types of phase transitions involving random tiling systems were discussed. One such transition, from a quasicrystalline to microcrystalline state, is known to occur reversibly in real quasicrystals.

Many open questions remain in the field of random tilings. For example, is there a proof that all quasicrystalline random tiling systems have an entropy density that is (to lowest order) harmonic in phason strain squared, or are there any counterexamples? Are exact solutions possible for those random tiling system for which no exact solution has yet been found? Can a more rigorous development of random tiling theory within the mathematical framework of probability spaces lead to any new insights or results? Perhaps the most important open question is what role tiling entropy plays in the stability of real quasicrystals. The near future holds promise for the development of realistic effective tiling Hamiltonians for such systems. Such models should allow quantitative determination of the role of energy *vs.* tiling entropy *vs.* other sources of entropy in real quasicrystals as a function of temperature.

REFERENCES

[1] Elser V., *Phys. Rev. Lett.* **54** (1985) 1730.

[2] Shechtman D., Blech I., Gratias D. and Cahn J.W., *Phys. Rev. Lett.* **53** (1984) 1951.

[3] Henley C.L., "Random Tiling Models", in Quasicrystals: The State of the Art, edited by D.P. DiVincenzo and P.J. Steinhardt (World Scientific, Singapore, 1991) p. 429.

[4] Grünbaum B. and Shephard G.C., Tilings and Patterns, edited by W.H. Freeman and Company (New York, 1986) p. 16.

[5] Steinhardt P.J. and Ostlund S., The Physics of Quasicrystals (World Scientific, Singapore, 1987) p. 12.

[6] Donnadieu P., Harmelin M., Su H.L., Seifert H.J., Effenberg G. and Aldinger F., *Z. Metallk.* **88** (1997) 33.

[7] Hahn T., International Union of Crystallography, International Tables for X-ray Crystallography (Reidel, Dordrecht, 1983) p. 774.

[8] Fristedt B. and Gray L., A Modern Approach to Probability Theory (Birkhäuser, Boston, 1997) p. 6-8.

[9] Berthé V., in Beyond Quasicrystals, edited by F. Axel and D. Gratias (Les Éditions de Physique, Les Ulis, France/Springer, Berlin, 1995) p. 441.

[10] Katz A., in Beyond Quasicrystals, edited by F. Axel and D. Gratias (Les Éditions de Physique, Les Ulis, France/Springer, Berlin, 1995) p. 141.

[11] Jarič M., *Phys. Rev. B* **34** (1986) 4685.

[12] Widom M., Strandburg K.J. and Swendsen R.H., *Phys. Rev. Lett.* **58** (1987) 706.

[13] Cockayne E., *J. Phys. A: Math. Gen.* **27** (1994) 6107.

[14] Mihalkovič M., in Proceedings of the International Conference on Aperiodic Crystals, Aperiodic 94, edited by G. Chapuis and W. Paciorek (World Scientific, Singapore, 1995) p. 552.

[15] Cockayne E., *Phys. Rev. B* **51** (1995) 14958.

[16] Stampfli P., *Helv. Phys. Acta* **59** (1986) 1260.

[17] Smith A.P., *J. Non-cryst. Solids* **153** (1993) 258.

[18] Gähler F., Doctoral dissertation (Swiss Federal Institute of Technology, 1988).

[19] Baake M., Klitzing R. and Schlottmann M., *Physica* **191A** (1992) 554.

[20] Janssen T., in Beyond Quasicrystals, edited by F. Axel and D. Gratias D. (Les Éditions de Physique, Les Ulis, France/Springer, Berlin, 1995) p. 128.

[21] Oxborrow M. and Henley C.L., *Phys. Rev. B* **48** (1993) 6966.

[22] Henley C.L., *Phys. Rev. B* **43** (1991) 993.

[23] Adler R.L., Kohneim A.G. and McAndrew M.H., *Trans. Amer. Math. Soc.* **114** (1965) 309.

[24] Kalugin P.A., Kitayev A.Yu., Levitov L.S., *J. Phys. Lett.* **46** (1985) L601.

[25] Van Smaalen S., *Phys. Rev. B* **39** (1989) 3850.

[26] Elser V., *Acta Cryst. A* **42** (1986) 36.

[27] Boudard M., de Boissieu M., Audier M., Kycia S., Goldman A.J., Hennion B., Bellissent R., Quilichini M. and Janot C., in Proceedings of the Fifth International Conference on Quasicrystals, edited by C. Janot and R. Mosseri (World Scientific, Singapore, 1995) p. 172.

[28] Capitan M.J., Bessiere M., Lefebvre S., Calvayrac Y., Quivy A. and Gratias D., in Proceedings of the Fifth International Conference on Quasicrystals, edited by C. Janot and R. Mosseri (World Scientific, Singapore, 1995) p. 652.

[29] de Boissieu M., personal communication.

[30] MacMahon P.A., Combinatory Analysis (The University Press, Cambridge, 1984) p. 1.

[31] Mosseri R., Bailly F. and Sire C., *J. Non-cryst. Solids* **153** (1993) 201.

[32] Destainville N., Mosseri R. and Bailly F., *J. Stat. Phys.* **87** (1997) 697.

[33] Widom M., Destainville N., Mosseri R. and Bailly F., in Quasicrystals, edited by S. Takeuchi and T. Fujiwara T. (World Scientific, Singapore, 1998) p. 83.

[34] Joseph D. and Baake M., *J. Phys. A: Math. Gen.* **29** (1996) 6709.

[35] Li W.X., Park H. and Widom M., *J. Stat. Phys.* **66** (1992) 1.

[36] Newman M.E.J. and Henley C.L., *Phys. Rev. B* **52** (1995) 6386.

[37] Peyrière J., in Beyond Quasicrystals, edited by F. Axel and D. Gratias (Les Éditions de Physique, Les Ulis, France/Springer, Berlin, 1995) p. 465.

[38] Widom M., *Phys. Rev. Lett.* **70** (1993) 2094.

[39] Kalugin P.A., *J. Phys. A: Math. Gen.* **27** (1994) 3599.

[40] de Gier J. and Nienhuis B., *Phys. Rev. Lett.* **76** (1996) 2918.

[41] de Gier J. and Nienhuis B., *J. Stat. Phys.* **87** (1997) 415.

[42] de Gier J. and Nienhuis B., *J. Phys. A: Math. Gen.* **31** (1998) 2141.

[43] Krajčí M., Hafner J. and Mihalkovič M., *Phys. Rev. B* **55** (1997) 843.

[44] Steurer W., Haibach T., Zhang B., Kek S. and Lück R., *Acta Cryst. B* **49** (1993) 661.

[45] Cockayne E., Phillips R., Kan X.B., Moss S.C., Robertson J.L., Ishimasa T. and Mori M., *J. Non-cryst. Solids* **153** (1993) 140.

[46] Elser V., *Philos. Mag. B* **73** (1996) 641.

[47] Elser V. and Henley C.L., *Phys. Rev. Lett.* **55** (1985) 2883.

[48] Cockayne E. and Widom, M., *Phys. Rev. Lett.* **81** (1998) 598.

[49] Cockayne E. and Widom M., *Philos. Mag. A* **77** (1998) 593.

[50] Lançon F. and Billard L., *J. Phys. France* **49** (1988) 249.

[51] Burkov S.E., *J. Stat. Phys.* **65** (1991) 395.

[52] Li X.Z., *Acta Cryst. B* **51** (1995) 265.

[53] Li X.Z, Frey F., Steurer W. and Kuo K.H., in Proceedings of the Fifth International Conference on Quasicrystals, edited by C. Janot and R. Mosseri (World Scientific, Singapore, 1995) p. 202.

[54] Ritsch S., Beeli C., Nissen H.-U. and Lück R., *Philos. Mag. A* **71** (1995) 671.

[55] Steurer W. and Kuo K.H., *Acta Cryst. B* **46** (1990) 703.

[56] Mihalkovič M., Zhu W.-J., Henley C.L. and Phillips R., *Phys. Rev. B* **53** (1996) 9021.

[57] Hiraga K., *J. Non-cryst. Solids* **153** (1993) 28.

[58] Duneau M. and Audier M., Figures 17 and 18 in Lectures in Quasicrystals, edited by F. Hippert and D. Gratias (Les Éditions de Physique, Les Ulis, France, 1994) p. 283.

[59] Cockayne E., Widom M., Launois P., Fettweis M. and Dénoyer F., in Proceedings of the International Conference on Aperiodic Crystals, Aperiodic 94, edited by G. Chapuis and W. Paciorek (World Scientific, Singapore, 1995) p. 578.

[60] de Bruijn N.G., *Math. Proc.* **A84** (1981) 39.

[61] Kalugin P., *JETP Lett.* **49** (1989) 467.

[62] Dotera T. and Steinhardt P.J., *Phys. Rev. Lett.* **72** (1994) 1670.

[63] Gähler F., in Proceedings of the Fifth International Conference on Quasicrystals, edited by C. Janot and R. Mosseri R. (World Scientific, Singapore, 1995) p. 236.

[64] Fettweis M., Launois P., Dénoyer F., Reich R. and Lambert M., *Phys. Rev. B.* **49** (1994) 15573.

[65] Kalning M., Kek S., Krane H.G., Dorna V., Press W. and Steurer W., *Phys. Rev. B.* **55** (1997) 187.

[66] Fettweis M., Launois P., Reich R., Wittmann R. and Dénoyer F., *Phys. Rev. B.* **51** (1995) 6700.

[67] Cockayne E., in preparation.

Model Sets: A Survey*

R.V. Moody

*Department of Mathematical Sciences, University of Alberta,
Edmonton, Alberta, T6G 2G1, Canada*

---

1. INTRODUCTION

Even when reduced to its simplest form, namely that of point sets in euclidean space, the phenomenon of genuine quasi-periodicity appears extraordinary. Although it seems unfruitful to try and define the concept precisely, the following properties may be considered as representative:

- discreteness;
- extensiveness;
- finiteness of local complexity;
- repetitivity;
- diffractivity;
- aperiodicity;
- existence of exotic symmetry (optional).

---

* Dedicated to the memory of Richard (Dick) Slansky The spirit of the universe is subtle and informs all life. Things live and die and change their forms, without knowing the root from which they come. Abundantly it multiplies; eternally it stands by itself. The greatest reaches of space do not leave its confines, and the smallest down of a bird in autumn awaits its power to assume form.

— Chuang Tzu (tr. Lin Yutang)

The purpose of this paper is to give an overview of the mathematics of the cut and project method which not only provides a very rich harvest of point sets (called model sets) satisfying these properties, but also provides a very natural way to link these ideas with many other structures in mathematics.

A subset Λ of $\mathrm{I\!R}^d$ is called a *Delone (Delaunay)* set if it is uniformly discrete and relatively dense. This means that there are radii $r, R > 0$ so that each ball of radius r (resp. R) contains at most (resp. at least) one point of Λ. Although this is a fairly strong version of the first two items on our list, it is the most commonly used one and coincides well with the primitive atomic picture of a (ideally infinite) piece of material.

The set Λ has *finite local complexity* if for each $r > 0$ there are, up to translation, only finitely many point sets (called *patches of radius* r) of the form $\Lambda \cap B_r(v)$. Here $B_r(v)$ is the ball of radius r about the point $v \in \mathrm{I\!R}^d$. So, on each scale, there are only finitely many different patterns of points. This condition can be expressed topologically: Λ has finite local complexity iff the closure of $\Lambda - \Lambda$ is discrete. It is conceivable to replace "translation" by "isometry" in this definition, but the theory would change considerably and, with the notable exception of the pinwheel tiling [33], little has been said so far on this more general situation.

Repetitivity means loosely that any finite patch that appears, appears infinitely often. More precisely, given any patch of radius r there is an R so that within each ball of radius R, no matter its position in $\mathrm{I\!R}^d$, there is at least one translate of this patch. A stronger form of this requires in addition that each type of patch of radius r should appear with a well-defined frequency. The sets that we deal with here normally have this additional property (see Sect. 3).

From the very beginning, diffractivity has been the hallmark of aperiodic order. Physically it is the most visible of its manifestations. Mathematically it is one of the most subtle and least visible! Very roughly we are asking (mathematically) that the Fourier transform of the autocorrelation density that arises by placing a delta peak on each point of Λ, should contain a part that looks discrete and point-like. Later on we will make this very vague prescription precise. The amazing thing is that in the context of model sets we obtain perfect diffractiveness, in the sense that the diffraction is purely point-like, under the fairly mild hypothesis of regularity. One of the goals of this survey is to show how this comes about.

Lack of periodicity speaks for itself. Lattices and unions of cosets of lattices are the basis of the most prevalent forms of long-range order (crystallography). Point sets based on them satisfy all the previous properties. But of course, that is the trivial part of the theory! The objective is to move into new territory.

Exotic symmetry usually means non-crystallographic symmetry. Although not mathematically essential, certainly the existence of physical structures with "forbidden" icosahedral symmetry was instrumental in the rapid development of this field.

Even with the rather strong interpretations on the various properties listed above, we still do not know how to characterize sets that satisfy them. For an extensive discussion of these problems see [26]. However there is one very general method of construction which relies on controlled projection from a discrete group located in some auxiliary "embedding" space. In its original form this so-called cut and project method is based on projection from lattices in higher dimensional spaces. Many people have written about this starting, in physics, with the work of Kramer [23] and including the very useful article of Meyer given in the previous edition of this School [28]. Meyer had already thought about sets formed by projection from the view point of harmonic analysis long before the discovery of quasi-crystals [27]. Even though it is convenient to somewhat rearrange the main components of his original construction, nonetheless he created a formalism which is ideal for creation of points sets with the desired properties of long-range aperiodic order. These are the *model sets*.

Some people object to the terminolgy "model set" prefering "cut and project set" which sounds more serious and professional. However, we prefer to interpret "model" as meaning exemplary and think that in terms of both of its priority and its greater generality the term deserves to be adopted.

The main purpose of this article is first to give some idea of the scope of the relevant examples that arise as model sets (this scope surely not yet fully realized) and then to show how the model sets are poised between a number of quite different areas of mathematics. It is the satisfying way in which they connect many diverse parts of mathematics that makes model sets so intriguing and offers to the imagination so many tantalizing prospects for future work. For the reader interested in more on the tiling side of quasiperiodicity we recommend the survey paper [1], which also provides a complementary source of some of the material presented here.

2. MODEL SETS

Let us launch ourselves directly into the notion of a model set. By definition, a cut and project scheme consists of a collection of spaces and mappings

$$\mathbb{R}^d \xleftarrow{\pi_1} \mathbb{R}^d \times G \xrightarrow{\pi_2} G$$
$$\cup$$
$$\tilde{L}$$
$$(1)$$

where $\mathbb{R}^d$ is a real euclidean space and G is some locally compact abelian group, π_1 and π_2 are the projection maps onto them, and $\tilde{L} \subset \mathbb{R}^d \times G$ is a lattice, *i.e.*, a discrete subgroup such that the quotient group $(\mathbb{R}^d \times G)/\tilde{L}$ is compact. We assume that $\pi_1|_{\tilde{L}}$ is injective and that $\pi_2(\tilde{L})$ is dense in G. We call $\mathbb{R}^d$ (resp. G) the *physical* (resp. *internal*) space. The product $\mathbb{R}^d \times G$ is the *embedding* space. We write $L = \pi_1(\tilde{L})$. It is very convenient to define the mapping

$$* : L \longrightarrow G \; : \mapsto \pi_2(\pi_1|_L^{-1}(x)). \tag{2}$$

Given any subset $W \subset G$, we define a corresponding set $\Lambda(W) \subset \mathbb{R}^d$ by

$$\Lambda(W) = \{\pi_1(x) \,|\, x \in \tilde{L},\, \pi_2(x) \in W\} = \{u \in L \,|\, u^* \in W\} \cdot \qquad (3)$$

We call such a set Λ (or more generally any translate of such a set) a *model set* (or *cut and project set*) if the following condition [W1] is fulfilled:

W1: W is nonempty and $W = \overline{\mathrm{int}(W)}$ is compact.

For some of the deeper results we need more precise assumptions, of which the following are the most relevant:

W2: The model set Λ is *generic* if the boundary ∂W of its window W satisfies $\partial W \cap \pi_2(\tilde{L}) = \emptyset$.

W3: The model set Λ is *regular*[1] if ∂W is of (Haar) measure 0.

The definition formalizes the notion of a point set in $\mathbb{R}^d$ constructed by projecting selected points from a lattice in some "super–space". The points selected for projection are those which fall into some bounded region when they are projected into the complementary internal space G. The notion of lattice, familiar in a real space as the $\mathbb{Z}$-span of a basis of that space, is replaced here by the more general definition that can be applied to any topological group. In condition [W1] the equality could be replaced by "$\subset$", but it is convenient to have this additional hypothesis since then $\overline{\Lambda^*} = W$.

We assume that the reader is familiar with the most common, and only easily visualized examples of this, that are based on a pair of orthogonal axes, at irrational slopes (*i.e.* axes, through the origin of the standard lattice $\mathbb{Z}^2$ in $\mathbb{R}^2$ which are taken to be the physical and internal spaces, and a window which is an interval on the internal space *e.g.* [1, 39]).

There are four different view points to the diagram above, which we can picture as follows. In the first place we have the physical space $\mathbb{R}^d$ and the point set Λ in it whose geometric properties are those that we wish to understand and describe. Lattices are discrete groups inside larger continuous groups, and so may be thought of as *arithmetic* in origin. We will see later how in many interesting cases the arithmetic aspect is quite central. Why the internal side should be thought of as having to do with *analysis* will also emerge in later, but an initial way to think of it is that on the internal side, the set of points of Λ appear in a totally different arrangement so that their closure is a very nice region of space. Finally, by definition, $T := (\mathbb{R}^d \times G)/\tilde{L}$ is a compact abelian group. In the usual situation of a real internal space, T is a torus, whence the notation. In any case, T has a totally natural action of $\mathbb{R}^d$ on it and it is this action that gives rise to a *dynamical system*. In the end we will have a second,

---

[1] The terminology here is not standardized. Sometimes what we call generic is called regular. Nonetheless, the generic situation is justifiably "generic" as we point out below. Our use of regular is close to the one used in [37, 38].

related, dynamical system which plays an important role in questions around diffraction.

$$\begin{array}{ccccc} & & \text{dynamical systems side} & & \\ & & T & & \\ & & \uparrow & & \\ \mathbb{R}^d & \longleftarrow & \mathbb{R}^d \times G & \longrightarrow & G \qquad (4)\\ \text{physical side} & & \uparrow & & \text{analytical side} \\ & & \tilde{L} & & \\ & & \text{arithmetic side} & & \end{array}$$

In fact this picture can be dualized, thereby producing yet another four pictures! This dualization plays quite an important role in Meyer's theory which we touch on only most briefly here. However we will use one part of the dual picture. In the dual picture it is $\hat{T}$ that is the lattice and we have

$$\begin{array}{ccccc} \widehat{\mathbb{R}^d} & \xleftarrow{\hat{\pi}_1} & \widehat{\mathbb{R}^d \times \widehat{G}} & \xrightarrow{\hat{\pi}_2} & \widehat{G} \\ & & \uparrow & & \\ & & \hat{T} & & \end{array} \qquad (5)$$

Here we have identified the dual of the direct product $\mathbb{R}^d \times G$ with the direct product of the duals, and have chosen to single out the canonical projections as the important maps, designating them by $\hat{\pi}_1$ and $\hat{\pi}_2$ respectively.

3. GEOMETRIC SIDE

The geometric properties of model sets $\Lambda = \Lambda(W)$ have been described in detail elsewhere (for instance [29,37]). We will restrict ourselves to pointing out a few of the most important features here. In the first place, model sets are Delone sets and have the property of finite local complexity. In fact they satisfy a very strong form of finite local complexity:

$$\Lambda - \Lambda \text{ is uniformly discrete.} \qquad (6)$$

A Delone set satisfying (6) is called a *Meyer set*. There are a remarkable number of ways describing Meyer sets [25,28,29] which link them strongly with harmonic analysis. Though the Meyer property is considerably weaker than that of a model set, we nonetheless have

Theorem 1 *[27] Any Meyer set is a Delone subset of some model set.*

The situation regarding repetitivity is complicated by the boundary of the window W. Λ is repetitive if it is generic. If we are allowed to modify a model set by moving its window around then it is straightforward using the fact that G is a Baire space to see that the window can be moved to make the resulting model set generic. A proof of this can be found in [9]. Furthermore, in the regular case, the frequency of repetition of each patch is well-defined in the sense that for each finite patch P the number of occurences of the patch P (up to translation) per unit of volume in the ball $B_r(0)$ of radius r approaches a positive limit as $r \to \infty$. This is actually not hard to prove once one has established uniformity of projection (Th. 2).

Lack of periodicity is automatic for model sets as long as the mapping $*$ is injective. Otherwise, the kernel of $*$ is the translation group of Λ.

The comprehensive paper of Lagarias [26] is the most extensive study to date of the geometry of point sets in the context of quasi-periodic structures.

4. ARITHMETIC SIDE

Although the requirement in the definition of a model set of the existence of a lattice is not in itself particularly arithmetic, nonetheless the interesting and important examples all have strong arithmetic aspects. In the usual cases where the internal space is a real space, the arithmetic arises through the standard inner product on the embedding space and the nature of the two projections.

We will illustrate here the typical arithmetic input into the theory with two very different examples.

4.1. The icosian model sets

It was M. Baake *et al.* [2] who first pointed out that the root and weight lattices of types A_4 and D_6 could be used as the lattices for projection in cut and project schemes for dihedral $\mathcal{D}_5$ and icosahedral symmetries in 2 and 3 dimensions. The fact that these two groups are Coxeter groups (finite reflection groups) and form the first two of the the series H_2, H_3, H_4 of non-crystallographic finite Coxeter groups[2] (of which H_3 and H_4 are the only examples of rank larger than 2) suggests that H_4 should also appear in this context. This was first pointed out in [15] and elaborated in more detail in [30]. We do nothing more than outline this here. It is not necessary to know anything about root systems to follow this example.

The elements of norm 1 of the usual quaternion ring $H = \mathbb{R} + \mathbb{R}i + \mathbb{R}j + \mathbb{R}k$ form a group isomorphic to SU(2). Since this group is a 2-fold cover of the orthogonal group SO(3), in particular it contains 2-fold covers of the icosahedral

---

[2] The group H_2 is the dihedral group of order 10. Usually it is fitted into the series $I_2(k)$ of dihedral Coxeter groups, but it is also completely natural to think of it in the icosahedral series, as we do here. In fact, we could go a step further and include H_1 which is simply the reflection group of order 2.

group. One such example is the following list I of 120 vectors:

$$\frac{1}{2}(\pm 1, \pm 1, \pm 1, \pm 1), \quad (\pm 1, 0, 0, 0) \quad \text{and all permutations,}$$

$$\frac{1}{2}(0, \pm 1, \pm \tau', \pm \tau) \qquad \qquad \text{and all even permutations}$$

(7)

where $\tau = (1 + \sqrt{5})/2$ is the Golden ratio and $'$ indicates the conjugation map $\sqrt{5} \mapsto -\sqrt{5}$.

The subring I generated by this group is called the *icosian ring*. Of course it depends on our particular choice of I, though it is straightforward to see that I is unique up to inner automorphisms of H. The form of the points of I makes it clear that I is a $\mathbb{Z}[\tau]$-module. We let $*$ denote the mapping on I that conjugates each of the coordinates with respect to the unique Galois non-trivial automorphism on $\mathbb{Z}[\tau]$ (defined by sending $\sqrt{5} \mapsto -\sqrt{5}$). Note that $I^* \neq I$.

The ring I is of rank 4 over $\mathbb{Z}[\tau]$ and rank 8 over $\mathbb{Z}$. We make an explicit embedding of I as a lattice $\tilde{I}$ in $\mathbb{R}^8$ by the mapping $x \mapsto (x, x^*)$.

This already provides the framework of a cut and project scheme:

$$\mathbb{R}^4 \xleftarrow{\pi_1} \mathbb{R}^4 \times \mathbb{R}^4 \xrightarrow{\pi_2} \mathbb{R}^4$$
$$\cup$$
$$\tilde{I}$$

(8)

with the projections being given by the first and second components of (x, x^*).

Remarkably the lattice $\tilde{I}$ has an entirely natural interpretation as the root lattice of type E_8 (see for instance [13] which underscores its arithmetic nature). This is explained in [12, 15, 30].

Now we wish to show that this cut and project scheme respects the symmetry that is inherent in its construction. Geometrically the points of I form the vertices of a regular polytope P in 4-space and also form the vectors of a root system Δ_4 of type H_4. The Coxeter group H_4 is none other than the group of automorphisms of P (and also of Δ_4), and is in fact very easily described: it is the set of all (14400) maps

$$x \mapsto uxv \quad ; \quad x \mapsto u\bar{x}v$$

(9)

where $u, v \in I$. The subgroup of these transformations in which $v = u^{-1}$ is obviously a copy of the icosahedral group and this subgroup stabilizes the 3-dimensional space $\mathbb{R}i + \mathbb{R}j + \mathbb{R}k$ of *pure* quaternions.

These maps provide automorphisms of the rings I and, via conjugation, on I^* too, and thus give rise to an action of I as automorphisms on the entire cut and project scheme. If the window W is chosen to be invariant under I then the resulting model set is also I-invariant.

Restricting everything to the pure quaternions we get a new cut and project scheme based on the 6-dimensional root lattice D_6 and an icosahedral symmetry. Restricting further to the planes orthogonal to the 5-fold axes brings us back to A_4 and the related dihedral $\mathcal{D}_5$ symmetry. A step further, and we arrive at the Fibonacci chain in 1 dimension. Thus all three families as well as the fundamental Fibonacci model sets fit together in this quaternionic model. Not only is this very pretty, it also essentially encompasses the generic situation for icosahedral symmetry in model sets: the only other relevant lattices in 6-space are the D_6 weight lattice and the lattices lying between the root and weight lattices. For more on this see [12,36].

4.2. p-adic model sets

Until recently, little thought had been given to the situation in which the internal group is something different than another real space, or at worst a real space crossed with a torus. However, there is a whole series of very natural locally compact abelian groups that are not euclidean in nature, namely the p-adic groups. Since these may not be familiar in this context let us recall the basic ideas.

Let p be a prime number in the integers $\mathbb{Z}$. Using p we can define a metric on the rational numbers Q, and by restriction on $\mathbb{Z}$, in the following way. For each $a \in \mathbb{Z}$, we define its p-value, $\nu_p(a)$, as the largest exponent k for which p^k divides a (with $\nu_p(0) := \infty$). This function is extended to the p-adic valuation $\nu_p : Q \longrightarrow \mathbb{Z}$ by $\nu_p(a/b) := \nu_p(a) - \nu_p(b)$ for all rational numbers a/b. We now define the "distance" between two rational numbers x, y as $d(x, y) = p^{-\nu_p(y-x)}$.

It is not hard to see that this does define a metric on Q, in which closeness to 0 is equivalent to high divisibility by the prime p. The completion of the rationals under this topology is the field of p-adic numbers Q_p and the completion of the subring $\mathbb{Z}$ is the subring of p-adic integers, $\mathbb{Z}_p$. Each non-negative p-adic integer can be given the more concrete representation as a series in the form $\sum_{n=0}^{\infty} a_n p^n$ where the a_n are integers in the range $0 \leq a_n < p$. Note that convergence here is automatic, even though there are infinitely many terms in the sum, because of the nature of the p-adic topology. The topologies defined by such metrics have other counter-intuitive properties. For example, for each non-negative integer k, the set $p^k \cdot \mathbb{Z}_p$, of elements of $\mathbb{Z}_p$ divisible by p^k, is the ball of radius p^{-k} and is clopen, *i.e.* both open and closed, as too are all its cosets, $a + p^k \cdot \mathbb{Z}_p$.

Seen as a topological space, $\mathbb{Z}_p$ is both compact and totally disconnected (but not discrete). In particular, Q_p and $\mathbb{Z}_p$ are locally compact abelian groups under addition. Thus, we can use $\mathbb{Z}_p$ to construct interesting cut and project schemes for $\mathbb{R}^d$ simply by taking $G := (\mathbb{Z}_p)^d$ and $L = \mathbb{Z}^d$ embedded diagonally into $\mathbb{R}^d \times \mathbb{Z}_p^d$ (based on the natural embedding of $\mathbb{Z}$ in $\mathbb{Z}_p$). For more on p-adic numbers and other totally disconnected groups, the reader may consult [11,31].

In [9] it was shown that a number of interesting substitution systems and tilings can be interpreted in a p-adic setting, including the well-known chair tiling. Rather than repeat these, let us give a different example, mentioned in [9] but not elaborated upon. One of the earliest classes of aperiodic tilings to be discovered was the class of Raphael Robinson's square tilings [35]. The title of his paper recalls that the mathematical interest in aperiodic structures

Fig. 1. — Pattern of increasing squares in a Robinson square tiling.

had a totally different (and earlier) origin than the physical one, namely the interest in decidability problems in the tiling of the plane with tiles of finitely many different types. The Robinson tilings are tilings of the plane by equally sized squares, in the usual fashion, with the twist that the square tiles come in 6 types (up to rotational and reflectional symmetries), distinguished by the markings of their edges, and the tiling is required to respect these edge markings by having the edges of adjacent tiles properly matched. Pictures of the tiles may be found in [35] and [18]. What is important for our discussion is that there is another set of markings by lines of these tiles and the correct tilings are those for which these lines arrange themselves into a pattern of squares of increasing scales $1, 2, 4, 8, \ldots$ (see Fig. 1). This picture is the one that Robinson used to prove the aperiodicity, for evidently no translation can map the squares of all scales onto themselves simultaneously. The same idea was used by Penrose in his recent hexagonal tiling [32].

Now the point is that the centres of the tiles of each of the six types form a model set based on an internal space which is 2-adic. Very briefly the argument is as follows.

Starting with Figure 1 and *ignoring the actual square tiles themselves* we have patterns of interlocking squares of increasing scale. Let's call these the *pattern-squares* to distinguish them from the actual tiling squares. The vertices of the smallest scale pattern-squares (order 1) form the vertices of a lattice L' once one of them, say $c_1 = (0,0)$, has been chosen as the origin. It is convenient to introduce the larger lattice $L = \frac{1}{2} \cdot L'$ which we can identify with $\mathbb{Z}^2$. The locations of the vertices of the increasingly scaled pattern-squares determine two sequences, $\{\alpha_k\}$ and $\{\beta_k\}$, composed out of the two numbers $\{\pm 1\}$ as follows: the vertices of the pattern-squares of order k (side-length 2^k) are the points of a coset

$$L_k := c_k + 2^k L \tag{10}$$

where

$$c_k = (\alpha_0 + \alpha_1 2 + \ldots, +\alpha_{k-2} 2^{k-2}, \beta_0 + \beta_1 2 + \ldots, +\beta_{k-2} 2^{k-2}) \tag{11}$$

for $k = 2, 3, \ldots$. Conversely, given two sequences α, β of ± 1's we can use (11) and (10) to define the vertices of a suitable pattern of squares.

The pattern-squares themselves are determined by the condition that the points of L_k are the centres of the squares of order $k - 1$ for all $k > 1$. This then establishes a coordinatization of the pattern-squares. Both the actual pattern of the pattern-squares and their coordinatization depend on the choice of our two sequences (more below).

Now looking again at the tiling squares, we distinguish the 6 types of tiles according to their location in the pattern-squares. We list these here together with a description of the coordinates of the centres of their tiles:

(1) the "corner tiles" of the squares of order 1;
coordinates $L_1 = 2L$; density $\frac{1}{4}$;

(2) the "corner tiles" of all squares of all higher orders;
coordinates $\bigcup_{k \geq 2} c_k + 2^k L$; density $\frac{1}{12}$;

(3) "cross tiles" where edges of two different orders of pattern squares meet (actually these orders always differ by exactly 1);
coordinates:
$$\left(\bigcup_{k \geq 3} c_k + (\pm 2^{k-2}, \pm 2^{k-3}) + 2^k L \right) \cup \left(\bigcup_{k \geq 3} c_k + (\pm 2^{k-3}, \pm 2^{k-2}) + 2^k L \right);$$
density $\frac{1}{6}$;

(4) "edge " squares, which contain part of a single edge of a pattern-square, except those in which are exactly in the middle of an edge; density $\frac{1}{6}$;

(5) "edge " squares, which contain part of a single edge of a pattern-square and which are exactly in the middle of an edge; density $\frac{1}{6}$;

(6) blank tiles, with no part of any edge in them; density $\frac{1}{6}$.

Observe that each of these sets is a countable union w_j of cosets of the form $a + 2^k L$. Let us replace each of these by the corresponding 2-adic clopen set $a + 2^k \cdot \mathbb{Z}_2^2$. In this way we get 6 open sets W_j, $j = 1, \ldots, 6$, whose closures, being closed subsets of the compact group $\mathbb{Z}_2^2$, are compact with non-empty interiors. Finally we can describe the centres of the squares of type j as

$$\{ x \in \mathbb{Z}^2 \mid x \in W_j \} \tag{12}$$

which, unlikely as it appears, is a model set under the scheme

$$\mathbb{R}^2 \xleftarrow{\pi_1} \mathbb{R}^2 \times \mathbb{Z}_2^2 \xrightarrow{\pi_2} G$$
$$\cup$$
$$\mathbb{Z}^2$$
(13)

where $\mathbb{Z}^2$ is embedded into $\mathbb{R}^2 \times \mathbb{Z}_2^2$ diagonally: $x \mapsto (x, x)$.

The entire tiling is determined by the vertices of the various squares and hence by the window

$$W := \bigcup_{k=1}^{\infty} c_k + 2^k \mathbb{Z}_2^2.$$
(14)

Evidently W is an open subset of $\mathbb{Z}_2^2$. Let $w \in \overline{W} \backslash W$. Then for each $m \in \mathbb{Z}_+$, $w + 2^m \mathbb{Z}_2^2$ meets W, so $w \equiv c_k \bmod 2^{\min\{m,k\}} \mathbb{Z}_2^2$ for some $k = k(m)$. If $k \leq m$ then $w \in c_k + 2^k \mathbb{Z}_2^2 \subset W$. Thus $k > m$ and $w \equiv c_{k(m)} \bmod 2^m \mathbb{Z}_2^2$. It follows that w is the limit of some subsequence of the $\{c_k\}$. Since the entire sequence evidently converges (in the p-adic topology, of course!) to some $c = (a, b)$, where $a = \sum \alpha_k 2^k$ and similarly for b, we see that $w = c$ and so $\partial W = \{c\}$. Thus the model set of all vertices is regular, and even generic provided that $c \notin \mathbb{Z}^2$.

More generally, one may expect these p-adic topologies to arise whenever there is a self-similarity $\theta : L \to L$ for which $\theta(L) \subset L$, but $\theta(L) \neq L$.

In [9] we also see the appearance of mixed p-adic and real spaces as the internal spaces. Beyond these types we are not aware of any interesting examples, though they may well exist.

5. ANALYTIC SIDE

The transition from an inherently discrete picture on the physical side to something inherently smooth on the internal side is made *via* Weyl's theory of uniform distribution (3). In the generality of the setting here it is derived in [6] based on Theorem 2 below.

Let us assume that we have a model set Λ. Now consider the following question. Suppose that we take a ball $B_R(0)$ of radius R about the origin in $\mathbb{R}^d$ and look at $\Lambda_R := \Lambda \cap B_R(0)$. Then we can ask how Λ_R^* is distributed over W. We say that the sets Λ_R^* are *uniformly distributed* if for each open set $U \subset W$ we have

$$\lim_{R \to \infty} \frac{\text{card}(\Lambda_R^* \cap U)}{\mu(W)} = \mu(U)/\mu(W)$$
(15)

where μ is Haar measure on G.

Theorem 2 *[21, 37] If Λ is regular then the sets Λ_R^* are uniformly distributed over W.*

Let $f^* : G \longrightarrow C$ be any function. We define $f : L \longrightarrow C$ by $f(x) = f^*(x^*)$. If f^* is supported on the window W then evidently f is supported on the model set Λ. If f^* is continuous (which is the case of interest) then this is iff.

Theorem 3 *(Weyl) [6, 41] If Λ is regular and f^* is continuous then*

$$\lim_{R \to \infty} \frac{1}{\mathrm{card}(\Lambda_R)} \sum_{x \in \Lambda_R} f(x) = \frac{1}{\mathrm{vol}(W)} \int_W f^*(u) \mathrm{d}\mu(u). \tag{16}$$

Since W has boundary of measure zero, it is not necessary to insist that f^* (which is supported on W) be continuous on all of internal space, only on the window W.

In this way discrete averaging on the model set is transformed into integration on the window. This process was used in [4, 6] to determine the existence of invariant measures on internal space in the presence of self-similarity on the quasi-crystal. We briefly explain this. We assume here that internal space is $\mathbb{R}^n$ for some n. For the general situation see [6].

A *self-similarity* of Λ is an affine linear mapping $t = t_{Q,v}$

$$t_{Q,v} : \quad x \mapsto Qx + v \tag{17}$$

on $\mathbb{R}^d$ that maps Λ into itself, where Q is a (linear) similarity and $v \in \mathbb{R}^d$. Thus $Q = qR$, *i.e.* it is made up of an orthogonal transformation R and an *inflation factor* q.

Let $t_{Q,v}$ be a self-similarity of Λ. Since Λ is uniformly discrete, we must have $|q| \geq 1$. We will assume $|q| > 1$ and that $QL = L$. We are interested in the *entire* set of affine inflations with the same similarity factor Q.

Note that Q naturally gives rise to an automorphism $\tilde{Q}$ of the lattice $\tilde{L}$, *i.e.* an element of $\mathrm{GL}_{\mathbb{Z}}(\tilde{L})$, and a linear mapping Q^* of $\mathbb{R}^n$ that maps W into itself. From the arithmetic nature of $\tilde{Q}$ we deduce that the eigenvalues of Q and Q^* are algebraic integers and from the compactness of W that Q^* is contractive.

Define

$$W_Q := \{u \in \mathbb{R}^n \mid Q^*W + u \subset W\} \cdot \tag{18}$$

We say that Q is *compatible* with Λ if $\mathrm{int}(W_Q) \neq \emptyset$. Assuming that this is the case (not a strong assumption), then the set $\mathcal{T}_Q$ of affine inflations with the same similarity Q is the set of mappings $t_{Q,v} : x \mapsto Qx + v$, where v runs through the set

$$T = T_Q := \{v \in L \mid v^* \in W_Q\} \cdot \tag{19}$$

Theorem 4 *If Q is a self-similarity and the above assumptions on Q apply then there is a unique absolutely continuous positive measure μ on internal space, supported on W, satisfying:*

- $\mu(W) = 1$;

- μ is invariant in the sense that, if we define $t_v^* \cdot \mu_f$ by $t_v^* \cdot \mu(Y) = \mu((t_v^*)^{-1}(Y))$, then

$$\mu = \lim_{s \to \infty} \frac{1}{\#\left(T \cap B_s(0)\right)} \sum_{v \in T \cap B_s(0)} t_v^* \cdot \mu. \tag{20}$$

The similarity of this measure to Hutchison measures in the context of iterated function systems is not coincidental. In fact, if we restrict to the ball $B_s(0)$ then the $\{t_v^*\}$ form a finite set of contractions which is indeed an iterated function system. For more on this and invariant density functions on model sets see [4].

The type of limit averaging involved here is a very natural one from the point of view of physical situations, representing the transition from the world of sets finite in extent to the ideal world of infinitely extended point sets.

Although no one to our knowledge has made any use of it, it is interesting to use Weyl's theorem to transfer the structure of $L^2(W)$ to a space of similar objects on Λ. Namely, the space of continuous functions on W leads to a space

$$\mathcal{C} = \mathcal{C}(\Lambda) \tag{21}$$

of funtions on Λ via the mapping $*$. Then the usual inner product $\langle f, g \rangle = \int_W \overline{f^*(u)} g^*(u) \mathrm{d}u = \langle \overline{f}, g \rangle_W$ defines an inner product on $\mathcal{C}$ and we can complete this space in order to get a Hilbert space $\overline{\mathcal{C}}$ isomorphic to $L^2(W)$. Of course the elements of $\overline{\mathcal{C}}$ can no longer be interpreted as functions on Λ since functions on Λ that differ by a function whose absolute square has limit average sum equal to 0 are identified.

6. DYNAMICAL SYSTEMS SIDE

So far we have looked at one model set in isolation. Now we move on to consider families of model sets. We start with a number of definitions and results. All of these may be found in the paper of Schlottmann [38] on which we have relied heavily here. Many are well-known in the context of tilings for which a recent reference with a good bibliography is [40]. In this section all point sets under discussion are assumed to be Delone sets in $\mathbb{R}^d$.

Two Delone sets S, S' in $\mathbb{R}^d$ are *locally isomorphic* (or some people say *locally indistinguishable*) if, up to translations, every patch of either of them occurs in the other. Thus on any finite scale, up to translation, the two sets are indistinguishable. Given a Delone set $S \subset \mathbb{R}^d$ we can look at its local isomorphism class (LI class) $\mathrm{LI}(S)$, namely all point sets locally isomorphic to it.

We denote by $\mathcal{X}(r)$ the set of all Delone sets of $\mathbb{R}^d$ for which the minimum separation between distinct points is at least r. We assume in the rest of this section that $r > 0$ has been fixed.

We define a Hausdorff topology on $\mathcal{X}(r)$ as follows: two sets $S, S' \in \mathcal{X}(r)$ are "close" if for some large compact set $K \subset \mathbb{R}^d$ and some small ϵ we have

$$(v + S) \cap K = S' \cap K \text{ for some } v \in \mathbb{R}^d \text{ with } |v| < \epsilon. \qquad (22)$$

More precisely we define a uniformity $\mathcal{U}$ on $\mathcal{X}(r)$ using as the sets $U(K, \epsilon)$ of uniformity the set of pairs (S, S') satisfying (22).

Theorem 5 *[34, 38, 40] With respect to this topology $\mathcal{X}(r)$ is a complete Hausdorff space.*

Let $S \in \mathcal{X}(r)$. Then $\mathbb{R}^d$ obviously acts on $\mathrm{LI}(S)$ by translation and in particular the entire orbit $[S]$ of S lies in $\mathrm{LI}(S)$. This action is continuous and hence extends also to an action on the closure $\overline{\mathrm{LI}(S)}$ of $\mathrm{LI}(S)$. The relationship between orbits and LI classes can be summed up by

$$S \in [S] \subset \mathrm{LI}(S) \subset \overline{[S]} = \overline{\mathrm{LI}(S)}. \qquad (23)$$

The second inclusion follows easily from the definitions. The inclusions may, according to the situation, be strict or actual equalities. For a lattice there is only one orbit in its LI class. For general model sets the situation is very different, as we shall see.

Recall that a Delone set S is said to be of *finite local complexity* if the closure of $S - S$ is discrete. Finite local complexity is a property that is inherited by whole LI classes.

Theorem 6 *An LI class is pre-compact (i.e. its completion is compact) iff it has finite local complexity.*

Thus, if S is a Delone set of finite local complexity, $\overline{[S]}$ is compact and we obtain a dynamical system $\mathcal{D}(S)$:

$$\mathbb{R}^d \times \overline{[S]} \longrightarrow \overline{[S]}. \qquad (24)$$

In the sequel we will use the symbols like $\mathcal{D}(S)$ to denote both the dynamical system itself and the corresponding defining space $\overline{[S]}$.

Theorem 7 *Let S be a Delone set of finite local complexity. The following are equivalent:*

(i) S is repetitive;

(ii) $\overline{[S]} = \mathrm{LI}(S)$ is closed;

(iii) The dynamical system $\mathcal{D}(S)$ is minimal.

We recall that minimal means that every $\mathbb{R}^d$ orbit is dense.

Since generic model sets are repetitive and also of finite local complexity, this leads to a very nice result:

Theorem 8 *Let Λ be a generic model set. Then its LI class $\mathrm{LI}(\Lambda)$ is a compact Hausdorff space and under the action of translation by $\mathbb{R}^d$ it becomes a minimal dynamical system, $\mathcal{D}(\Lambda)$.*

This is the first of the dynamical systems that we wish to consider. Its rather abstract form is better understood by relating it to a more accessible dynamical system.

To this end, let $\Lambda = \Lambda(W) = \{x \in L \mid x^* \in W\}$ be a model set. Each element (u, v) of the group $\mathbb{R}^d \times G$ can be used to form a new model set

$$\Lambda(W, u, v) := u + \{x \in L \mid x^* \in -v + W\} \, . \tag{25}$$

If $(u, v) \in \tilde{L}$ then $v = u^*$ and we can rewrite this as $\{u + x \in L \mid (u+x)^* \in W\}$, which is just Λ again. Thus we get a whole family of model sets parametrized by $T := (\mathbb{R}^d \times G)/\tilde{L}$ with $\mathbb{R}^d \times G$ acting on it. This is the second dynamical system. Its points correspond to the model sets $\Lambda(W, u, v)$. This is the so-called *torus parametrization* introduced by Baake *et al.* in [3]. We use the same terminology in the more general context here, although in general T is not a torus!

The action of $\mathbb{R}^d$ on T, $(x, y + \tilde{L}) \mapsto x + y + \tilde{L}$, is a faithful transcription of the operation of translation in physical space, so the orbits of $\mathbb{R}^d$ on T correspond to model sets that differ only by translation. The action of G on T corresponds to translating the window around.

Theorem 9 *Let Λ be a generic model set. Then*

$$\mathbb{R}^d \times T \longrightarrow T \tag{26}$$

is a minimal uniquely ergodic dynamical system $\mathcal{D}_{\mathrm{tor}}$. The unique invariant probability measure is normalized Haar measure. The set of points of $\mathcal{D}_{\mathrm{tor}}$ corresponding to generic model sets is dense and indeed the set of points corresponding to the non-generic model sets is of the first category.

It is noteworthy that this dynamical system is independent of W but the actual parametrization of model sets is clearly dependent on it.

So now given a generic model set Λ, there are two dynamical systems for the group $\mathbb{R}^d$, one $\mathcal{D}(\Lambda)$ coming from the closure of the orbit of Λ under action of $\mathbb{R}^d$ and another $\mathcal{D}_{\mathrm{tor}}$ coming from the torus parametrization. Not surprisingly they are related, but rather surprisingly this relation is somewhat subtle. All the elements of $\mathcal{D}(\Lambda)$ are, by definition, in the same LI class. The same is not the case for the model sets parametrized by $\mathcal{D}_{\mathrm{tor}}$. Indeed, Λ is generic, but translating the window around is bound to produce model sets that are not generic. These non-generic model sets are not locally isomorphic to the regular ones, because they have certain special local configurations of points that are related to the boundaries of their windows.

Theorem 10 *[38] Let Λ be a generic model set. Then there is a continuous surjective mapping*

$$\beta : \mathcal{D}(\Lambda) \longrightarrow \mathcal{D}_{\text{tor}} \qquad (27)$$

which is $\mathbb{R}^d$-equivariant and which maps Λ onto the point 0 of the torus. Furthermore, for each of the points of $\mathcal{D}_{\text{tor}}$ which parametrize generic model sets, the preimage in $\mathcal{D}(\Lambda)$ consists of a unique point.

This mapping comes about as follows: Let $\Lambda' \in \mathcal{D}(\Lambda)$. First suppose that $\Lambda' \subset L$. Then it is not hard to see that $\bigcap_{x \in \Lambda'}(W - x^*)$ is a single point, call it $b(\Lambda')$. Furthermore, for all $u \in L$, $b(\Lambda' - u) = b(\Lambda') + u^*$. Now for arbitrary $\Lambda' \in \mathcal{D}(\Lambda)$, we can always find $v \in \mathbb{R}^d$ with $\Lambda' - v \subset L$. This v is nothing like unique but it follows from what we have just said that the pair $\beta(\Lambda') := (v, b(\Lambda' - v))$ is unique mod $\tilde{L}$, and this is the mapping that we require.

Using these facts it can be established that

Theorem 11 *[38] Assume that Λ is a regular and generic model set. Then $\mathcal{D}(\Lambda)$ is uniquely ergodic and furthermore $L^2(\mathcal{D}(\Lambda))$ and $L^2(\mathcal{D}_{\text{tor}})$ are isometrically isomorphic as $\mathbb{R}^d$-spaces.*

The importance of this is that it shows that from the spectrum of $\mathcal{D}_{\text{tor}}$ being discrete, which surely is since T is a compact abelian group, it follows that the spectrum of $\mathcal{D}(\Lambda)$ is discrete. It is from this that the pure point diffractivity of Λ can be deduced. In the final section we briefly describe how this happens.

7. DIFFRACTION

The theoretical framework for the discussion of diffraction has been very well described in several places. The two papers of Hof [19, 20] are standards and there are also good descriptions in [8, 17]. Here we just quickly formulate the definitions.

Let Λ be a regular model set and define the (tempered) distribution

$$\delta_\Lambda := \sum_{x \in \Lambda} \delta_x, \qquad (28)$$

where δ_x is the Dirac measure at x. For each $s > 0$ we calculate the *autocorrelation* of δ_Λ restricted to the ball of radius s:

$$\delta_{\Lambda \cap B_s(0)} * \tilde{\delta}_{\Lambda \cap B_s(0)} = \sum_{x,y \in \Lambda \cap B_s(0)} \delta_{x-y}, \qquad (29)$$

where, as usual, the over-tilde indicates changing the sign of the argument. The limit as s goes to infinity of the volume-averaged auto-correlation of this measure, which exists for model sets, is the *auto-correlation measure* of Λ (its so-called *Patterson function*):

$$\gamma = \lim_{s \to \infty} \frac{1}{\text{vol}(B_s(0))} \sum_{x,y \in \Lambda_s} \delta_{x-y}.$$

This limit, taken in the vague topology, converges to a tempered distribution (*i.e.* this limit exists when taken against rapidly decreasing test functions). Its Fourier transform is a positive measure $\hat{\gamma}$ (a result of Bochner's theorem applied to the positive definite distribution γ) which is the *diffraction pattern* of Λ. The measure decomposes into a point part and a continuous part. The point part of this measure is the *Bragg spectrum* of Λ. The model set has *pure point spectrum* if the continuous part is the trivial 0-measure.

The complexity of the definition makes it hard to discover the nature of the diffraction pattern, in particular whether or not we have pure-point diffraction or not. One approach has been to use the ergodic theory outlined above, and indeed it is able to give the main result:

Theorem 12 *[38] Any regular model set has pure point spectrum. Furthermore this spectrum is supported on the projection into Fourier space on the physical side of the dual of the compact group T (5), i.e. it has the form*

$$\hat{\gamma} = \sum_{k \in \hat{T}} w(k) \delta_{\hat{\pi}_1(k)}. \tag{30}$$

The proof of this is based on an idea of Dworkin [14]. The argument is spelled out in [20] and we repeat it here since otherwise it is difficult to see the connection between dynamical systems and diffraction.

We can assume that Λ is generic since translation of the window does not alter the qualitative nature of the diffraction. The next step is to replace δ_Λ by a smooth approximation to it. To this end, let $b : \mathbb{R}^d \longrightarrow \mathbb{R}_{\geq 0}$ be a smooth function whose support is contained in the ball $B_r(0)$ of radius r, where $B_{2r}(0) \cap (\Lambda - \Lambda) = \{0\}$. Define a function $\psi : \mathcal{D}(\Lambda) \longrightarrow \mathbb{R}$ by

$$\psi(\Lambda') = \int_{\mathbb{R}^d} b(-u) \delta_{\Lambda'}(u) du, \tag{31}$$

which is continuous on $\mathcal{D}(\Lambda)$. The action of $\mathbb{R}^d$ on $[\Lambda]$ gives rise to a corresponding action $x \mapsto T_x$ on the space of functions on the orbit $[\Lambda]$ of Λ under translation.

For each $x \in \mathbb{R}^d$ we have

$$T_x \psi(\Lambda) = \int_{\mathbb{R}^d} b(-u) \delta_{-x+\Lambda}(u) du = \int_{\mathbb{R}^d} b(-u) \delta_\Lambda(x+u) du = b * \delta_\Lambda(x), \tag{32}$$

which shows that the function $\sigma^{(b)}$: $\mathbb{R}^d \longrightarrow C$ defined by $x \mapsto T_x(\psi)(\Lambda)$ is obtained by centering a copy of b at each point of Λ.

Now consider the autocorrelation of $\sigma^{(b)}$:

$$
\begin{aligned}
\gamma^{(b)}(x) &= \lim_{s \to \infty} \tfrac{1}{\text{vol}(B_s(0))} \int_{B_s(0)} T_{x+y}(\psi)(\Lambda) T_y(\psi)(\Lambda) \, dy \\
&= \int_{\mathcal{D}(\Lambda)} T_x(\psi) \psi \, d\mu = \int_{\mathcal{D}(\Lambda)} \overline{T_x(\psi)} \psi \, d\mu = (T_x \psi, \psi).
\end{aligned}
\tag{33}
$$

The main point here is use of the *Birkhoff ergodic theorem* and the ergodicity of the action of $\mathbb{R}^d$ on $\mathcal{D}(\Lambda)$ to replace the integral over $\mathbb{R}^d$ by an integral over $\mathcal{D}(\Lambda)$. Note that the *uniqueness* of ergodicity and the continuity of ψ is needed here to guarantee the statement *for all* x rather than a.e. ([16], Sect. 3.2).

In view of Theorem 11, we have a Fourier expansion of ψ in terms of the eigenfunctions for the action of $\mathbb{R}^d$: $\psi = \sum a_\lambda \phi_\lambda$, where ϕ_λ is the eigenfunction for the character $x \mapsto e^{2\pi i \lambda \cdot x}$ on $\mathbb{R}^d$. Thus $(T_x \psi, \psi) = \sum |a_\lambda|^2 e^{2\pi i \lambda \cdot x}$ and taking Fourier transforms we have

$$
\hat{\gamma}^{(b)} = \sum |a_\lambda|^2 \delta_\lambda
\tag{34}
$$

which is a pure point measure on $\mathbb{R}^d$.

On the other hand we know that $\sigma^{(b)} = b * \delta_\Lambda$ whose autocorrelation can be calculated directly as $b * \tilde{b} * \gamma$ and so $\hat{\gamma}^{(b)} = |\hat{b}|^2 \hat{\gamma}$. Finally, taking a sequence of bump functions $\{b\}$ converging in the vague topology to δ_0 we obtain the required pure-point nature of the diffraction pattern.

Theorem 12 is qualitative in nature. The quantitative counterpart is this:

Theorem 13 *[28] Let $k \in \hat{T}$ and let χ denote the characteristic (or indicator) function of W. Then $w(k) = |\hat{\chi}(-\hat{\pi}_2(k))/\text{vol}(W)|^2$.*

There are a number of variations on this theme that are worthwhile mentioning. First we may imagine replacing the simple sum (28) by a weighted sum

$$
\delta_\Lambda^\omega := \sum_{x \in \Lambda} \omega(x) \, \delta_x,
\tag{35}
$$

where ω: $\Lambda \longrightarrow C$ is some function.

Theorem 14 *If Λ is a regular model set and if $\omega \in C(\Lambda)$ (see (21)) then the weighted point distribution is pure point diffractive.*

Next we consider the case that our points of Λ are considered stochastically:

$$
\delta_{\text{stochastic}} := \sum_{x \in \Lambda} \eta(x) \, \delta_x,
\tag{36}
$$

where the $\eta(x)$ form a collection of independent identically distributed random variables that take the values $1, 0$ (indicating occupancy or not of the respective model set sites) with the probability of occupancy being p:

Theorem 15 *[7] Let Λ be a regular model set and let η be as above with the mean and second moment equal to m_1 and m_2 respectively. Then the autocorrelation of Λ and that of its stochastic version are, with probability one, related by*

$$\gamma_{\text{stochastic}} = (m_1)^2\,\gamma + d\,(m_2 - (m_1)^2)\delta_0 \tag{37}$$

with Fourier transforms

$$\hat{\gamma}_{\text{stochastic}} = (m_1)^2\,\hat{\gamma} + d\,(m_2 - (m_1)^2), \tag{38}$$

where d is the density per unit volume of the Λ.

Thus the pure point nature of Λ is affected by at most the addition of a constant continuous background. More on this stochastic approach may be found in [7].

Yet a different variation is to allow the points of Λ to be moved in some regular way.

Theorem 16 *[20] Let Λ be a regular model set and let $f : x \mapsto (f_1(x), \ldots, f_d(x))$ be some mapping of Λ into $\mathbb{R}^d$, where each $f_i \in C(\Lambda)$ (see (21)). Then the set $\Lambda_f := \{x + f(x) \mid x \in \Lambda\}$ is pure point diffractive.*

These variations can be combined in the obvious ways.

The problems of determining which point sets are pure point diffractive is a fascinating and challenging one which is still wide open. The examples above show how much model sets can be modified without serious damage to their diffractive properties. Obviously adding or removing points whose average density is 0 also does not alter the diffraction. But there are point sets that are even more remote that are diffractive. One example is the set of visible points of a lattice. Given a lattice L in $\mathbb{R}^d$ its *visible* points are those points $x \in L$, $x \neq 0$, satisfying $Qx \cap L = \mathbb{Z}\,x$. What is interesting about the visible points is that they do *not* form a Delone set (they are not relatively dense in $\mathbb{R}^d$). In fact, for each $r > 0$, the set of holes of radius exceeding r has positive density. However,

Theorem 17 *[8] The set of visible points of any lattice of rank at least 2 is pure point diffractive.*

7.1. Comments

In this paper we have considered point sets in $\mathbb{R}^d$ that are constructed through the method of projection from an embedding group and a lattice. We have assumed that the embedding group is of the form $\mathbb{R}^d \times G$ where G is a locally compact abelian group, which, in view of Theorem 1, is fairly natural. However, it is possible to study model sets in the situation where the "physical space" has been replaced by an arbitrary locally compact abelian group without losing many of the most interesting properties. In particular the diffraction results of Theorem 12 have formulations in this generality [38].

We have not touched here the notion of *multi-component* model sets. These arise completely naturally in the context of tilings, for instance. Many tilings have vertices that can be described in terms of model sets, but usually only after distinguishing the different types of vertices and treating each type individually. For example the Penrose tilings are of this type, with 4 different types of vertices. There are several ways incorporate this into our model set formalism, either by taking more complicated internal spaces (in the Penrose case we can use $\mathbb{R}^2 \times \mathbb{Z}/5\mathbb{Z}$ and a 4 component window) or by taking products of model sets. We refer the reader to [5], and more particularly to [6] for a discussion of these ideas.

Acknowledgments

It is a pleasure to thank Martin Schlottmann for his edifying insights into this material.

REFERENCES

[1] M. Baake, A guide to mathematical quasicrystals, in: *Quasicrystals*, edited by J.B. Luck, M. Schreiber and P. Häussler (Springer, 1998).

[2] M. Baake, P. Kramer, M. Schlottmann and D. Zeidler, Planar patterns with five-fold symmetry as sections of periodic structures in 4-space, *Int. J. Mod. Phys. B* **4** (1990) 2217-2268.

[3] M. Baake, J. Hermisson and P. Pleasants, The torus parametrization of quasiperiodic LI-classes, *J. Phys. A: Math. Gen.* **30** (1997) 3029-3056.

[4] M. Baake and R.V. Moody, Self-similarities and invariant densities for model sets, in: *Algebraic Methods and Theoretical Physics*, edited by Y. St. Aubin (Springer, New York, 1997) in press.

[5] M. Baake and R.V. Moody, Multi-component model sets and invariant densities, in: *Aperiodic '97*, edited by M. de Boissieu, J.-L. Verger-Gaugry and R. Currat (World Scientific, Singapore, 1998) 9-20.

[6] M. Baake and R.V. Moody, Compact sets of contractions, Weyl's theorem, and invariant densities for multi-component model sets, in: *Directions in Mathematical Quasicrystals*, edited by M. Baake and R.V. Moody, CRM Monograph Series (AMS, Rhode Island, 2000) in preparation.

[7] M. Baake and R.V. Moody, Diffractive Point Sets with Entropy, *J.Phys. A: Math. Gen.* **31** (1998) 9023-9039.

[8] M. Baake, R.V. Moody and P. Pleasants, Diffraction from visible lattice points and k-th power free integers, *Journal of Discrete and Computational Geometry* (1999) in press.

[9] M. Baake, R.V. Moody and M. Schlottmann, Limit-periodic point sets as quasicrystals with p-adic internal spaces, *J. Phys. A: Math. Gen.* **31** (1998) 5755-5765.

[10] M. Baake and M. Schlottmann, Geometric Aspects of Tilings and Equivalence Concepts, in: *Proc. of the 5th Int. Conf. on Quasicrystals*, edited by C. Janot and R. Mosseri (World Scientific, Singapore, 1995) 15–21.

[11] N. Bourbaki, *Topology 1* (Addison-Wesley, Reading, 1966).

[12] L. Chen, R.V. Moody and J. Patera, Non-crystallographic root systems, *Quasicrystals and Discrete Geometry*, edited by J. Patera, *Fields Institute Monographs* **10** (AMS, Rhode Island, 1998).

[13] J.H. Conway and N.J.A. Sloane, *Sphere packings, lattices and groups* 2nd Ed. (Springer, New York, Berlin, 1998).

[14] S. Dworkin, Spectral theory and X-ray diffraction, *J. Math. Phys.* **34** (1993) 2965-2967.

[15] V. Elser and N.J. Sloane, A highly symmetric quasicrystal, *J. Phys. A: Math. Gen.* **20** (1987) 6161.

[16] H. Furstenberg, *Recurrence in ergodic theory and combinatorial number theory* (Princeton University Press, Princeton, New Jersey, 1981).

[17] F. Gähler and R. Klitzing, The diffraction pattern of self-similar tilings, in: *The Mathematics of Long-Range Aperiodic Order*, edited by R.V. Moody, *NATO ASI Series C* **489** (Kluwer, Dordrecht, 1997) 141–74.

[18] B. Grünbaum and G.C. Shephard, *Tilings and Patterns* (Freeman, New York, 1987).

[19] A. Hof, On diffraction by aperiodic structures, *Commun. Math. Phys.* **169** (1995) 25-43.

[20] A. Hof, Diffraction by aperiodic structures, in: *The Mathematics of Long-Range Aperiodic Order*, edited by R.V. Moody, *NATO ASI Series C* **489** (Kluwer, Dordrecht, 1997) 239–68.

[21] A. Hof, Uniform distribution and the projection method, in: *Quasicrystals and Discrete Geometry*, edited by J. Patera, *Fields Institute Monographs* **10** AMS (1998).

[22] A. Katz and M. Duneau, Quasiperiodic patterns and icosahedral symmetry, *J. Phys. France* **47** (1986) 181–96.

[23] P. Kramer, Non-periodic central space filling with icosahedral symmetry using copies of seven elementary cells, *Acta Cryst. A* **38** (1982) 257–64.

[24] P. Kramer and R. Neri, On periodic and non-periodic space fillings of E^m obtained by projection, *Acta Cryst. A* **40** (1984) 580–7; and *Acta Cryst. A* **41** (1985) 619 (*Erratum*).

[25] J.C. Lagarias, Meyer's concept of quasicrystal and quasiregular sets, *Comm. Math. Phys.* **179** (1996) 365-376.

[26] J.C. Lagarias, Mathematical Quasicrystals, in: *Directions in Mathematical Quasicrystals*, edited by M. Baake and R.V. Moody, *CRM Monograph series, AMS* (Rhode Island, 2000) in preparation.

[27] Y. Meyer, *Algebraic numbers and harmonic analysis* (North Holland, Amsterdam, 1972).

[28] Y. Meyer, *Quasicrystals, Diophantine approximation, and algebraic numbers*, in: Quasicrystals and Beyond, edited by F. Axel and D. Gratias (Les Éditions de Physique, Springer-Verlag, 1995).

[29] R.V. Moody, Meyer sets and their duals, in: *The Mathematics of Long-Range Aperiodic Order*, edited by R.V. Moody, *NATO ASI Series C* **489** (Kluwer, Dordrecht, 1997) 403–41.

[30] R.V. Moody, J. Patera, Quasicrystals and Icosians, *J. Phys. A: Math. Gen.* **26** (1993) 2829-2853.

[31] J. Neukirch, The *p*-adic numbers, in: *Numbers*, edited by H.-D. Ebbinghaus *et al.* (Springer, New York, 1990) 155–178.

[32] R. Penrose, Remarks on tiling: Details of a $(1 + \epsilon + \epsilon^2)$-aperiodic set, in: *The Mathematics of Long-Range Aperiodic Order*, edited by R.V. Moody, *NATO ASI Series C 489* (Kluwer, Dordrecht, 1997) 467–97.

[33] C. Radin, The pinwheel tilings of the plane, *Annals of Mathematics* **139** 661-702.

[34] C. Radin and M. Wolff, Space tilings and local isomorphism, *Geometriae Dedicata* **42** (1992) 355-360.

[35] R.M. Robinson, Undecidability and nonperiodicity of tilings of the plane, *Inv. Math.* **44** (1971) 177–209.

[36] D.S. Rokshar, D.C. Wright and N.D. Mermin, Scale equivalence of quasicystallographic space groups, *Phys. Rev. B* **37** (1988) 8145–8149.

[37] M. Schlottmann, Cut-and-project sets in locally compact abelian groups, in: *Quasicrystals and Discrete Geometry*, edited by J. Patera, *Fields Institute Monographs* **10** (AMS, Rhode Island, 1998).

[38] M. Schlottmann, Generalized model sets and dynamical systems, to appear in: *Directions in Mathematical Quasicrystals*, edited by M. Baake and R.V. Moody, CRM Monograph Series (AMS, Rhode Island, 2000) in preparation.

[39] M. Senechal, Quasicrystals and geometry (Cambridge University Press, 1995).

[40] B. Solomyak, Dynamics of self-similar tilings, *Ergod. Th. & Dynam. Syst.* **17** (1997) 695–738.

[41] H. Weyl, Uber die Gleichungverteilung von Zahlen mod. Eins, *Math. Ann.* **77** (1916) 313–352.

Course N° 7

Acceptance Windows Compatible with a Quasicrystal Fragment

Given by J. Patera

Written by Z. Masáková[1], J. Patera[2] and E. Pelantová[1]

[1] *Department of Mathematics, Czech Technical University, Faculty of Nuclear Sciences and Physical Engineering, Trojanova 13, 12000 Prague 2, Czech Republic*
[2] *Centre des Recherches Mathématiques, Université de Montréal, C.P. 6128 Succursale Centre-Ville, Montréal, Québec, H3C 3J7, Canada*

1. INTRODUCTION

Crystals and quasicrystals are physical materials distinguished clearly by the presence or absence of periodicity in their discrete internal structure. They have as well important similarities which are finding expression in the commonly used term "aperiodic crystal" [9,20]. Their similarities stem from the fact that both are deterministic structures: every atom of an ideal periodic or aperiodic crystal has its place: it neither can be removed nor an additional atom can be inserted into the structure without creating a defect in it.

The results of this paper underline a fundamental difference between the real life periodic and aperiodic crystals by providing a specific content to their distinction. Their different nature is revealed when one considers a fragment of each of them: a piece of a crystal determines its repetitive structure everywhere in the space, a quasicrystal sample can be extended ("grown") into a space filling ideal quasicrystal in (uncountably) many different ways. A finite size crystal is deterministic in this way, while a finite quasicrystal is not.

The range of the order which is left undetermined by a given quasicrystal fragment can be measured in a number of ways. One way is to introduce the entropy E of the disorder, compatible with the fragment. It can be defined in terms of the volumes of the smallest and the largest acceptance windows capable to reproduce precisely the points of the quasicrystal in the fragment. The windows are denoted in the article $\Omega_{\min}$ and $\Omega_{\max}$ respectively. In the case of an infinite quasicrystal, one has $\Omega_{\min} = \Omega_{\max}$ by definition of the cut and project quasicrystal. In the case of a fragment, one always has $\Omega_{\min} \subset \Omega_{\max}$. Therefore we put $E = -r \ln r$, where r is the ratio of volumes of the acceptance windows $\Omega_{\min}$ and $\Omega_{\max}$, (see (17)).

Quasicrystal data may be given by the bright spots of a diffraction pattern, or as the atom positions of a direct experiment, or taken from a computer generated model [21]. In all cases it is a finite quasicrystal sample. The purpose of this article is to develop well defined and rigorously justified tools for analysis of the quasicrystal data, provided the quasicrystals are of the cut and project type [5, 9, 20]. Most quasicrystal definitions can be (re)interpreted in this way [17]. In this article we take an experimental quasicrystal diffraction pattern to be a fragment of just another (centrally symmetric) cut and project quasicrystal. A rigorous justification of this should be appropriate, given the complication arising from various sensitivities of measuring instruments. We shall deal with this question elsewhere. An implicit justification of our assumption here is found for example in [8].

The position of every point of an ideal (*i.e.* infinite size) cut and project quasicrystal in any dimension is determined by its "acceptance window". It is therefore interesting to ask the opposite question: what can we learn about the acceptance window from a finite size fragment of a quasicrystal? Surprisingly, no solution of such a natural problem appears to be found in the literature, although the question is evidently important. In it one is asking for a confrontation of the well known model with quasicrystal data.

Our analysis is based on a rather general definition of (cut and project) quasicrystals [5]. It is assumed that the quasicrystal points have coordinates, relative to an appropriate basis, of the $\mathbb{Z}[\tau]$ type (see Eq. (4)), where $\tau = \frac{1}{2}(1 + \sqrt{5})$. The definition (5) of a quasicrystal $\Sigma(\Omega)$ involves a bounded region Ω, called the acceptance window, which together with a mapping (star map) selects the points of $\Sigma(\Omega)$ from a dense set.

In this article we consider only the 1-dimensional problem. First of all, it is an indispensable tool for the study of higher dimensional analogues of our problems. Secondly 1-dimensional subquasicrystals are plentiful and everpresent in multidimensional quasicrystals. Every straight line, through two points in a multidimensional quasicrystal, contains a 1-dimensional quasicrystal. Thus information about acceptance windows-intervals of 1-dimensional subquasicrystals provides information about linear sizes of higher dimensional acceptance windows.

Many properties of 1-dimensional quasicrystals are generally known, some of them are part of the folklor of the field. But it proved rather difficult to find where they are actually demonstrated. Most recently the structure of 1-dimensional quasicrystals was therefore studied again in [11]. Some of the properties rigorously established in [11] and much used in this paper, are the following (see also three examples in Fig. 1). The symbols used to denote quasicrystal point sets are defined in equation (5).

- One dimensional quasicrystals (with connected acceptance window-interval) split into two classes: 2-tile and 3-tile quasicrystals.

- The 2-tile ones are singled out by the length of their acceptance windows which have to be equal to integer powers of τ: $\mathrm{vol}(\Omega) = \tau^k$, $k \in \mathbb{Z}$. Lengths of their tiles are in the ratio $1 : \tau$. All quasicrystals with certain fixed length of the acceptance window form one LI class [1]. However, among them there are still uncountably many non-equivalent quasicrystals, in the sense that any pair of them is not related by an affine mapping.

- Among all the 2-tile quasicrystals, the so called Fibonacci chains, form an infinite discrete subset. Their acceptance windows are bounded by points of $\mathbb{Z}[\tau]$. A general Fibonacci chain $\Sigma([c, c + \tau^k))$, $c \in \mathbb{Z}[\tau]$ differs from $\Sigma([c, c + 1))$ by uniform rescaling by the factor $(-\tau)^{-k}$. In most problems it suffices to consider only $\Sigma([0, 1))$ as representing them all. A choice of the origin is rarely an issue. In this case a curious phenomena may occur. Adding or removing the boundary points c, or $c + \tau^k$ of the acceptance interval, one may create an exceptional third tile, so that if the acceptance window is closed, the exceptional tile is smaller than the others, if the window is open, it is larger. Always the ratio of all tiles is $1 : \tau : \tau^2$.

- Quasicrystals with $\mathrm{vol}(\Omega) \neq \tau^k$ have three tiles with lengths in the ratio $1 : \tau : \tau^2$. The adjacent tiles are either of the same length, or in the ratio $1 : \tau$, which follows from the property of τ-inflation invariance. Three tiles of the same length are never adjacent in a $\mathbb{Z}[\tau]$-quasicrystal.

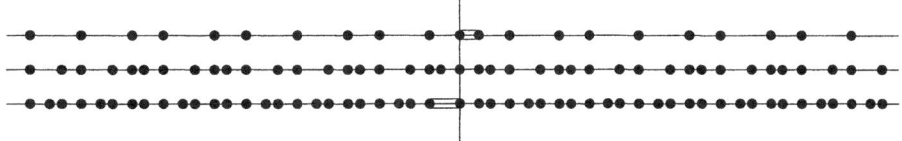

Fig. 1. — Vertically aligned fragments of three 1-dimensional quasicrystals $\Sigma([0, 1])$, $\Sigma([-1, 1])$, and $\Sigma((-1, \tau))$. For the definition see (5). The first and the third one are generally 2-tile quasicrystals with one exceptional tile, which is marked by the triple line. The second quasicrystal has three tiles. Vertical line indicates alignement at the point 0.

From now on we assume that a finite 1-dimensional point set $S \subset \mathbb{Z}[\tau]$ is given. We denote its convex hull by $\langle S \rangle$. The first problem we must face is to recognize whether a given fragment $\langle S \rangle$ is a piece of a quasicrystal. More precisely, we seek to determine, whether there exists a quasicrystal $\Sigma(\Omega)$ whose intersection with the fragment consists precisely of points of S,

$$\Sigma(\Omega) \cap \langle S \rangle = S. \tag{1}$$

We say that $\langle S \rangle$ is a quasicrystal fragment if (1) holds for some Ω.

In the given fragment we are given only relative distances between points and no absolute scale, while the points of $\Sigma(\Omega)$ are given on the real axis. The intersection between the two implies that a scale and an origin are chosen for S. The choices are not unique, but the resulting ambiguity does not lead to complications.

The property of S, which allows us to decide whether there exists a $\Sigma(\Omega)$, containing $\langle S \rangle$ as its fragment, is the τ-inflation (or quasiaddition) invariance [12]. More precisely, we introduce the local invariance as the restriction of the τ-inflation invariance property of $\Sigma(\Omega)$ to a finite set S (Def. 3.1).

Suppose that the first question was answered positively, *i.e.* there are no other quasicrystal points in the volume $\langle S \rangle$ than those of S. Thus we have the relation (1). It is well known that a quasicrystal cannot be specified precisely by its finite subset S. Any finite S which is contained in a quasicrystal is found in (infinitely) many non-equivalent ones. Therefore our aim is to determine all quasicrystals containing the given S. Hence our problem is to find all Ω verifying (1).

The smallest among the acceptance windows satisfying (1) is denoted $\Omega_{\min}$. Determination of $\Omega_{\min}$ is rather natural. It is given as the convex hull $\langle S' \rangle = \Omega_{\min}$ of the star map image of S. Exact determination of the largest interval $\Omega_{\max}$, verifying (1), is one of the results of this paper. It exploits the duality of S and its star map image S'. Indeed, in the same way that $\langle S' \rangle$ is an acceptance window for the points of $S \subset \Sigma(\langle S' \rangle)$, $\langle S \rangle$ can be also taken as an acceptance window for points S' of the (dual) quasicrystal $\Sigma(\langle S \rangle)$.

In order to facilitate the use of the results in this article, we now formulate a step-by-step procedure. Justification of the steps is the content of the paper. The first three steps of the procedure below are dealt with in Section 3, and step 4 is justified in Section 4.

1. We are given a finite point set S. In order that S be a fragment of a quasicrystal, it should be locally invariant under τ-inflation. It is a property which is practically verified by visual inspection.

2. We choose the scale of the picture so that the coordinates of points are in $\mathbb{Z}[\tau]$. This is possible because of the local τ-inflation invariance of the set.

3. Use the star map operation on each point of the set and find the smallest interval $\Omega_{\min}$ containing the star map images of the points of S. Then $\Omega_{\min}$ is the lower bound for the acceptance window Ω.

4. The upper limit $\Omega_{\max}$ for the acceptance window is determined by the number of the given points. The larger the number of points, the closer the upper bound is to the acceptance window Ω. In fact, our conclusion is more curious than that: $\Omega_{\max}$ is determined by the number of largest distances-tiles between adjacent points in S, rather than by the total number of points in it. Enlarging S so that no new largest tile appears in it, does not change the interval $\Omega_{\max}$.

In Section 5 the method is illustrated by examples of two types, both of them being 2-dimensional quasicrystal fragments. In each case we use several 1-dimensional sub-quasicrystals in order to determine bounds on the acceptance window in the appropriate direction.

In the first example our method is explained on a computer generated quasicrystal fragment. In the example of the second type we analyze quasicrystal diffraction data of [2].

Let us emphasize that the bounds on the acceptance window in the 1-dimensional case, which we find here, are not estimates but exact values: adding to $\Omega_{\max}$ or removing from $\Omega_{\min}$ even a single point would prevent the equality in (1). Such a result cannot be repeated in any dimension > 1. Only in dimension 1, any connected Ω is automatically convex. In higher dimensions the minimal window can be enlarged to a convex window satisfying (1) in a number of different ways. However, the union of such enlargements is not necessarily convex. Nevertheless, one can learn a lot about bounds on Ω along any direction. We illustrate how to use such possibilities in the 2-dimensional examples in Section 5. It is conceivable that the acceptance windows of physical quasicrystals are well behaved: convex with rather smooth boundaries. In that case the analysis of 1-dimensional samples can provide good estimates of the bounds on the higher dimensional acceptance window. Only after the analysis suggested here has been applied to sufficiently many diverse experimental data, will we be in position to decide whether such an expectation is justified.

2.   NOTATION AND AUXILIARY FACTS

Let τ and τ' be the roots of the algebraic equation $x^2 = x + 1$,

$$\tau = \frac{1 + \sqrt{5}}{2}, \qquad \tau' = \frac{1 - \sqrt{5}}{2}. \tag{2}$$

The "golden mean" τ is a Pisot number. The two important relations satisfied by τ and τ' follow from the fact that they are solutions of the mentioned quadratic equation.

$$\tau + \tau' = 1, \qquad \tau' = -\frac{1}{\tau}. \tag{3}$$

The quadratic extension $Q[\sqrt{5}] \equiv Q[\tau]$ of the rational numbers Q is the algebraic number field which is crucial in our consideration. The correspondence between the two roots τ and τ' defines an automorphism on $Q[\tau]$, denoted by $'$, by:

$$(a + b\tau)' = a + b\tau' = a - \frac{b}{\tau} \quad \text{for all} \ a, b \in Q.$$

In analogy with more dimensional structures the automorphism is sometimes called the star map.

This automorphism $'$ is everywhere *discontinuous*, *i.e.* in any neighbourhood of some given element x, there is a point y such that the distance between x' and y' is arbitrarily large.

The ring of integers of $Q[\tau]$, is the set,

$$\mathbb{Z}[\tau] = \{a + b\tau \mid a, b \in \mathbb{Z}\}. \tag{4}$$

The ring $\mathbb{Z}[\tau]$ is dense in the set of all real numbers $\mathbb{R}$. It is a unique factorization domain, hence the greatest common divisor of a subset F of $\mathbb{Z}[\tau]$ is well defined. If $d = \gcd\{F\}$ is the greatest common divisor of a set $F \subset \mathbb{Z}[\tau]$, then d' is the greatest common divisor of the set F' of conjugated points to points of F.

The definition of a cut and project quasicrystal is given for simplicity only in the one dimensional case, which is studied in this article. However, the generalization for higher dimensions is straightforward; it can be found for example in [5,11]. Such a generalization to quasicrystals in $\mathbb{R}^n$, $n > 1$, involves a choice of basis in $\mathbb{R}^n$, $\mathbb{Z}[\tau]$-span of the basis and determination of the star map images of the basis vectors.

Definition 2.1. *Let $\Omega \subset \mathbb{R}$ be a bounded interval. A* **quasicrystal** *$\Sigma(\Omega)$ is a set of points*

$$\Sigma(\Omega) = \{x \in \mathbb{Z}[\tau] \mid x' \in \Omega\}. \tag{5}$$

Ω *is called the* **acceptance window** *of the quasicrystal.*

It is known [17], that the boundedness of the acceptance interval (if it is non degenerated) assures the Delone property of $\Sigma(\Omega)$. Since Ω is convex, $\Sigma(\Omega)$ has many inflation symmetries [14].

The structure of one dimensional quasicrystals is studied in [11]. Some of their properties are mentioned in the Introduction and illustrated on examples in Figure 1.

Throughout the article proofs of our statements are based on the notion of τ-expansion [6,7,19]. A τ-*expansion* is an infinite sequence $(x_i)_{k \geq i \geq -\infty}$, satisfying

$$x = \sum_{i=-\infty}^{k} x_i \tau^i, \qquad x_i \in \{0, 1\}, \tag{6}$$

for certain integer $k \geq 0$. The digits x_i of τ-expansion can be computed by the Rényi algorithm [6,19]. In order to have the τ-expansion unique, the sequence x_i has to fulfill $x_i x_{i-1} = 0$ for any $i \leq k$, which means that two 1 are never adjacent in the sequence $(x_i)_{k \geq i \geq -\infty}$. This follows from the identity $\tau^j + \tau^{j+1} = \tau^{j+2}$, for any $j \in \mathbb{Z}$. For negative real numbers we put $x_i = -|x|_i$, negatives of τ-expansion coefficients for $|x|$.

If an expansion ends in infinitely many zeros, it is said to be *finite*, and the zeros at the end are omitted. The set $Fin(\tau)$ of all numbers with finite τ-expansion coincides with the ring $\mathbb{Z}[\tau]$. In particular, the τ-expansion of any $x \in \mathbb{Z}[\tau] \cap (0, 1)$ contains only negative powers of τ,

$$x = \sum_{i=1}^{k} \alpha_i \tau^{-i}, \qquad \alpha_i \in \{0, 1\}; \, \alpha_i \alpha_{i+1} = 0. \tag{7}$$

An interesting property of numbers with finite τ-expansions is shown in [4].

Lemma 2.2. *Let* $x, y \in \mathbb{Z}[\tau]$, *with* τ-*expansions*

$$x = \sum_{i=j}^{k} x_i \tau^i, \qquad y = \sum_{i=l}^{m} y_i \tau^i.$$

Then the sum $x + y$ *has the* τ-*expansion*

$$z := x + y = \sum_{i=\min\{j,l\}-2}^{n} z_i \tau^i, \qquad \text{for some } n \in \mathbb{Z}.$$

A criterium to decide whether a given finite point set can be a fragment of a quasicrystal, which is given in the next section, hinges on an important property of quasicrystals called inflation invariance. An operation of s-inflation is the mapping $\vdash_s \colon \mathbb{R} \times \mathbb{R} \to \mathbb{R}$, defined by

$$x \vdash_s y := sx + (1 - s)y. \tag{8}$$

Considering x, y, and s to be elements of $\mathbb{Z}[\tau]$, one can define the operation $\dashv$ which corresponds to $\vdash$ in the following sense,

$$x' \dashv_s y' := s'x' + (1 - s')y' = (x \vdash_s y)'.$$

A quasicrystal with convex acceptance window Ω is invariant under s-inflation, if $s \in \mathbb{Z}[\tau]$ and $s' \in [0, 1]$. Indeed, taking $x, y \in \Sigma(\Omega)$, we have $(x \vdash_s y)^* = s'x^* + (1 - s')y^*$, which is a convex combination of elements of convex Ω, therefore $x \vdash y$ belongs to $\Sigma(\Omega)$. All inflation properties of cut and project quasicrystals in general dimension were studied in [10].

Among all the values of s, from the definition (8), $s = -\tau$ plays an exceptional role. If a Delone set is closed under $(-\tau)$-inflation, it is closed under s-inflations for any $s \in \mathbb{Z}[\tau]$, such that $s' \in [0, 1]$. Any such set can be identified with a cut and project quasicrystal [12]. The operation (8) for $s = -\tau$ was introduced by Berman and Moody in [3] and is called τ-*inflation* or *quasiaddition*. In this case we omit the subscript s in the notation, using simply $\vdash$. The corresponding $\dashv$ operation is called τ'-inflation.

The operation of quasiaddition can be used to generate the quasicrystal [12]. In the next section we use the following two lemmas from [12]. For that we need the following notation. Let S be a point set in $\mathbb{R}$. The smallest set closed under $\vdash$ is called the inflation closure of S. It is denoted by $S^\vdash$; similarly one defines $S^\dashv$.

Lemma 2.3. *All points of $\mathbb{Z}[\tau]$ belonging to the interval $[0,1]$ can be generated using the operation of τ'-inflation starting from 0 and 1,*

$$\mathbb{Z}[\tau] \cap [0,1] = \{0,1\}^\dashv.$$

Lemma 2.4. *Let S be a subset of $\mathbb{Z}[\tau]$ containing 0 and denote $\alpha = \gcd\{S'\}$. Then*

$$(S')^\dashv = \langle S' \rangle \cap \alpha\mathbb{Z}[\tau].$$

Moreover, if $\alpha = 1$, then

$$S^\vdash = \Sigma(\langle S' \rangle).$$

3. LOCAL INVARIANCE AND THE FORWARD GROWTH

Our aim is to associate to the given point set S cut and project quasicrystals in such a way that (i) the points of S are contained in the quasicrystals, and (ii) no other points of the quasicrystals would be found in between them. More precisely, we are looking for a convex acceptance window Ω such that the intersection of the quasicrystal $\Sigma(\Omega)$ with the convex hull $\langle S \rangle$ of the set S contains precisely the points of S, see (1).

Determination of all such windows Ω is the aim of the four steps of the procedure in Introduction. In this Section we deal with the first three steps. In the step one and two we have to make precise the starting assumptions concerning the given point set. Then we can demonstrate some of its properties. The step three of the procedure is crucial. In order to justify it, we have to prove that the points in S are all in $\Sigma(\Omega)$, and that no new points are added to $\langle S \rangle$ by the requirement of the closure of $\Sigma(\Omega)$ under the quasiaddition (*i.e.* global τ-inflation invariance of $\Sigma(\Omega)$), see Lemma 3.2, and Proposition 3.3.

Given a finite point set S which may be given by the bright spots of a diffraction pattern, or as the atom positions of a direct experiment, or taken from a computer generated model. In all the cases the set is finite. In order that the given set is a part of a quasicrystal, it has to display some quasicrystal atributes. The pertinent property of all quasicrystals with convex acceptance window is the τ-inflation invariance. Its restriction to a finite set is called local τ-inflation invariance (Def. 3.1).

The global τ-inflation invariance of a set Λ implies that for any $x, y \in \Lambda$ also $x \vdash y \in \Lambda$. Geometrically it means that to each pair $x, y \in \Lambda$ we find the point

$x \vdash y$ at a distance $\tau|y - x|$ from x, on the straight line fixed by x, y. Clearly, global τ-inflation invariance requires that, Λ containing more than one point, contains infinitely many of them. Therefore a finite set S cannot verify the requirement. In a finite point set S, which should be a one piece fragment of a τ-inflation invariant infinite one, we have to admit that $x \vdash y$ may fall outside of the set.

Suppose that a 1-dimensional set S satisfies the restricted condition of inflation invariance, *i.e.* for each x, y we either find $x \vdash y$ in S, or it falls outside of the fragment (*i.e.* the convex hull of S denoted by $\langle S \rangle$), one can deduce (see [11,12]) that the ratio of adjacent distances is τ, 1, or τ^{-1}. Therefore we can assume without loss of generality that the length of the tiles are 1, τ, and τ^2. In the case where only two different tiles occur, we consider length τ and τ^2. This fact implies that we have set the scale and the origin in S in such a way that $S \subset \mathbb{Z}[\tau]$ and the greatest common divisor of the set S is equal to 1, which accomplishes the step 2 of Introduction. Hence we have the local τ-inflation invariance (or local invariance for simplicity).

Definition 3.1. *A set $S = \{a_1 < \ldots < a_n\} \subset \mathbb{Z}[\tau]$ containing zero is said to be* **locally invariant** *under quasiaddition if it satisfies the following conditions:*

(i) *the ratio $(a_{i+1} - a_i)/(a_i - a_{i-1}) \in \{1, \tau, \tau^{-1}\}$, for any $i = 2, \ldots, n-1$.*

(ii) $\gcd\{S\} = 1$.

(iii) *For any $x, y \in S$, either $x \vdash y \in S$, or $x \vdash y$ does not belong to $\langle S \rangle$.*

Our aim is to show that a locally invariant set is complete in the following sense. The τ-inflation closure $S^{\vdash}$ of S does not bring new points into $\langle S \rangle$, (Prop. 3.3). Since the closure $S^{\vdash}$ is a quasicrystal with convex acceptance window (see the main theorem of [12]), we then conclude that S is a fragment of the quasicrystal.

There still remains an ambiguity because a finite size fragment does not determine the quasicrystal uniquely. We deal with that problem in the next Section.

The proof of Proposition 3.3 requires a property established as Lemma 3.2. Namely that the τ-inflation growth of a point set from two seeds proceeds outwards.

Lemma 3.2. *Let $x \in \{0, 1\}^{\vdash}$, $x > 1$. Then there exist $y, z \in \{0, 1\}^{\vdash}$ such that*

$$0 \leq y < z < x \quad and \quad x = y \vdash z.$$

Proof. Let $x \in \{0, 1\}^{\vdash}$. Due to Lemma 2.3, one has $x' \in [0, 1] \cap \mathbb{Z}[\tau]$. To prove the statement, it suffices to show that there exist $y, z \in \mathbb{Z}[\tau]$, $y', z' \in [0, 1]$, $0 \leq y, z < x$, such that $x' = \frac{1}{\tau}y' + \frac{1}{\tau^2}z'$.

Any $w' \in [0, 1] \cap \mathbb{Z}[\tau]$ can be written in its τ-expansion as

$$w' = \frac{a_1}{\tau} + \frac{a_2}{\tau^2} + \ldots + \frac{1}{\tau^k},$$

for some $k \in \mathbb{N}$. From the properties of τ-expansion, the condition $w = -a_1\tau + a_2\tau^2 - \ldots + (-\tau)^k > 0$ is satisfied if and only if k is even.

Let

$$x' = \frac{b_1}{\tau} + \frac{b_2}{\tau^2} + \ldots + \frac{1}{\tau^{2k}}$$

be the τ-expansion of an $x > 0$. Suppose at first that x' has only even powers in its expansion. If $x' = 1/\tau^2$, then $x = \tau^2 = 0 \vdash 1$ and the statement of the lemma is satisfied for $y = 0$, $z = 1$. Let now

$$x' = \frac{b_2}{\tau^2} + \frac{b_4}{\tau^4} + \ldots + \frac{1}{\tau^{2k}} = \frac{1}{\tau}\underbrace{\left(\frac{b_2}{\tau^2}\right)}_{y'} + \frac{1}{\tau^2}\underbrace{\left(\frac{b_2}{\tau^2} + \frac{b_4}{\tau^2} + \ldots + \frac{1}{\tau^{2k-2}}\right)}_{z'},$$

with $k > 1$. Clearly, $y' \in [0,1] \cap \mathbb{Z}[\tau]$. Further $0 \le y = b_2\tau^2 < x = b_2\tau^2 + \ldots + \tau^{2k}$. Since

$$z' = \frac{b_2}{\tau^2} + \left(\frac{b_4}{\tau^2} + \ldots + \frac{1}{\tau^{2k-2}}\right) < \frac{1}{\tau^2} + \sum_{i=1}^{\infty}\frac{1}{\tau^{2i}} = \frac{1}{\tau^2} + \frac{1}{\tau} = 1$$

and

$$x - z = (\tau^4 - \tau^2)b_4 + \cdots + (\tau^{2k} - \tau^{2k-2})\cdot 1 > 0,$$

the condition $z' \in [0,1] \cap \mathbb{Z}[\tau]$ and $0 \le z < x$ is satisfied.

Suppose now that the τ-expansion of x' contains also odd powers of τ. If $b_1 = 1$, we may write

$$x' = \frac{1}{\tau} + \frac{b_3}{\tau^3} + \ldots + \frac{1}{\tau^{2k}} = \frac{1}{\tau}\cdot 1 + \frac{1}{\tau^2}\underbrace{\left(\frac{b_3}{\tau} + \ldots + \frac{1}{\tau^{2k-2}}\right)}_{z'},$$

so that $x = 1 \vdash z = -\tau + \tau^2 z$, i.e. $z = (x+\tau)/\tau^2 < x$. Clearly, $z' \in [0,1] \cap \mathbb{Z}[\tau]$.

Now we assume that $b_1 = 0$ and that $2r - 1$ is maximal odd power in the τ-expansion of x', i.e. $b_{2r-1} = 1$ and $b_{2j-1} = 0$ for all $j > r$. Clearly, $r < k$.

$$\begin{aligned}
x' &= \frac{b_2}{\tau^2} + \ldots + \frac{1}{\tau^{2r-1}} + \frac{b_{2r+2}}{\tau^{2r+2}} + \ldots + \frac{1}{\tau^{2k}} = \\
&= \frac{1}{\tau}\underbrace{\left(\frac{b_2}{\tau} + \ldots + \frac{1}{\tau^{2r-2}}\right)}_{y'} + \frac{1}{\tau^2}\underbrace{\left(\frac{b_{2r+2}}{\tau^{2r}} + \ldots + \frac{1}{\tau^{2k-2}}\right)}_{z'}.
\end{aligned}$$

Obviously both $y', z' \in [0,1] \cap \mathbb{Z}[\tau]$. We have to show that $0 < y, z < x$. From the τ-expansion of y' and z' it is clear that $0 < y < z$. Therefore we have $x = -\tau y + \tau^2 z > -\tau z + \tau^2 z = z$. This completes the proof of the lemma. □

The above Lemma states that all points of the quasicrystal $\Sigma[0,1] = \{0,1\}^\vdash$, greater than 1, can be generated by τ-inflation from other non-negative points of the quasicrystal, which are closer to the origin, and which have been already generated. Therefore, while generating the set $\{0,1\}^{\vdash}$ by quasiaddition from 0 and 1, it suffices to proceed forward and avoid quasiadding points: for which the result would be closer to origin than the generators. This is of course valid for all quasicrystals $\{a,b\}^{\vdash}$, where $b - a = \tau^k$, for some $k \in \mathbb{Z}$. Similarly one builds the quasicrystal backwards from 0 and 1 into negative elements.

In the following proposition, a criterium is given so that a point set S could be a fragment of a quasicrystal. In the same time, the step 3 of the Introduction is accomplished.

Proposition 3.3. *Let $S \subset \mathbb{Z}[\tau]$ be a set locally invariant under quasiaddition. Then*

$$S^{\vdash} \cap \langle S \rangle = S.$$

In particular, S is a fragment of a cut and project quasicrystal with acceptance window $\langle S' \rangle$,

$$\Sigma\left(\langle S' \rangle\right) \cap \langle S \rangle = S.$$

Proof. Due to Lemma 2.4, it suffices to show that we have the implication

$$x \in \langle S \rangle \text{ and } x' \in \langle S' \rangle \implies x \in S. \tag{9}$$

We proceed recursively on the number of elements in S.

1. Clearly, if S contains only two elements, $S = \{a_1, a_2\} \subset \mathbb{Z}[\tau]$, $a_1 < a_2$, one can assume without loss of generality that $a_2 - a_1 = 1$. For such numbers we know [12] that the tiles in the quasicrystal $\{a_1, a_2\}^{\vdash}$ are of length τ, τ^2, except for the unique one between a_1 and a_2. Therefore no additional point falls in between a_1 and a_2. Therefore (9) holds.

2. Suppose $S = \{a_1, a_2, a_3\} \subset \mathbb{Z}[\tau]$, $a_1 < a_2 < a_3$. Such a set has only two tiles. Without loss of generality we choose the scale and the origin in S, such that $a_1 = 0$ and the distances in S are either 1, τ. One has the following non equivalent possibilities for S: $S = \{0, 1, \tau^2\}$, or $S = \{0, 1, 2\}$. The corresponding star map images are then $S' = \{0, \frac{1}{\tau^2}, 1\}$, $S' = \{0, 1, 2\}$. In the first case the acceptance interval has the length 1, therefore the tiles in the quasicrystal $\Sigma\left(\langle S' \rangle\right)$ are τ, and τ^2, except of the one between 0 and 1. For the second case, the length of $\langle S' \rangle = [0, 2]$ is 2. Since $2 \in (\tau, \tau^2)$, the distances in $\Sigma[0, 2]$ are $\frac{1}{\tau}$, 1, and τ, therefore no additional point can occur in $S \cap \langle S \rangle$.

3. Suppose S contains n elements $a_1 < a_2 < \ldots < a_{n-1} < a_n$. In the star map the order of elements is not preserved. Consider the sequence $i_1, \ldots, i_n$ such that $a'_{i_1} < a'_{i_2} < \ldots < a'_{i_{n-1}} < a'_{i_n}$. To show the implication (9), take $x \in [a_1, a_n]$, such that $x' \in [a'_{i_1}, a'_{i_n}]$.

(a) Suppose that $x' \in [a'_{i_2}, a'_{i_{n-1}}]$. Therefore we have both $x' \in [a'_{i_1}, a'_{i_{n-1}}]$ and $x' \in [a'_{i_2}, a'_{i_n}]$. Then necessarily either $x \in \langle a_{i_1}, \ldots a_{i_{n-1}} \rangle$, or $x \in \langle a_{i_2}, \ldots a_{i_n} \rangle$. Thus we can use the induction hypothesis to say that $x \in \{a_{i_1}, \ldots, a_{i_{n-1}}\} \subset S$, or $x \in \{a_{i_2}, \ldots, a_{i_n}\} \subset S$.

(b) Suppose that $x' \in [a'_{i_1}, a'_{i_2}] \subset [a'_{i_1}, a'_{i_1+1}]$. Then

$$x \in \Sigma[a'_{i_1}, a'_{i_1+1}] = \{a_{i_1}, a_{i_1+1}\}^{\vdash},$$

because $a_{i_1+1} - a_{i_1} = 1, \tau$, or τ^2. According to Lemma 3.2, each element x of $\{a_{i_1}, a_{i_1+1}\}^{\vdash}$ can be constructed step by step, starting from a_{i_1} and a_{i_1+1} proceeding uniquely forwards, or uniquely backwards. Since a_{i_1}, a_{i_1+1}, and x are contained in $\langle S \rangle$ and S is locally invariant, one has $x \in S$.

(c) One proceeds similarly, if $x' \in [a'_{i_{n-1}}, a'_{i_n}]$. $\square$

4. THE MAXIMAL ACCEPTANCE WINDOW

In Proposition 3.3, we have found just one window $\Omega_{\min} = \langle S' \rangle$ (the smallest one), verifying our requirement (1). However, since a finite fragment $\langle S \rangle$ does not determine the cut and project quasicrystal uniquely, there are other windows Ω, verifying the condition (1). Obviously, they have to contain the minimal window $\langle S' \rangle$.

The aim of this Section is to find the upper bound $\Omega_{\max}$ for the acceptance interval. Then any Ω between $\Omega_{\min}$ and $\Omega_{\max}$ satisfies (1), (step 4). Clearly, the greater is the given fragment, the narrower bounds for Ω we can expect.

Simultaneously with S we are given as well S'. Therefore we have also the convex hulls $\langle S \rangle$ and $\langle S' \rangle$. Our strategy now is to reverse the role of S and S'. The acceptance window $\langle S \rangle$ then determines all the points of $\Sigma(\langle S \rangle)$, in particular the nearest one to S' on each side. In the same way as we took $\langle S' \rangle$ as the (minimal) acceptance window for the infinite quasicrystal $\Sigma(\langle S' \rangle)$ containing S, we can consider $\langle S \rangle$ as the (minimal) acceptance window for the infinite quasicrystal $\Sigma(\langle S \rangle)$ containing S'. We want now to enlarge maximally the minimal window $\langle S' \rangle$ so that the new quasicrystal with enlarged window would still have the same intersection with $\langle S \rangle$ as in (1). That is equivalent to the requirement that the extension $\Omega_{\max}$ of $\langle S' \rangle$ contains the same points of $\Sigma(\langle S \rangle)$ as $\langle S' \rangle$,

$$\Sigma(\langle S \rangle) \cap \Omega_{\max} = \Sigma(\langle S \rangle) \cap \langle S' \rangle = S'.$$

Practically one simply has to find both nearest points to $\langle S' \rangle$ in $\Sigma(\langle S \rangle)$. Their position is the strict upper bound on $\Omega_{\max}$.

The explicit formulation of the result is given in the following theorem.

Theorem 4.1. *Let $S = x_1, x_2, \ldots, x_n \subset \mathbb{Z}[\tau]$ be a set locally invariant under τ-inflation, such that $x_1 < \ldots < x_n$. Without loss of generality assume that $x_{i+1} - x_i \in \{1, \tau, \tau^2\}$ and that there exists $i_0 \in \{1, \ldots, n\}$ such that $x_{i_0+1} - x_{i_0} = \tau^2$. Denote the following.*

$$c = \max\{ x'_i \mid x_{i+1} - x_i = \tau^2 \},$$
$$d = \min\{ x'_{i+1} \mid x_{i+1} - x_i = \tau^2 \}.$$

Then the interval $\left(c - \frac{1}{\tau}, d + \frac{1}{\tau}\right)$ *is the maximal acceptance window satisfying the relation (1). In particular, if* $a = \min\{x'_i\}$ *and* $b = \max\{x'_i\}$, *then* $\Sigma(\Omega) \cap [x_1, x_n] = \{x_1, \ldots, x_n\}$ *if and only if*

$$[a, b] \subset \Omega \subset \left(c - \frac{1}{\tau}, d + \frac{1}{\tau}\right).$$

Proof. We have to show following two facts. First, that the acceptance interval $\left(c - \frac{1}{\tau}, d + \frac{1}{\tau}\right)$ satisfies the property (1). Second, the mentioned interval is maximal with this property, *i.e.* adding a point to it, a new point would occur in the convex hull $\langle S \rangle$ which would in that case become the closed interval $[x_1, x_n]$.

From the definition of c and d one has

$$d \leq c + \frac{1}{\tau^2}. \tag{10}$$

It implies that the interval $\left(c - \frac{1}{\tau}, d + \frac{1}{\tau}\right)$ has the length

$$d - c + \frac{2}{\tau} \leq \frac{1}{\tau^2} + \frac{2}{\tau} = \tau.$$

A quasicrystal with acceptance window of the length smaller than τ has distances at least 1.

Let us show that the quasicrystal $\Sigma\left(c - \frac{1}{\tau}, d + \frac{1}{\tau}\right)$ satisfies (1). Suppose the opposite, that there exists $y \in \Sigma\left(c - \frac{1}{\tau}, d + \frac{1}{\tau}\right) \cap [x_1, x_n]$, such that $y \notin \{x_1, \ldots, x_n\}$. Necessarily there exists an index i, such that $x_i < y < x_{i+1}$, and $x_{i+1} - x_i = \tau^2$. Otherwise there would be a distance smaller than 1 in $\Sigma\left(c - \frac{1}{\tau}, d + \frac{1}{\tau}\right)$, which contradicts the assumption. Hence there are two possibilities for y:

$$y = \begin{cases} x_i + 1 & = x_{i+1} - \tau \\ x_i + \tau & \end{cases}$$

implying

$$y' = \begin{cases} x'_{i+1} + \frac{1}{\tau} & \geq d + \frac{1}{\tau} \\ x'_i - \frac{1}{\tau} & \leq c - \frac{1}{\tau} \end{cases}$$

respectively. Therefore $y' \notin \left(c - \frac{1}{\tau}, d + \frac{1}{\tau}\right)$, which gives the contradiction. Thus the interval $\left(c - \frac{1}{\tau}, d + \frac{1}{\tau}\right)$ satisfies (1).

Let us show the maximality of the interval. Take its extreme points $c - \frac{1}{\tau}$ and $d + \frac{1}{\tau}$. We show that their $'$ preimages fall into $[x_1, x_n]$, but that they are not points of S. Denote by $c = x'_{i_c}$, $d = x'_{i_d}$, $1 \leq i_c, i_d \leq n$, the points where the maximum and minimum in the definition of c and d were achieved. We have $x_{i_c+1} - x_{i_c} = \tau^2$ and $x_{i_d} - x_{i_d-1} = \tau^2$. Therefore

$$\left(c - \frac{1}{\tau}\right)' = \left(x'_{i_c} - \frac{1}{\tau}\right)' = x_{i_c} + \tau,$$

$$\left(d + \frac{1}{\tau}\right)' = \left(x'_{i_d} + \frac{1}{\tau}\right)' = x_{i_d} - \tau.$$

It is clear that $x_{i_c} < x_{i_c} + \tau < x_{i_c+1}$, and $x_{i_d-1} < x_{i_d} - \tau < x_{i_d}$, which means that $x_{i_c} + \tau$ and $x_{i_d} - \tau$ belong to the convex hull $[x_1, x_n]$ of S. However, they are not contained in S. They split the longest tile between x_{i_c} and x_{i_c+1}, (x_{i_d-1} and x_{i_d} respectively). This completes the proof.   □

It may happen that there is only one distance τ^2 between adjacent points of S. The equality in (10) is then satisfied. Therefore the maximal window is of the form

$$\left(c - \frac{1}{\tau}, c + \frac{1}{\tau^2} + \frac{1}{\tau} \right) = \left(c - \frac{1}{\tau}, c + 1 \right) = (\alpha, \alpha + \tau) ,$$

which is the case of a 2-tile quasicrystal, with only one exceptional longest tile. Example of such a quasicrystal is $\Sigma(-1, \tau)$, see Figure 1.

Another interesting thing to notice is that the determination of the maximal window in fact does not depend directly on the number of points in the locally invariant set S. Decisive is the occurence of longest tiles. Suppose we have several locally invariant sets ordered by inclusion. The estimate of the maximal acceptance window would be the same for all of them, if they have the same number of long tiles, whatever is the total number of points contained in the respective samples.

5. EXAMPLE: ANALYSIS OF TWO-DIMENSIONAL QUASICRYSTAL DATA

In this section we consider two 2-dimensional examples of quasicrystals with decagonal symmetries. The first example is computer generated fragment of a quasicrystal, therefore it is perfect: all the required properties for our method to be applicable are verified. The second example are quasicrystal diffraction data [2]. The data display 10-fold symmetry of Bragg peaks and can be therefore taken as a quasicrystal with central symmetry. Although the interpretation of the acceptance window is far from clear, our aim is just to illustrate the applicability of the method in this situation and to discuss the complications which may arise in comparison with the ideal computer generated model.

Both our examples are quasicrystals with H_2 symmetry. It means that their points can be viewed as points of the complex plane with $\mathbb{Z}[\tau]$ coordinates in the basis 1, and $e^{4\pi i/5}$.

The first example which we study in this Section comes from Figure 2. We start with a computer generated two dimensional quasicrystal of the type H_2 (for definition see [5]) with circular acceptance window of radius 0.7.

In these lectures we have defined only 1-dimensional quasicrystals. Quasicrystal of general dimension n can be defined in a similar way (see for example [10]). Just to explain the example here, let us point out that the dots in Figure 2 can be given as points in the complex plane with coordinates in $\mathbb{Z}[\tau]$ relative to the basis $\{1, e^{2\pi i/5}\}$. Origin is the center point and the real axis

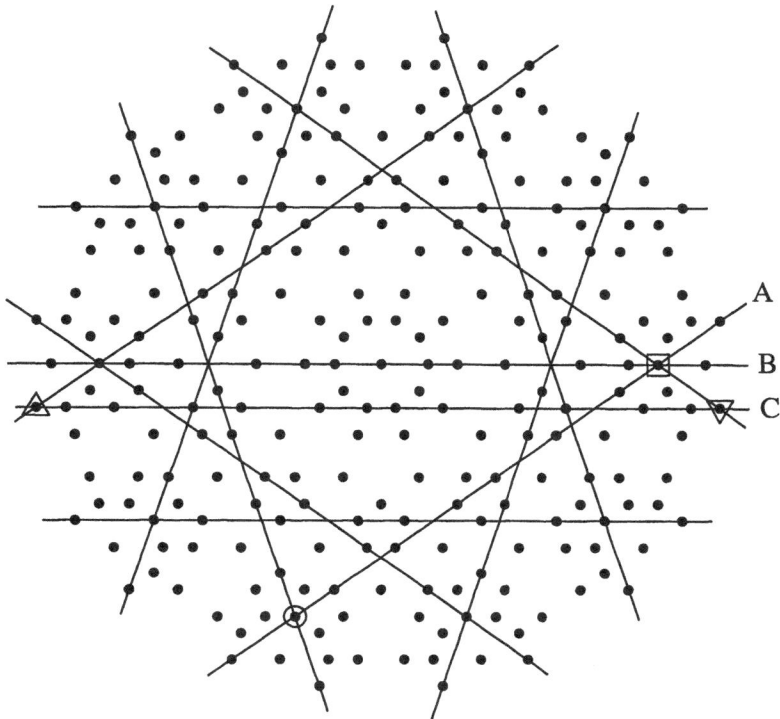

Fig. 2. — A circular view of a 2-dimensional H_2 quasicrystal is shown. The points on the line segments A, B, C constitute the examples of locally invariant sets, considered in Section 5. The acceptance window of the 2-dimensional quasicrystal is a disk of radius 0.7, centered at origin (Fig. 3). Therefore the displayed quasicrystal fragment has 10-fold rotational symmetry with respect to the origin.

is along the horizontal direction. A point $X = x_1 + x_2 e^{2\pi i/5}$, x_1, $x_2 \in \mathbb{Z}[\tau]$, is then drawn in Figure 2, provided it satisfies the acceptance condition

$$|X^*| = \left| x_1' + x_2' e^{\frac{4\pi i}{5}} \right| = \left| a + b\tau' + (c + d\tau') e^{\frac{4\pi i}{5}} \right| \leq 0.7, \qquad a, b, c, d \in \mathbb{Z}.$$
(11)

The size of the drawing and the scale chosen in Figure 2 allow one to show only the points with

$$|X| = \left| a + b\tau + (c + d\tau) e^{\frac{2\pi i}{5}} \right| \leq 12.$$
(12)

Separately, the inequalities (11) and (12) admit infinitely many points. Together they allow only 251 points shown Figure 2 and also in Figure 3.

In our first example we consider simultaneously three 1-dimensional samples of quasicrystal fragments and determine the bounds $\Omega_{\min}$, and $\Omega_{\max}$ on the size of the acceptance windows Ω for each of them. First we explain the samples as they are shown in Figures 2 and 3. Then we calculate the bounds $\Omega_{\min}$, and $\Omega_{\max}$ following the procedure described in the Introduction. Then we measure the lack of determinism which is due to the finite size of the quasicrystal data. In particular, we find the value of the entropy (17) for the three cases.

The three quasicrystal fragments chosen here have three special features.

- Firstly, the data are taken from a 2-dimensional computer generated quasicrystal we know all about. In particular, we know its acceptance window. Therefore the results of our analysis of the 1-dimensional data can be compared with the known parameters of the model.

- Secondly, the three 1-dimensional examples, being taken from the same 2-dimensional quasicrystal, provide partial information on bounds of the 2-dimensional acceptance window.

- Third, since we have chosen a 2-dimensional fragment with the 10-fold symmetry around its center, each of our 1-dimensional examples has 10 (or 5) equivalent copies along different directions through the 2-dimensional quasicrystal.

All that put together give us more restrictions on the 2-dimensional acceptance window.

In Figure 2 we have three 1-dimensional examples on the straight lines denoted by A, B, and C, containing respectively 10, 15, and 14 points. In Figure 3 we show the star map images of the points from Figure 2. Under the star map, a straight line is transformed into a straight line oriented in general differently; only the horizontal lines do not change their orientation.

Example A and its star map A^* are identified in Figure 2 and Figure 3 by the dots marked by a square and circle. The coordinates of the points are respectively

$$
\begin{aligned}
\square &\iff X = \tau^4 + \tau^2, & X^* &= \frac{1}{\tau^4} + \frac{1}{\tau^2}, \\
\bigcirc &\iff X = -(\tau^4 + \tau^2)e^{\frac{2\pi i}{5}}, & X^* &= -\left(\frac{1}{\tau^4} + \frac{1}{\tau^2}\right)e^{\frac{4\pi i}{5}}.
\end{aligned}
\tag{13}
$$

In Figure 2 we have also drawn the ten equivalent lines arising by the 10-fold rotations of A around the origin. The example A differs from the other two by having tiles of only 2 distinct length.

The example B is a 1-dimensional quasicrystal fragment containing the origin. In our coordinate choice it is found on the real axis of the complex plane. Its star map B^* is identified in Figure 3 by the origin, $0^* = 0$, and by the dot in the square (13). The global 10-fold symmetry transforms B into 5 equivalent 1-dimensional examples with the origin in common. We do not show the corresponding lines on our figures.

The example C is given by the points

$$\triangle \iff X = \tau^5 + 1 - \tau e^{\frac{2\pi i}{5}}, \qquad X^* = 1 - \frac{1}{\tau^5} + \frac{1}{\tau} e^{\frac{4\pi i}{5}},$$

$$\triangledown \iff X = -\tau^5 - 1 - \tau e^{\frac{2\pi i}{5}}, \qquad X^* = -1 + \frac{1}{\tau^5} + \frac{1}{\tau} e^{\frac{4\pi i}{5}},$$

in Figures 2 and 3. Again we have refrained from drawing the ten equivalent examples which are due to the global 10-fold symmetry.

All together we have 10 copies of A and C, and 5 copies of B as our examples. The result of our analysis are bounds on the acceptance intervals in the 25 examples. For each of them we show in Figure 3 the range of variability of the corresponding Ω as the two short line-segments on the corresponding straight line. The line-segments start at the limits of $\Omega_{\min}$ and end at the boundary points of $\Omega_{\max}$. Note that the line-segments of the two families A and C happen to have common end points.

In the main part of this section we calculate the bounds on Ω in the three types of examples. For that we can ignore their relative position in the 2-dimensional quasicrystal. We consider each one separately as if it is situated on the real axis. Let us proceed according to steps mentioned in the Introduction.

1. The first step consists in checking the local invariance under τ-inflation of the given finite point sets $S^{(A)}$, $S^{(B)}$, and $S^{(C)}$. It is satisfied, since the examples originate as fragments of 1-dimensional subquasicrystals of a 2-dimensional quasicrystal with convex acceptance window. In general the assumption of local invariance would have to be justified.

2. In the second step, one has to assign coordinates to points, which are for the moment given only by the picture. In doing that we choose the origin and set the scale of the picture. The goal is to have the coordinates in $\mathbb{Z}[\tau]$ with 1 as their gratest common divisor. The origin can be chosen anywhere, preferably as one of the fragment points. If the set is centrally symmetric like $S^{(B)}$, it is convenient to choose the symmetry center for the origin.

   For the example B, let us fix the origin in the point of the symmetry. For examples A and C the origin is chosen as one of the extreme points of $S^{(A)}$, $S^{(C)}$, respectively. Without loss of generality, let the tiles in S be of length 1, τ and τ^2. In the example $S^{(A)}$ the fragment has tiles of only 2 lengths. In order to apply Theorem 4.1 without modification, we take the two tiles to be of length τ and τ^2. We thus have the following point sets as our quasicrystal data:

$$S^{(A)} = \{0, \tau^2, \tau^3, \tau^4, \tau^4 + \tau^2, \tau^5, \tau^5 + \tau^2, \tau^5 + \tau^3 + 1, \tau^6, \tau^6 + \tau^2\},$$

$$S^{(B)} = \{0, \pm\tau, \pm\tau^2, \pm\tau^3, \pm\tau^4, \pm(\tau^4 + \tau), \pm(\tau^4 + \tau^2), \pm\tau^5\}. \qquad (14)$$

$$S^{(C)} = \{ \, 0,\, 1,\, \tau^2,\, \tau^3 + 1,\, \tau^4,\, \tau^4 + \tau^2,\, \tau^5,\, \tau^5 + 1,\, \tau^5 + \tau^2,\, \tau^5 + \tau^3 + 1,$$
$$\tau^6,\, \tau^6 + \tau^2,\, \tau^6 + \tau^3,\, \tau^6 + \tau^3 + 1 \, \}.$$

3. In the third step of the procedure, we are to determine the minimal acceptance window $\Omega_{\min} = \langle S' \rangle$ for a quasicrystal in order to satisfy (1). For that we first need to calculate the star map images of points. One easily finds

$$S^{(A)'} = \Big\{ \, 0,\, \frac{1}{\tau^2},\, -\frac{1}{\tau^3},\, \frac{1}{\tau^4},\, \frac{1}{\tau^4} + \frac{1}{\tau^2},\, -\frac{1}{\tau^5},\, \frac{1}{\tau^2} - \frac{1}{\tau^5},\, 1 - \frac{1}{\tau^3} - \frac{1}{\tau^5},\, \frac{1}{\tau^6},$$
$$\frac{1}{\tau^6} + \frac{1}{\tau^2} \, \Big\}.$$

$$S^{(B)'} = \Big\{ \, 0,\, \mp\frac{1}{\tau},\, \pm\frac{1}{\tau^2},\, \mp\frac{1}{\tau^3},\, \pm\frac{1}{\tau^4},\, \pm\Big(\frac{1}{\tau^4} - \frac{1}{\tau}\Big),\, \pm\Big(\frac{1}{\tau^4} + \frac{1}{\tau^2}\Big),\, \mp\frac{1}{\tau^5} \, \Big\}.$$

$$S^{(C)'} = \Big\{ \, 0,\, 1,\, \frac{1}{\tau^2},\, 1 - \frac{1}{\tau^3},\, \frac{1}{\tau^4},\, \frac{1}{\tau^4} + \frac{1}{\tau^2},\, -\frac{1}{\tau^5},\, 1 - \frac{1}{\tau^5},\, \frac{1}{\tau^2} - \frac{1}{\tau^5},$$
$$1 - \frac{1}{\tau^3} - \frac{1}{\tau^5},\, \frac{1}{\tau^6},\, \frac{1}{\tau^6} + \frac{1}{\tau^2},\, \frac{1}{\tau^6} - \frac{1}{\tau^3},\, 1 + \frac{1}{\tau^6} - \frac{1}{\tau^3} \, \Big\}.$$

In order to find the minimal interval containing S', *i.e.* convex hull $\langle S' \rangle$, we determine the minimal and maximal elements in S'. After a simple computation one finds

$$\Omega_{\min}^{(A)} = \langle S^{(A)'} \rangle = \Big[-\frac{1}{\tau^3},\, 1 - \frac{1}{\tau^3} - \frac{1}{\tau^5} \Big] \approx [-0.236, 0.674],$$

$$\Omega_{\min}^{(B)} = \langle S^{(B)'} \rangle = \Big[-\frac{1}{\tau},\, \frac{1}{\tau} \Big] \approx [-0.618, 0.618], \tag{15}$$

$$\Omega_{\min}^{(C)} = \langle S^{(C)'} \rangle = \Big[-\frac{1}{\tau^3} + \frac{1}{\tau^6},\, 1 \Big] \approx [-0.180, 1].$$

All acceptance windows Ω, satisfying (1), contain $\Omega_{\min}$.

4. The determination of the maximal acceptance window is the subject of the last step of the procedure. It is based on Theorem 4.1. First we find the numbers c, d of the theorem. For that, one has to examine the set (14) and/or Figure 2, and find all longest tiles, *i.e.* distances τ^2 between adjacent points. The longest tiles are clearly visible in Figure 2. The definition of c requires one to consider lower boundary points of the tiles of length τ^2. The number c is then equal to maximum of their star map images,

$$c^{(A)} = \max\Big\{ \, 0,\, -\frac{1}{\tau^3},\, \frac{1}{\tau^4},\, -\frac{1}{\tau^5},\, \frac{1}{\tau^2} - \frac{1}{\tau^5},\, \frac{1}{\tau^6} \, \Big\} = \frac{1}{\tau^2} - \frac{1}{\tau^5},$$

$$c^{(B)} = \max\left\{-\frac{1}{\tau^4}, -\frac{1}{\tau^3}\right\} = -\frac{1}{\tau^4},$$

$$c^{(C)} = \max\left\{0, \frac{1}{\tau^2}, \frac{1}{\tau^4}, \frac{1}{\tau^2} - \frac{1}{\tau^5}, \frac{1}{\tau^6}\right\} = -\frac{1}{\tau^2}.$$

Similarly, one determines d as the minimum of star map images of points, which are the upper boundary points of the tiles τ^2, *i.e.*

$$d^{(A)} = \min\left\{\frac{1}{\tau^2}, -\frac{1}{\tau^4}, \frac{1}{\tau^4} + \frac{1}{\tau^2}, \frac{1}{\tau^2} - \frac{1}{\tau^5}, 1 - \frac{1}{\tau^3} - \frac{1}{\tau^5}, \frac{1}{\tau^6} + \frac{1}{\tau^2}\right\} = \frac{1}{\tau^4},$$

$$d^{(B)} = \min\left\{\frac{1}{\tau^3}, \frac{1}{\tau^4}\right\} = \frac{1}{\tau^4},$$

$$d^{(C)} = \min\left\{1 - \frac{1}{\tau^3}, \frac{1}{\tau^4} + \frac{1}{\tau^2}, 1 - \frac{1}{\tau^3} - \frac{1}{\tau^5}, \frac{1}{\tau^6} + \frac{1}{\tau^2}\right\} = \frac{1}{\tau^6} + \frac{1}{\tau^2}.$$

Having found c and d for the three examples, we get the maximal acceptance window according to Theorem 4.1,

$$\Omega_{\max} := \left(c - \frac{1}{\tau}, d + \frac{1}{\tau}\right).$$

For the three examples one has

$$\Omega_{\max}^{(A)} = \left(-\frac{1}{\tau^3} - \frac{1}{\tau^5}, \frac{1}{\tau} + \frac{1}{\tau^4}\right) \approx (-0.326, 0.764),$$

$$\Omega_{\max}^{(B)} = \left(-\frac{1}{\tau^4} - \frac{1}{\tau}, \frac{1}{\tau^4} + \frac{1}{\tau}\right) \approx (-0.764, 0.764), \tag{16}$$

$$\Omega_{\max}^{(C)} = \left(-\frac{1}{\tau^3}, 1 + \frac{1}{\tau^6}\right) \approx (-0.236, 1.056).$$

Let us now underline that we have chosen the scale and origin in the three 1-dimensional examples arbitrarily. Therefore the position of $\Omega_{\min}$ and $\Omega_{\max}$ is determined up to an arbitrary translation by a distance in $\mathbb{Z}[\tau]$, and the length of the windows are found up to a multiplication by a power τ^k, $k \in \mathbb{Z}$. Our three examples come from a single 2-dimensional quasicrystal. In each of them the long tile is of the same length, which we set to be τ^2. Therefore the scales in the examples are the same. Thus we can compare the length of acceptance windows for A, B, and C,

$$\text{vol}\left(\Omega_{\min}^{(A)}\right) = 1 - \frac{1}{\tau^5} \approx 0.910, \qquad \text{vol}\left(\Omega_{\max}^{(A)}\right) = 1 + \frac{1}{\tau^2} + \frac{1}{\tau^5} \approx 1.090,$$

$$\text{vol}\left(\Omega_{\min}^{(B)}\right) = \frac{2}{\tau} \approx 1.236, \qquad \text{vol}\left(\Omega_{\max}^{(B)}\right) = \frac{2}{\tau} + \frac{2}{\tau^4} \approx 1.528,$$

$$\text{vol}\left(\Omega_{\min}^{(C)}\right) = 1 + \frac{1}{\tau^3} - \frac{1}{\tau^6} \approx 1.180, \quad \text{vol}\left(\Omega_{\max}^{(C)}\right) = 1 + \frac{1}{\tau^3} + \frac{1}{\tau^6} \approx 1.292.$$

The short line segments in Figure 3 determine the extension of the linear acceptance windows of any of the three families of straight lines. The line segment shows the direction of the window and ends exactly at the boundary point of the corresponding $\Omega_{\max}$.

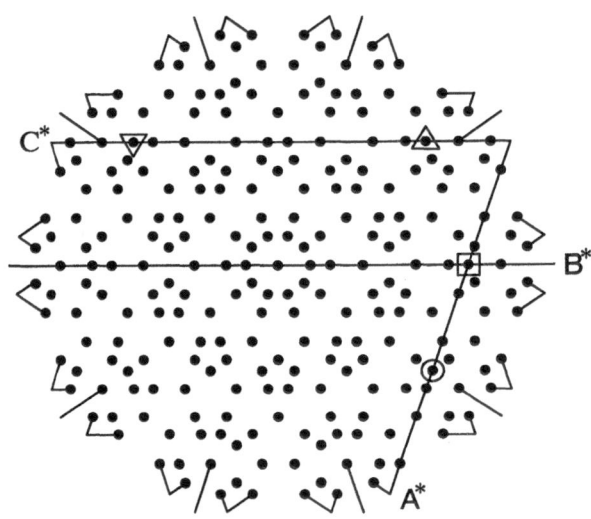

Fig. 3. — Star map images of points in Figure 2. Marked dots are images of the dots from Figure 2 with the same markings; A^*, B^*, C^* are the images of A, B, C. The short lines indicate the direction and the range of the uncertainty in the corresponding acceptance interval Ω.

Let us now consider the precision of the estimates constructed by the above procedure. It can be measured by an entropy

$$E = -\frac{\mathrm{vol}(\Omega_{\min})}{\mathrm{vol}(\Omega_{\max})} \ln\left(\frac{\mathrm{vol}(\Omega_{\min})}{\mathrm{vol}(\Omega_{\max})}\right). \qquad (17)$$

The values of the entropy of the fragments of straight lines a, b, and c, according to (17), is

$$E^{(A)} = -\frac{5\tau + 2}{5\tau + 4} \ln\left(\frac{5\tau + 2}{5\tau + 4}\right) \approx 0.151,$$

$$E^{(B)} = -\frac{2\tau + 1}{2\tau + 2} \ln\left(\frac{2\tau + 1}{2\tau + 2}\right) \approx 0.171,$$

$$E^{(C)} = -\frac{10\tau + 5}{10\tau + 7} \ln\left(\frac{10\tau + 5}{10\tau + 7}\right) \approx 0.082.$$

Note that the case A with lowest number of points (ten) does not imply the highest value of entropy. This is caused by the higher number of longest tiles τ^2 in it. The highest of three values of entropy is achieved by the example B with 15 points.

In order to appreciate the difference between $\Sigma\left(\Omega_{\min}\right)$ and $\Sigma\left(\Omega_{\max}\right)$, let us choose the example B and compare (Fig. 4) larger fragments of these quasicrystals than in Figures 2 and 3 with the corresponding size fragment of the original quasicrystal $\Sigma(-0.7, 0.7)$. Since the diameter of the original acceptance window is 1.4, which is greater than 1 and less than τ, we know from [11] that the distances between adjacent points are exactly 1, τ, and τ^2. So that our chosen scale coincides with that of the computed 2-dimensional example.

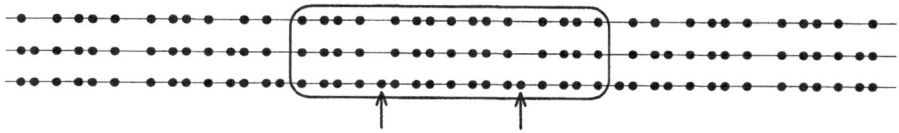

Fig. 4. — Three aligned fragments of quasicrystals are shown. They coincide in the marked region, except for the two dots indicated by arrows. The two points added in the third example are due to inclusion of the extreme points to the maximal acceptance window. Had we taken the true $\Omega_{\max}$, which is open, the marked points would be absent.

In Figure 4, three quasicrystals are shown,

$$\Sigma\left(\Omega_{\min}^{(B)}\right) = \Sigma\left(\left[-\frac{1}{\tau}, \frac{1}{\tau}\right]\right) \subset \Sigma(-0.7, 0.7) \subset \Sigma\left(\overline{\Omega}_{\max}^{(B)}\right) = \Sigma\left(\left[-\frac{1}{\tau^3}, 1+\frac{1}{\tau^6}\right]\right).$$

Here $\overline{\Omega}_{\max}^{(B)}$ denotes the closed acceptance window, unlike the $\Omega_{\max}^{(B)}$ from (16), which is open. By taking the closed interval, we have added the two (boundary) points to the maximal acceptance window, which is open. The preimages of the two points appear as two new dots in the convex hull $\langle S^{(B)} \rangle$. Hence the open window is really maximal (Fig. 4). The three quasicrystals are ordered by inclusion. They coincide in the convex hull $\langle S^{(B)} \rangle$ marked in Figure 4. The arrows indicate the two points which would be absent, had we taken the open acceptance interval $\Omega_{\max}^{(B)}$.

Our second example is to analyze the experimental diffraction data [2] shown in Figure 5.

The majority of quasicrystal data originate from X-ray diffraction experiments. It is known that an ideal diffraction pattern would be densely covered

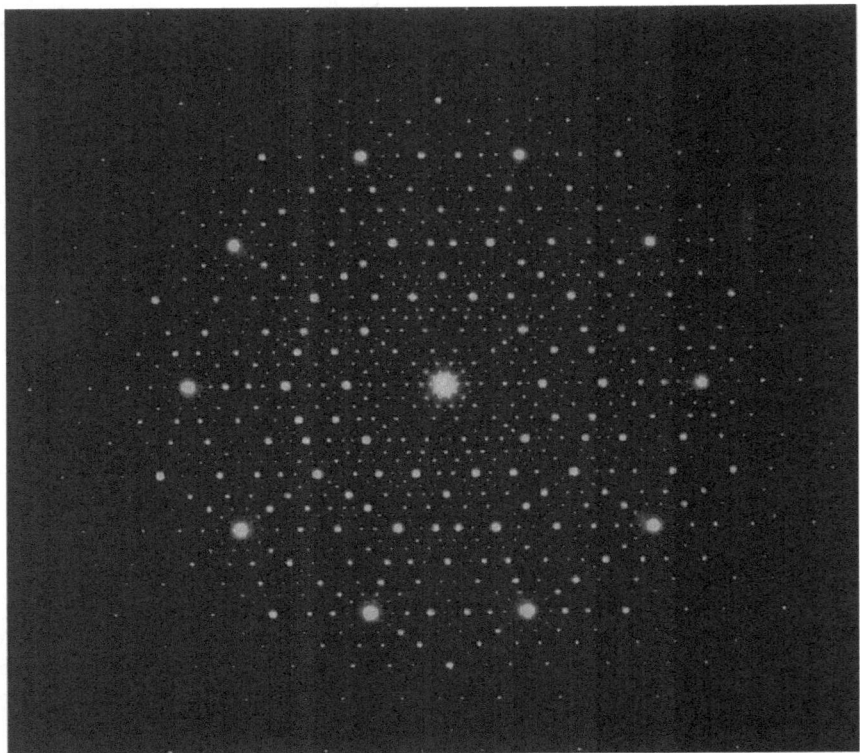

Fig. 5. — Electron diffraction pattern of the Ni-rich basic decagonal state of Al-Co-Ni along the tenfold projection. Taken from [2], p. 102, Figure 1a.

by diffraction peaks of various intensities [8, 9, 20]. Fortunately the limited sensitivity of detectors saves the situation: only a discrete set of peaks gets detected. It turns out that the centrally symmetric (Delone) point set which arises in this way has virtually all the properties of quasicrystals. Namely, one can associate to its points coordinates in $\mathbb{Z}[\tau]$, the lengths between adjacent points along any direction are in the ratio τ, there occur only 2 or 3 distinct tiles on any straight line, etc.

The first striking feature of the data is the different intensity of various points, which is further complicated by the fact that the density of points and their intensity is not uniform accross the fragment. For our analysis we have to cut off the exterior part of the data. Even then one faces a difficulty in deciding which points are still visible on the figure. It is not difficult, although somewhat laborious to assign to each point its two coordinates in $\mathbb{Z}[\tau]$, similarly as in (13).

It naturally involves setting up the scale and choosing the origin. We take the center of the figure to be the origin of the coordinate system and the minimal distance between visible points to be 1.

Due to the 10-fold symmetry around the center it is sufficient to assign the coordinates to points within a segment with angle $2\pi/10$ at the vertex in the origin. The remaining points of the figure may be obtained by reflection with respect to sides of the segment. In fact, a detailed comparison of the ten segments reveals certain discrepancy. Some of the most faint points may be missing in several fragments, apparently due to non uniform sensitivity accross the picture.

As in the computer generated example before, we now consider 1-dimensional sub-quasicrystals in the figure.

First let us select the five straight lines with origin as their common point, which are the most densely occupied by the Bragg peaks. We consider only segments of these lines, containing 19 points on each side of the center, (limited by the bright spots of the exterior decagon of the radius τ^7 in our scale). Making no distinction between the faint and bright points, one can verify the presence of the local invariance under the quasiaddition. As before we determine bounds on the linear sizes of the acceptance window along the five directions.

Next we choose another fragment of a 1-dimensional sub-quasicrystal containing relatively many points. Our choice is the line segment bounded by the complex numbers τ^7 and $\tau^7 e^{3\pi i/5}$, (real axis in the horizontal direction). There are ten copies of this line symmetrically placed around the origin. They should be populated by the same 26 points. In fact, a detailed comparison reveals that the copies differ by one or two of the faintest points. The requirement of local invariance under quasiaddition is not satisfied in this case.

Thus we have two options. First we may conclude that the whole figure will not have a convex acceptance window. In this case we continue by choosing other line segments of the figure, which are locally invariant, in order to find more information about the shape of the non-convex acceptance window.

The second option is open if we can add/remove a small number of the faintest points to assure the local invariance of the chosen samples. Then we can continue the determination of the bounds on the acceptance window compatible with the given fragment.

Either option is not difficult to pursue, only a few dozens of peaks are to be analyzed. We refrain from further analyzing the options because meaningful conclusions would require a greater insight into the experimental setup and most likely also additional experiments.

6. COMMENTS AND REMARKS

1. The results of these lectures are the bounds $\Omega_{\min}$ and $\Omega_{\max}$ on the intervals $\Omega \in \mathbb{R}$,

$$\Omega_{\min} \subseteq \Omega \subseteq \Omega_{\max}.$$

Within these bounds we find all the Ω which would reproduce the given locally τ-inflation invariant point set S, interpreted as a fragment of the cut and project quasicrystal $\Sigma(\Omega)$ with the acceptance window Ω.

Our result can be considered as a rigorously defined and quantitative illustration of the well known difficulty in the current understanding of the physics of quasicrystals: any finite size sample of the quasicrystal does not uniquely determine how such a quasicrystal would grow if it were allowed to do so in ideally favorable physical conditions.

The difficulty is, of course, inherent to the cut and project model of quasicrystals. It suffices to introduce a random agent into the growth process and the difficulty is eliminated. However, in doing so, one loses at the same time much of the intrinsic beauty of the aperiodic ordered structures as defined by the cut and project procedure, their enormous flexibility (quasicrystals in any dimension, any irrationality not only $\sqrt{5}$, etc.), many local symmetries, uniformity of their properties, and their relation to other basic and far more general structures and processes in mathematics [15, 16].

Perhaps the most important result of this paper is an argument in favor of a different conclusion. Namely that we are much too demanding of Nature in requiring the strictest locality in mathematical sense from its growth process. Indeed, in the example of Section 5, we determined $\Omega_{\max}$ using a few largest tiles of the quasicrystal samples (14) and still the bounds came quite close to each other. Samples that small will hardly be ever called quasicrystals in practice. More realistic samples S, with many largest tiles, would set the bounds on Ω so close that any distinction between various quasicrystals grown from that sample would be very small. There would be a missing point here or there between any two such quasicrystals, but far apart. Moreover, a real physical growth would necessarily be subject to other interactions with its environment, which may significantly affect its development, so that the miniscule differences between quasicrystals due to the slightly distinct acceptance windows would be indistinguishable from the effects of other interactions on the process.

2. Interesting is to notice the outcome of our procedure in the extreme cases when the given sample S is too small. It may happen that the length of $\Omega_{\max}$ is τ times bigger than that of $\Omega_{\min}$. Then the bounds $\Omega_{\max}$ and $\Omega_{\min}$ fix only the scale of $\Sigma(\Omega)$, which corresponds to our choice of the distances in S. The conclusion in this case is that the sample S is so small that it can be found (up to a scale) in any quasicrystal.

3. In the ideal case where a cut and project quasicrystal fills the entire physical space, there is a big difference between the quasicrystal and the content of its acceptance window. The latter is merely a bounded region (in an "unphysical" space) which is densely filled by the star map images of the quasicrystal points.

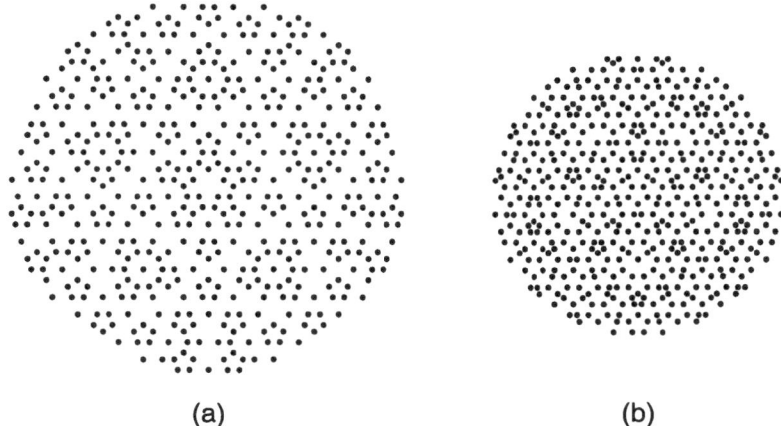

(a)   (b)

Fig. 6. — A quasicrystal fragment and its dual. (a) An H_2 quasicrystal $\Sigma(\Omega)$ visible through a circular window of diameter 8. (b) The (star map) image of the points from (a) in the acceptance window Ω. The window is a disk of diameter 6.

The case considered in this paper is much closer to the real world. Here the quasicrystal S is of finite size containing a finite number of points. The star map images of its points form a discrete set S' which has all the attributes of a quasicrystal. In particular, it is locally invariant under τ-inflation. If the complex hulls $\langle S \rangle$ and $\langle S' \rangle$ are identical as geometric figures, the two quasicrystals S and S' are indistinguishable. In general the two sets are different. They form a pair of dual quasicrystals. Either of the two can be taken for the "true" quasicrystal, while the other is its star map image. An example of such a pair is found in Figure 6 and also in Figures 2 and 3.

The duality between the two finite quasicrystals is crucial for the results, the importance of the pair for the study of real world quasicrystals was stressed already in [18].

4. So far we have considered quasicrystals with coordinates from $\mathbb{Z}[\tau]$. One may expect that similar conclusions concerning the finite size fragments and partial determination of linear sizes of the corresponding acceptance windows can be made for quasicrystals with other algebraic irrationalities. The present case (the golden mean τ) can be viewed as the common lowest member in two infinite families of quadratic unitary Pisot numbers allowing one to define quasicrystals with similar properties [13]. It is likely that a suitable approach for all the irrationalities would use binary operations analogous to quasiaddition here. However, the analogy may not go very far because in most of the new cases one would have to consider simultaneously more than one quasiaddition operation [13].

Acknowledgements

This work was partially supported by the National Science and Engineering Research Council of Canada, and FCAR of Quebec. J.P. is grateful for the hospitality of the Aspen Center for Physics where part of the work was done.

REFERENCES

[1] M. Baake, J. Hermisson and P. Pleasant, The torus parametrization of quasiperiodic LI-classes, *J. Phys. A: Math. Gen.* **30** (1997) 3029–3056.

[2] S. Ritsch, C. Beeli, H.-U. Nissen, T. Gödecke, M. Scheffer and R. Lück, Highly perfect decagonal Al-Co-Ni quasicrystals, *Philos. Mag. Lett.* **74** (1996) 99–106.

[3] S. Berman and R.V. Moody, The algebraic theory of quasicrystals with five-fold symmetry, *J. Phys. A: Math. Gen.* **27** (1994) 115–130.

[4] Č. Burdík, Ch. Frougny, J.P. Gazeau and R. Krejcar, Beta-Integers as Natural Counting Systems for Quasicrystals, *J. Phys. A: Math. Gen.* **31** (1998) 6449–6472.

[5] L. Chen, R.V. Moody, J. Patera, *Noncrystallographic root systems, in Quasicrystals and Discrete Geometry*, Fields Institute Monograph Series Vol. 10, edited by J. Patera (Amer. Math. Soc., Providence, RI, 1998) 145–192.

[6] Ch. Frougny and B. Solomyak, Finite β-expansions, *Ergodic Theory Dynamical Systems* **12** (1994) 713–723.

[7] J.P. Gazeau and J. Patera, τ-wavelets of Haar, *J. Phys. A: Math. Gen.* **29** (1996) 4549–4559.

[8] A. Hof, *Diffraction on aperiodic material*, in *Mathematics of Long Range Aperiodic Order*, Proc. NATO ASI, Waterloo, edited by R.V. Moody (Kluwer, 1996) 239–268.

[9] C. Janot, *Quasicrystals: A primer* (Oxford Univ. Press, Oxford, UK, 1994).

[10] Z. Masáková, J. Patera and E. Pelantová, Inflation centers of the cut and project quasicrystals, *J. Phys. A: Math. Gen.* **31** (1998) 1443–1453.

[11] Z. Masáková, J. Patera and E. Pelantová, Minimal distances in quasicrystals, *J. Phys. A: Math. Gen.* **31** (1998) 1539–1552.

[12] Z. Masáková, J. Patera and E. Pelantová, Selfsimilar Delone sets and quasicrystals, *J. Phys. A: Math. Gen.* **31** (1998) 4927–4946.

[13] Z. Masáková, J. Patera and E. Pelantová, s-convexity, model sets and their relation, CRM-2565 (Université de Montréal, 1998).

[14] Z. Masáková and E. Pelantová, Quasicrystals, tilings and scaling symmetries, Proceedings of the workshop on Self-similar Systems (Dubna, July 1998).

[15] Y. Meyer, *Nombres de Pisot, nombres de Salem et analyse harmonique*, Lecture Notes in Mathematics 117 (Springer, 1970).

[16] Y. Meyer, *Algebraic numbers and harmonic analysis* (North-Holland, 1972).

[17] R.V. Moody, *Meyer sets and their duals*, in *Mathematics of Long Range Aperiodic Order*, Proc. NATO ASI, Waterloo, edited by R.V. Moody (Kluwer, 1996) 403–441.

[18] J. Patera, *Noncrystallographic root systems and quasicrystals,* in *Mathematics of Long Range Aperiodic Order*, Proc. NATO ASI, Waterloo, edited by R.V. Moody (Kluwer, 1996) 443–465.

[19] A. Rényi, Representations for real numbers and their ergodic properties, *Acta Math. Acad. Sci. Hung.* **8** (1957) 477–493.

[20] M. Senechal, *Quasicrystals and Geometry* (Cambridge Univ. Press, Cambridge, UK, 1995).

[21] Proceedings of the 6-th International Conference on Quasicrystals (Tokyo, Japan, May 1997).

Course N° 8

Counting Systems with Irrational Basis for Quasicrystals

Given by J.P. Gazeau

Written by J.P. Gazeau[1] and R. Krejcar[2]

[1] *Laboratoire de Physique Théorique de la Matière Condensée,*
Université Paris 7 – Denis-Diderot,
2 place Jussieu, 75251 Paris Cedex 05, France
[2] *Department of Mathematics, Czech Technical University,*
Faculty of Nuclear Sciences and Physical Engineering, Trojanova 13,
12000 Prague 2, Czech Republic

1. INTRODUCTION

In the quasicrystalline context (see for instance [1–3]), quasilattices can be thought as mathematical discrete sets supporting atomic sites or Bragg peaks beyond a certain intensity. They play the same role as lattices do for crystals. Most of the proposed definitions are of geometrical nature, sticking to crystalline lattice theory through the celebrated Cut and Project method (see [4]), recently renamed model set method [5] or issued from involved packing construction in real space like the generalized dual method [6,7]. More "algebraic" approaches were also initiated, by several authors (see for instance [8,9]). Recent school or workshop proceedings give a comprehensive account of this original interactive field mixing number theory, lattices and experimental physics (see in Refs. [5, 10] for instance).

It has also been acknowledged that most of the algebraic and functional approaches to quasilattices, *e.g.* the Cut and Project method and involved Fourier analysis, should mention pioneer results obtained more than 25 years ago by Meyer [11, 12]. The notion of a quasilattice $\Lambda \subset \mathbb{R}^d$ proposed by

Meyer rests upon the idea that the quasilattice should be "almost" closed under subtraction

$$\Lambda - \Lambda \subset \Lambda + F, \qquad (1)$$

where F is some finite set. Such a requirement has important consequences on the diffraction pattern of Λ, and we refer to Lagarias for recent discussions on that aspect [13, 14]. More generally, one can deal with lattice internal laws within an equivalence class of quasilattices which differ from each other by the addition of finite sets.

An interesting algebraic definition of quasilattices [15,16] has been introduced more than five years ago by Moody and Patera, and their possible symmetry groups and semi-groups have been investigated [17–19]. More recently [20–24], it has been suggested to study algebraic models of quasilattices

$$\Lambda_\beta \stackrel{\text{def}}{=} \sum_{i=1}^{d} \mathbb{Z}_\beta e_i, \text{ where } \{e_i\} \text{ is a basis in } \mathbb{R}^d, \ d = 1, 2, 3, \qquad (2)$$

based on countable sets of numbers, denoted by $\mathbb{Z}_\beta$, and named β-*integers*, where β is some real number > 1. These quasilattices Λ_β are scaling invariant under dilation by β, and $\mathbb{Z}_\beta$ is precisely the counting system with origin, *i.e.* the numerical frame, in which we should think about structural properties of Λ_β, exactly like the first crystallographers did with lattices and ordinary integers.

As a matter of fact, these sets $\mathbb{Z}_\beta$ are natural candidates for coordinating quasicrystalline nodes in 1, 2 or 3 dimensions, and also the Bragg peaks in related diffraction patterns [25, 26]. In the observed cases :

$$\beta = \tau = \frac{1+\sqrt{5}}{2} = 2\cos\frac{\pi}{5} \qquad \text{(penta − or decagonal quasilattices)} \ (3)$$

$$\beta = \gamma = 1 + \sqrt{2} = 1 + 2\cos\frac{\pi}{4} \qquad \text{(octogonal case)} \qquad (4)$$

$$\beta = \delta = 2 + \sqrt{3} = 2 + 2\cos\frac{\pi}{12} \qquad \text{(dodecagonal case).} \qquad (5)$$

Generically, $\mathbb{Z}_\beta$ is obtained by means of a finite algorithm where β is a *Pisot-Vijayaraghavan number* or more simply PV or Pisot number (see [27]), *i.e.* an algebraic integer $\beta > 1$, which is solution to the irreducible polynomial of the form

$$X^m = a_{m-1}X^{m-1} + \cdots + a_1 X + a_0, \ a_i \in \mathbb{Z} \qquad (6)$$

such that all other solutions $\beta^{(i)}$ of (6) (Galois conjugates of β) have modulus strictly smaller than 1,

$$\beta^{(0)} = \beta \,, \ |\beta^{(i)}| < 1, \ i = 1, 2, \cdots, m - 1.$$

Therefore d-dimensional discrete sets of the form (2) can advantageously play the role of "grid frame" or ("millimeter paper" if $d = 2$) for labelling quasicrystalline atomic sites in real space, exactly as integer lattices ($\mathbb{Z}$-modules) are appropriate to real crystalline structures. Indeed, it seems that most of quasilattices obtained by Cut and Project [28] or by algebraic "filtering" [16] within the dense $\mathbb{Z}[\beta]$-module are supported by sets of the type Λ_β. Those β-integer quasilattices display nice algebraic and geometrical features, which straightforwardly generalize the ones for lattices. We already mentioned their similarity property under scaling by β:

$$\beta\Lambda \subset \Lambda.$$

This property holds because, on a more basic level and by construction, we have

$$\beta\mathbb{Z}_\beta \subset \mathbb{Z}_\beta.$$

Furthermore, it appears to be natural to study the additive and multiplicative properties of $\mathbb{Z}_\beta$:

$$\mathbb{Z}_\beta + \mathbb{Z}_\beta \subset \mathbb{Z}_\beta + X, \tag{7}$$

$$\mathbb{Z}_\beta \mathbb{Z}_\beta \subset \mathbb{Z}_\beta + Y, \tag{8}$$

where X and Y are to be determined in a non-ambiguous way, at least in the case of PV quadratic numbers.

This lecture is an introduction to the concept of β-integers and to the use it can be made of them. For pedagogical purposes, we shall mainly insist on examples involving $\tau = \frac{1+\sqrt{5}}{2}$. Indeed, the latter is in many senses the simplest one among the PV numbers. In the next section, we shall present the basic definitions and properties concerning $\mathbb{Z}_\tau$. In Sections 3 and 4, simple one-dimensional illustrations of τ-integer labelling will be given: the Fibonacci chain and its diffraction pattern. We shall consider two-dimensional examples in Section 5: a Penrose tiling and its diffraction pattern, and the diffraction pattern of a quasicrystalline structure. Section 6 is devoted to more general quadratic Pisot cases. More precisely, we shall give some insights on the arithmetics and algebra of the β-integers, keeping in mind the questions arising from (7) and (8).

2.   THE SET OF τ-INTEGERS

Let us first recall how the binary system is used to express real numbers as series in powers of 2:

$$\mathbb{R} \ni x = \pm(\xi_j 2^j + \xi_{j-1} 2^{j-1} + \cdots + \xi_l 2^l + \cdots), \tag{9}$$

where $j = j(x) \in \mathbb{Z}$ is the highest power of 2 such that $2^j \leq |x| < 2^{j+1}$, $\xi_j =$ integral part of $|x|/2^j \equiv [|x|/2^j] \in \{0,1\}$. The other expansion coefficients $\xi_l \in \{0,1\}$ are inductively defined by $\xi_l = [2r_{l+1}]$, $r_l =$ fractional part of $2r_{l+1} \equiv \{2r_{l+1}\}$, with $r_j = \{|x|/2^j\}$. In consequence, a positive x is a word, *e.g.* $\xi_j\xi_{j-1}\ldots\xi_1\xi_0 \cdot \xi_{-1}\xi_{-2}\ldots = 110 \cdot 100\ldots$, made with letters ξ_l in the alphabet $\{0,1\}$. The standard ordering of the positive real numbers corresponds to the lexicographical ordering of those words, and the natural numbers are those for which $\xi_l = 0$ for all $l < 0$. The same algorithm, called greedy algorithm (see [29–31]), can be employed to represent real numbers in a system based on an arbitrary real number $\beta > 1 : \mathbb{R} \ni x = \pm(\xi_j\beta^j + \xi_{j-1}\beta^{j-1} + \cdots + \xi_l\beta^l + \cdots) \equiv \pm\overline{\xi_j\xi_{j-1}}\ldots\xi_1\xi_0 \cdot \xi_{-1}\xi_{-2}\ldots$, where the "letters" assume their values in the alphabet $\{0,1,\ldots,\beta-1$ if β is natural , and $[\beta]$ if β is not$\}$. But if β is not natural, all words are not allowed. Let us see how things organize in the case $\beta = \tau$. More details on generic β will be given in Section 6. Since $\tau = 1.618\ldots$, the alphabet is $\{0,1\}$, and so any positive x is represented in "basis τ" by

$$x = \sum_{l=-\infty}^{j} \xi_l\tau^l \equiv \xi_j\xi_{j-1}\ldots\xi_1\xi_0 \cdot \xi_{-1}\xi_{-2}\ldots, \qquad (10)$$

where $\xi_l \in \{0,1\}$, and $\xi_{l+1}\xi_l = 0$. The latter "forbidding rule" expresses the algebraic fact that $\tau^{l+1} + \tau^l = \tau^{l+2}$. So, any 1 in the τ-expansion (10) has to be followed by 0 whereas 0 can be followed by 0 or 1. All allowed words can be lexicographically ordered following that constraint and this order corresponds to the standard order on the real line. By definition, the positive τ-integers are those real numbers which have only positive powers of τ in their τ-expansion (10). So, we shall denote them by

$$\mathbb{Z}_\tau^+ = \{x = \sum_{l=0}^{j} \xi_l\tau^l, \ \xi_l \in \{0,1\}, \ \xi_l\xi_{l+1} = 0\}. \qquad (11)$$

Accordingly, the set of τ-integers is defined by

$$\mathbb{Z}_\tau = \mathbb{Z}_\tau^+ \cup (-\mathbb{Z}_\tau^+).$$

For instance, the first positive τ-integers are given by

1	$\frac{1}{\tau}$	1	1	$\frac{1}{\tau}$

0	1	τ	τ^2	τ^2+1	τ^3
(0)	(1)	(10)	(100)	(101)	(1000).

We thus obtain a sequence of numbers strictly increasing in steps of length equal to 1 or $\frac{1}{\tau}$. In a certain sense, this quasiperiodic sequence with two incommensurable periods is the closest one to the periodic sequence with period equal to 1, namely the sequence of the integers. The countable set $\mathbb{Z}_\tau$ is naturally self-similar and symmetrical with respect to the origin:

$$\tau\mathbb{Z}_\tau \subset \mathbb{Z}_\tau, \ \mathbb{Z}_\tau = -\mathbb{Z}_\tau.$$

Any τ-integer is an element of the algebraic ring

$$\mathbb{Z}[\tau] = \{m + n\tau \mid m, n \in \mathbb{Z}\} = \mathbb{Z} + \mathbb{Z}\,\tau.$$

The latter is actually identical to the set of real numbers which have a finite τ-expansion (see [31]).

The fact that positive τ-integers are represented by a finite string of $0'$s and $1'$s, with condition that no run of two adjacent $1'$s occur, has to be related to the representation of natural numbers in the Fibonacci numeration system. Indeed, if the Fibonacci numbers are defined by

$$f_{n+2} \stackrel{\text{def}}{=} f_{n+1} + f_n, \ f_1 \stackrel{\text{def}}{=} 2, \ f_0 \stackrel{\text{def}}{=} 1, \tag{12}$$

then there is an explicit bijection between integers and the τ-integers (see for instance [32]),

$$n = \sum_{i=0}^{j_n} \xi_i f_i \longrightarrow x_n = \sum_{i=0}^{j_n} \xi_i \tau^i. \tag{13}$$

Note that properties of Fibonacci representations of natural numbers have been investigated by many people (see [33]). We shall come back to this important point in Section 6.

3. TAU-INTEGER LABELLING OF THE FIBONACCI CHAIN

The one-dimensional example which is usually presented as a toy model of quasicrystals is the so-called Fibonacci chain. It is a discrete quasiperiodic subset of the real line and is often given as an illustration of the Cut-and-Project method [1,28]. Consider the semi-open band $\mathcal{B}$ obtained by translating the unit square through the square lattice $\mathbb{Z}^2$ along the straight line E of slope $\alpha = \tan\theta$. $E = E^{\parallel}$ is referred to as a "cut" or "parallel" space or "physical space". Then project on E and along $E^{\perp}$ the lattice points lying in $\mathcal{B}$. Note that the latter belong to a unique path made of horizontal segments (A) and vertical segments (B). The resulting sequence of points lying in E are the nodes of the Fibonacci chain if $\alpha = \frac{1}{\tau}$. This chain is made of the projected paths and reads $\ldots ABAABABAABAA\ldots$. Note that a short link B is never adjacent to another B whereas two adjacent long links A can occur. We meet here the order already present in the τ-numeration. $E^{\perp}$ is called the perpendicular or "internal" space, and $\mathcal{B} \cap E^{\perp}$ is the "window" or "acceptance zone", or also "atomic surface".

The set of Fibonacci nodes is equivalently obtained as a subset of $\mathbb{Z}_\tau$ through a purely algebraic filtering procedure. Let us first consider the extension ring $\mathbb{Z}[\tau]$ of the algebraic integer τ. It can be obtained as the projection onto E and along $E^{\perp}$ of the whole square lattice $\mathbb{Z}^2$. There exists in this type of a ring an algebraic conjugation, called Galois automorphism, and defined by:

$$x = m + n\tau \longrightarrow x' = m - n\frac{1}{\tau}, \tag{14}$$

where $\tau' = -\frac{1}{\tau}$ is the other root of the golden mean equation $X^2 = X + 1$. This conjugation is an automorphism of the ring $\mathbb{Z}[\tau]$, which means that $(x_1 + x_2)' = x_1' + x_2'$ and $(x_1 x_2)' = x_1' x_2'$ for all $x_1, x_2 \in \mathbb{Z}[\tau]$. Note the Cut and Project interpretation of (14): if we choose $\overrightarrow{E_1} = \frac{1}{\tau + \frac{1}{\tau}}(1, \frac{1}{\tau})$ as a basis vector of the physical space E, and its conjugate $(\overrightarrow{E_1})' = \overrightarrow{E_2} = \frac{1}{\tau + \frac{1}{\tau}}(-1, \tau)$ as a basis of $E^\perp$, a point (n, m) in the lattice $\mathbb{Z}^2$ has coordinates $X_1 = m + n\tau = x$ and $X_2 = m - n\frac{1}{\tau} = x'$ in the (non-normalized) frame $(O, \overrightarrow{E_1}, \overrightarrow{E_2})$. Now let P be a bounded subset of $E^\perp$. There corresponds to it a quasilattice $\Sigma^P \subset E$ made up of the points $\overrightarrow{OM_1} = (m + n\tau)\overrightarrow{E_1}$ in E such that their conjugates $\overrightarrow{OM_2} = (m - n\frac{1}{\tau})\overrightarrow{E_2}$ lie inside the window P. But we could forget about this higher-dimensional detour by just using the internal sieving rule in the ring $\mathbb{Z}[\tau]$ itself [16]:

$$\Sigma^P = \left\{ x = m + n\tau \in \mathbb{Z}[\tau] \text{ such that } x' = m - n\frac{1}{\tau} \in P \right\} = (\mathbb{Z}[\tau] \cap P)'. \tag{15}$$

The Fibonacci set, say $\mathcal{F}$, with link lengths $A = \tau^2$, $B = \tau$, is precisely that set Σ^P with $P = [0, 1)$. Actually, and this is the deep relevance of the τ-integers in pentagonal and decagonal quasicrystallography, the set $\mathcal{F}$ is a underline{subset of $\mathbb{Z}_\tau$}. This fact results from a (non-trivial!) property of the latter:

Proposition 1. *Any number $m + n\tau$ in the ring $\mathbb{Z}[\tau]$ which has modulus ≤ 1 is the algebraic conjugate of a τ-integer. More precisely,*

$$\mathbb{Z}_\tau' \subset (-\tau, \tau) \cap \mathbb{Z}[\tau], \text{ and } \mathbb{Z}_\tau' \cap [-1, 1] = \mathbb{Z}[\tau] \cap [-1, 1]. \tag{16}$$

Instead of developing an algebraic proof of this proposition, let us a give a geometrical description of it.

In Figure 1 are plotted the lattice points (m, n) in $\mathbb{Z}^2$ such that $m + n\tau \in \mathbb{Z}_\tau$. They are clearly all the lattice points lying within the bands defined by

$$\tau x - \tau^2 < y < \tau x + \tau$$

in the first quadrant, and by

$$\tau x - \tau < y < \tau x + \tau^2$$

in the opposite sign quadrant. Note that the band width is $\tau^3 / \sqrt{1 + \tau^2}$. This "inverse" Cut and Project method leads to the following definition of the positive and negative τ-integers

$$\mathbb{Z}_\tau^+ = \left\{ x = m + n\tau \; ; \; m, n \in \mathbb{Z}, \; m, n \geq 0, \; -1 < x' = m - \frac{n}{\tau} < \tau \right\} \tag{17}$$

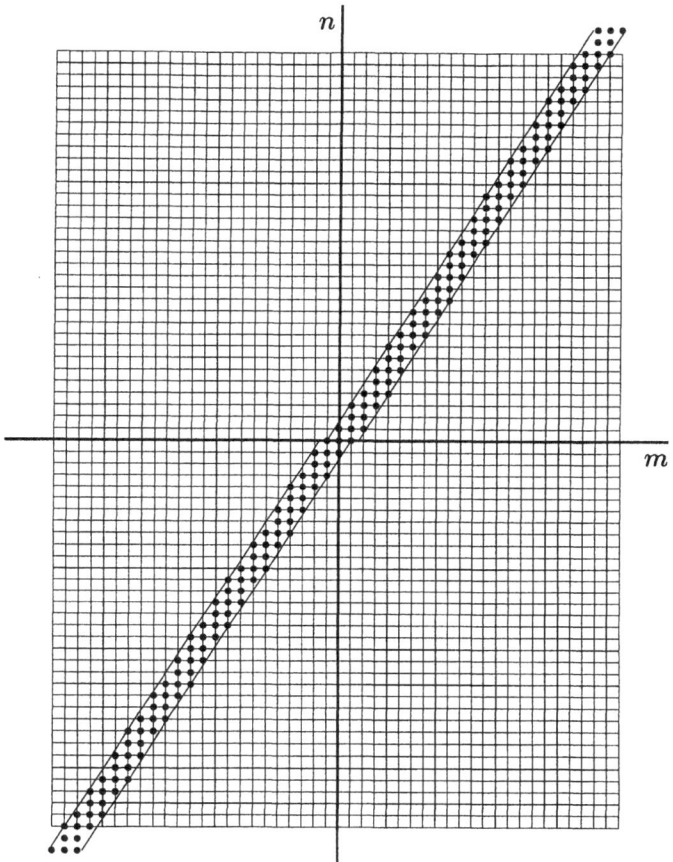

Fig. 1. — $\mathbb{Z}_\tau$-numbers.

$$\mathbb{Z}_\tau^- = \left\{ m + n\tau \; ; \; m, n \in \mathbb{Z}, \; m, n \leq 0, \; -\tau < x' = m - \frac{n}{\tau} < 1 \right\}. \quad (18)$$

Then we can deduce from the above that

$$\mathbb{Z}_\tau^+ = \text{positive part of } \Sigma^{(-1,\tau)}, \quad (19)$$

$$\mathbb{Z}_\tau^- = \text{negative part of } \Sigma^{(-\tau,1)}, \quad (20)$$

and

$$\Sigma^{(-1,1)} \subset \mathbb{Z}_\tau \subset \Sigma^{(-\tau,\tau)}. \quad (21)$$

The inclusion relations (21) also mean that it is sufficient to sieve from the discrete set $\mathbb{Z}_\tau$ in order to obtain $\Sigma^{(-1,1)}$

$$\Sigma^{(-1,1)} = \{ x \in \mathbb{Z}_\tau \; ; \; x' \in (-1,1) \; \}. \quad (22)$$

Scaling (22) with arbitrary powers of τ leads to the interesting chain [20, 22] of embeddings (see [18] and [34] for an algebraic proof)

$$\Sigma^{(-\tau^j, \tau^j)} \subset \mathbb{Z}_\tau / \tau^j \subset \Sigma^{(-\tau^{j+1}, \tau^{j+1})}, \ j \in \mathbb{Z}. \tag{23}$$

From these results the task of selecting and labelling nodes of the Fibonacci chain becomes straightforward:

$$\text{Fibonacci } \mathcal{F} \equiv \Sigma^{[0,1)} = \{x \in \mathbb{Z}_\tau \mid x' \in [0,1)\} \subset \mathbb{Z}_\tau. \tag{24}$$

This sieving procedure is illustrated by the following example involving a few numbers in $\mathbb{Z}_\tau$. The numbers belonging to $\mathcal{F}$ are underlined.

$$\ldots, \ -\tau^4 - \tau^2 - 1, \ -\tau^4 - \tau^2, \ \underline{-\tau^4 - \tau}, \ -\tau^4 - 1, \ -\tau^4, \ \underline{-\tau^3 - \tau}, \ -\tau^3 - 1, \ \underline{-\tau^3},$$
$$-\tau^2 - 1, \ -\tau^2, \ \underline{-\tau}, \ -1, \ \underline{0}, \ 1, \ \tau, \ \underline{\tau^2}, \ \tau^2 + 1, \ \tau^3, \ \underline{\tau^3 + 1}, \ \tau^3 + \tau, \ \underline{\tau^4}, \ \tau^4 + 1,$$
$$\tau^4 + \tau, \ \underline{\tau^4 + \tau^2}, \ \tau^4 + \tau^2 + 1, \ldots$$

At this point, we should notice a property of the Fibonacci chain $\mathcal{F}$ which makes it optimal among other subsets of $\mathbb{Z}_\tau$. Since by definition its Galois conjugate $\mathcal{F}'$ is equal to $\mathbb{Z}[\tau] \cap [0,1)$, we have $\mathcal{F}' - \mathcal{F}' = \mathbb{Z}[\tau] \cap (-1,1)$, and so, by (21) and (22),

$$\mathcal{F}' - \mathcal{F}' = \mathbb{Z}'_\tau \cap (-1,1) \Longrightarrow \mathcal{F} - \mathcal{F} \subset \mathbb{Z}_\tau. \tag{25}$$

This inclusion is optimal precisely from (16). In other words, one can assert the following.

Proposition 2. *Any node of the Fibonacci chain $\mathcal{F}$ may be chosen as the origin of its $\mathbb{Z}_\tau$ indexing set.*

Indeed, (25) can be rewritten $\mathcal{F} \subset \mathbb{Z}_\tau + x_0$ for an arbitrary choice of origin node $x_0 \in \mathcal{F}$. It is remarkable that this "democracy" property with respect to the tau-integer labelling system is shared by Penrose quasilattices in the two-dimensional case (see Sect. 5).

4.   TAU-INTEGER LABELLING OF DIFFRACTION PATTERN

It is well known that Bragg peaks in diffraction patterns of crystals form themselves crystal-like structures and so can be indexed by integers. Let us see how the τ-integers play exactly the same role for the diffraction pattern of our one-dimensional Fibonacci example $\mathcal{F}$. The Fourier intensity or Bragg spectrum in $q \in \mathbb{R}$ of $\mathcal{F}$ is defined as the following limit (see for instance [35]):

$$I(q) = \lim_{j \to \infty} \frac{1}{|\mathcal{F}_j|} \left| \sum_{\lambda \in \mathcal{F}_j} e^{i\lambda q} \right|^2, \tag{26}$$

where $\mathcal{F}_j$ is a finite approximant to $\mathcal{F}$ in the sense that $\mathcal{F}$ is the inductive limit $\mathcal{F} = \cup_{j \leq 0} \mathcal{F}_j$. $|\mathcal{F}_j|$ denotes the cardinal of $\mathcal{F}$. It can be shown that (26) is also equal to

$$I(q) = \lim_{j \to \infty} \left| \frac{1}{|\mathcal{F}_j|} \sum_{\lambda \in \mathcal{F}_j} e^{i\lambda q} \right|^2 . \tag{27}$$

A natural sequence $\mathcal{F}_j$ of approximants to $\mathcal{F}$ is defined through Galois conjugates (24) (in "internal space") obtained in a recursive way through successive subdivisions of the unit interval $[0, 1]$ into "τ-adic" subintervals [21, 36]. Let us explain this in full details. The algebraic nature of τ, as a solution to $X^2 = X + 1$, allows for the τ-adic property,

$$\frac{1}{\tau^j} = \frac{1}{\tau^{j+1}} + \frac{1}{\tau^{j+2}}, \ j \in \mathbb{Z}. \tag{28}$$

This equality provides a subdivision of the unit interval into two parts:

$$A \equiv [0, 1] = [0, 1/\tau] \cup [1/\tau, 1]. \tag{29}$$

Equation (29) is the starting point of an iterative sequence of subdivisions of A into intervals. At each step only the large intervals get divided. Thus at each stage we have two lengths of intervals:

$$A_{j,b} = \left[\frac{b}{\tau^j}, \frac{b+1}{\tau^j} \right] \tag{30}$$

$$A_{j,b} = A_{j+1,\tau b} \cup A_{j+2,\tau^2 b + \tau}. \tag{31}$$

At the j-th step we get the subdivision of A into $f_j = f_{j-1} + f_{j-2}$ subintervals, where the f_j's are Fibonacci numbers with $f_{-1} = 1$ and $f_{-2} = 0$ (the same as in (12)):

$$A = \left(\cup_{r=1}^{f_{j-1}} L_r^{(j)} \right) \cup \left(\cup_{s=1}^{f_{j-2}} S_s^{(j)} \right), \tag{32}$$

where the $L_r^{(j)} = A_{j,b_r}$ stand for large intervals whereas the $S_s^{(j)} = A_{j+1,b_s}$ do for small intervals. In the boundary values of these τ-adic intervals, we precisely find τ-integers $b \in \mathbb{Z}_\tau^+$ submitted to boundedness conditions depending on $j \in \mathbb{N}$. In order to make things more precise, let us consider the set $\mathbb{Z}_\tau^+$ of positive τ-integers as the inductive limit

$$\mathbb{Z}_\tau^+ = \cup_{N \geq 0} B_N, \tag{33}$$

$$\text{where } B_N = \{x \in \mathbb{Z}_\tau^+ \mid 0 \leq x < \tau^N\}. \tag{34}$$

Note that

$$B_N = \tau B_{N-1} \cup (1 + \tau^2 B_{N-2}) = B_{N-1} \cup (\tau^{N-1} + B_{N-2}), \tag{35}$$

with $B_0 = \{0\} \subset B_1 = \{0, 1\} \subset B_1 = \{0, 1, \tau\} \subset \cdots$. Now, we can assert:

Proposition 3. *The τ-adic interval $A_{j,b} = [\frac{b}{\tau^j}, \frac{b+1}{\tau^j}]$ appears at a certain step of the sequence of τ-adic subdivisions of $A = [0, 1]$ if b satisfies the condition*

$$b \in \tau B_{j-1}.$$

It is clear from the above and from (33) that the infinite set

$$\{b/\tau^j \mid b \in B_j, \ j \in \mathbb{N}\} \tag{36}$$

is identical to the set of all finite τ-expansions with negative powers of τ only:

$$\left\{ \sum_{l=-L}^{-1} \xi_l \tau^l, \ 0 < L < \infty, \ \xi_l \in \{0, 1\}, \ \xi_l \xi_{l+1} = 0 \right\}. \tag{37}$$

This set is dense in the interval $[0, 1)$, and is equal to $\mathbb{Z}[\tau] \cap [0, 1)$, since the ring $\mathbb{Z}[\tau]$ is made of all numbers with finite τ-expansion. Therefore (36) is equal to the Galois conjugate $\mathcal{F}'$ of the Fibonacci chain, according to (24). Hence we define

$$\mathcal{F}_j \equiv (B_j/\tau^j)'. \tag{38}$$

From (35) we infer the recurrence

$$\mathcal{F}_j = \mathcal{F}_{j-1} \cup ((-\tau)^j + \mathcal{F}_{j-2}). \tag{39}$$

The increasing sequence $\mathcal{F}_0 = \{0\} \subset \mathcal{F}_1 = \{0, -\tau\} \subset \mathcal{F}_2 = \{0, -\tau, \tau^2\} \subset \cdots$ tends to $\mathcal{F}$. We note that $|\mathcal{F}_j| = f_j$.

Now, we need to evaluate the limit (in the sense of Bohr) of the sequence of Fourier transforms of the finite point sets $\mathcal{F}_j$ appearing in (27). For that we consider again the τ-subdivision of the unit interval (29) which implies the following scaling equation for the corresponding characteristic functions:

$$1_{[0,1]}(x) = 1_{[0,1]}(\tau x) + 1_{[0,1]}(\tau^2 x - \tau). \tag{40}$$

The Fourier transform of (40) reads

$$\int_{-\infty}^{+\infty} 1_{[0,1]}(x) e^{-i\xi x}\, dx = e^{-i\xi/2} \frac{\sin \xi/2}{\xi/2} \equiv \phi(\xi) = \frac{1}{\tau} \phi\left(\frac{\xi}{\tau}\right) + \frac{e^{-i\xi/\tau}}{\tau^2} \phi\left(\frac{\xi}{\tau^2}\right). \tag{41}$$

Iterating (41) $j - 1$ times leads to

$$\phi(\xi) = \frac{f_{j-1}(\xi)}{\tau^j} \phi\left(\frac{\xi}{\tau^j}\right) + \frac{e^{-i\xi/\tau^j} f_{j-2}(\xi)}{\tau^{j+1}} \phi\left(\frac{\xi}{\tau^{j+1}}\right), \tag{42}$$

where the functions $f_j(\xi)$ obey the recurrence equation:

$$f_j(\xi) = f_{j-1}(\xi) + e^{-i\xi/\tau^j} f_{j-2}(\xi), \quad f_{-2} = 0, \ f_{-1} = 1. \tag{43}$$

Note that $\{f_j(0) = f_j\}$ is the sequence of Fibonacci numbers of (12). The solution to (43) is given by

$$f_j(\xi) = \sum_{b \in B_j} e^{-i\xi b/\tau^j}$$

which is precisely the Fourier transform of $\mathcal{F}'_j$. We thus obtain at the limit $j \to \infty$ the interesting summation formula [21, 23]

$$e^{-i\xi/2} \frac{\sin \xi/2}{\xi/2} = \lim_{j \to \infty} \frac{1 + \tau^2}{\tau^{j+3}} \sum_{\lambda' \in \mathcal{F}'_j} e^{-i\xi\lambda'}, \tag{44}$$

where we should notice that $\frac{\tau^{j+3}}{1+\tau^2} \approx f_j = |\mathcal{F}'_j|$ for large j. Now, it is well known (see for instance [3]) that the Bragg spectrum of the Fibonacci chain is supported by the dense set $q \in \frac{2\pi}{\tau^2+1} \mathbb{Z}[\tau]$. From the algebraic identity

$$\frac{1}{\tau^2 + 1} \lambda x + \frac{\tau^2}{\tau^2 + 1} \lambda' x' \in \mathbb{Z},$$

true for all λ, $x \in \mathbb{Z}[\tau]$, and from (27) and (44), we get the following exact expression for the Bragg spectrum:

$$I(q) = \left(\frac{\sin q'/2}{q'/2} \right)^2, \tag{45}$$

for $q' \equiv 2\pi \frac{\tau^2}{\tau^2+1}(m - n/\tau)$ if $q = 2\pi \frac{1}{\tau^2+1}(m + n\tau)$ (with abusive notations). Now we are in position to conclude about the relevance of the τ-integers in the labelling of the Bragg peaks (see Fig. 2):

Proposition 4. *The Bragg peaks of the Fibonacci chain $\mathcal{F}$ with intensity $I(q) > c \equiv (\frac{\sin \gamma/2}{\gamma/2})^2$ (cut-off) are located at all $q \in \mathbb{R}$ such that $q' \in (-\gamma, \gamma)$. If $\gamma < 2\pi \frac{\tau^2}{\tau^2+1}$, then $q \in \frac{2\pi}{\tau^2+1} \mathbb{Z}_\tau$.*

This assertion straightforwardly results from (22).

5. TAU-INTEGER LABELLING OF TWO-DIMENSIONAL STRUCTURES

It is well known that the condition $2 \cos \frac{2\pi}{n} \in \mathbb{Z}$ characterizes n-fold Bravais lattices in $\mathbb{R}^2$ (and in $\mathbb{R}^3$). Let us put $\zeta = e^{\frac{2\pi i}{n}}$, $\zeta^n = 1$. If we consider the $\mathbb{Z}$-module in the plane:

$$\mathbb{Z}[\zeta] = \mathbb{Z} + \mathbb{Z}\zeta + \mathbb{Z}\zeta^2 + \cdots + \mathbb{Z}\zeta^{n-1}, \tag{46}$$

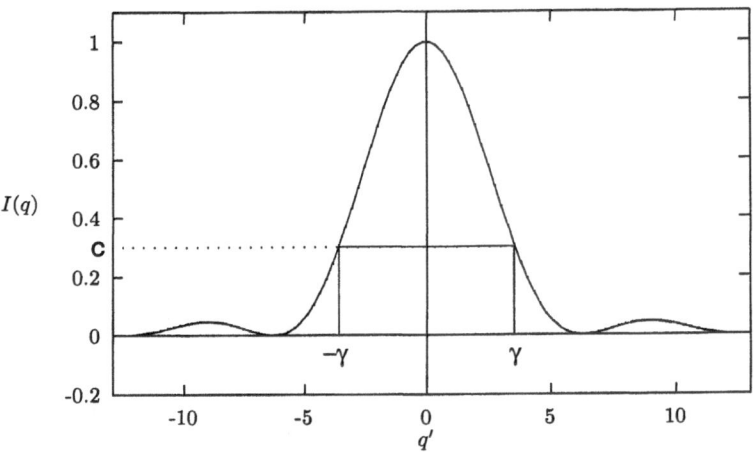

Fig. 2. — Intensity of the Bragg peaks of the Fibonacci chain.

we get the cyclotomic ring of order n. This n-fold structure is generically dense in $\mathbb{C}$, except precisely for the crystallographic cases. We indeed check that $\mathbb{Z}[\zeta] = \mathbb{Z}$ for $n = 1$ or 2, $\mathbb{Z}[\zeta] = \mathbb{Z} + \mathbb{Z}i$ for $n = 4$ (square lattice), and $\mathbb{Z}[\zeta] = \mathbb{Z} + \mathbb{Z}e^{\frac{i\pi}{3}}$ for the hexagonal cases $n = 3$ and $n = 6$.

What can we do for the non-crystallographic cases in order to preserve discreteness? Let us examine the simplest one, namely $n = 5$ or 10. Let us observe what it happens if we replace the ordinary integers by the τ-integers in (46). It is more convenient to introduce the root of unity $\xi = e^{\frac{i\pi}{5}}$, since $\tau = 2\cos\pi/5 = \xi + \xi^{cc}$. We obtain the set

$$\mathbb{Z}_\tau[\xi] \equiv \mathbb{Z}_\tau + \mathbb{Z}_\tau\xi + \mathbb{Z}_\tau\xi^2 + \mathbb{Z}_\tau\xi^3 + \mathbb{Z}_\tau\xi^4, \tag{47}$$

which we can call the *five-fold cyclotomic quasiring*. This discrete set $\mathbb{Z}_\tau[\xi]$ is displayed in Figure 3. If we now consider the following discrete sets in the plane,

$$\Gamma_q = \mathbb{Z}_\tau + \mathbb{Z}_\tau\xi^q, \quad q = 1, \, 2, \, 3, \text{or } 4, \tag{48}$$

we can easily guess that none of them is identical to (47). This fact is obvious by simple inspection of Figure 4 in which the set Γ_1 is displayed. On the other hand, the following inclusions hold true

$$\Gamma_q \subset \mathbb{Z}_\tau[\xi] \subset \frac{\Gamma_q}{\tau^4}. \tag{49}$$

One can understand from this illustrative example that the "τ-grids" Γ_q are "universal" labelling frames for a large class of planar pentagonal or decagonal

Fig. 3. — Five-fold cyclotomic quasiring.

quasilattices Λ of interest in physics. Universal means that it is always possible to embed Λ into a suitably deflated or inflated version of such "τ-square" papers:

$$\Lambda \subset \Gamma_q/\tau^j, \; j \in \mathbb{Z}. \tag{50}$$

This embedding is understood through algebraic filtering and labelling procedures which are analogous to those described in the two previous sections. We just have to define the algebraic dual operation extending the Galois conjugation. Due to the following algebraic relations based on the cyclotomic nature of $\tau = \xi + \xi^{cc} = 2 \cos \frac{\pi}{5}$, we have

Fig. 4. — τ-grid Γ_1.

$$\xi^2 = -1 + \tau\xi, \ \xi^3 = -\tau + \tau\xi, \ \xi^4 = -\tau + \xi. \tag{51}$$

It follows the equality: $\mathbb{Z}[\xi] = \mathbb{Z}[\tau] + \mathbb{Z}[\tau]\xi$. Consistently to the fact that $\tau' = 2\cos\frac{3\pi}{5} = \xi^3 + (\xi^{cc})^3$, we introduce the automorphism in $\mathbb{Z}[\xi]$,

$$z = m + n\tau + (p + q\tau)\xi \ \longrightarrow \ z^\star = m - n\frac{1}{\tau} + \left(p - q\frac{1}{\tau}\right)\xi^3, \tag{52}$$

and the filtering procedure (possibly involving a "phason" shift $\Phi \in \mathbb{C}$),

$$\Sigma_\Phi^P = \{z \in \mathbb{Z}[\xi] \text{ such that } z^\star - \Phi \in P\} = (\mathbb{Z}[\xi] \cap (P + \Phi))^\star, \tag{53}$$

where P is bounded in the plane. Then it is just a matter of choice of appropriate scale and origin in order to get $P + \Phi \subset \mathbb{Z}'_\tau + \mathbb{Z}'_\tau \xi^{3q} = \Gamma_q^\star$, with the notations of (48), and for a certain $q = 1, 2, 3, 4$. The embedding into a τ-grid Γ_q follows from (53) and we get the labelling of the quasilattice points with those of Γ_q:

$$\Sigma_\Phi^P = \{z \in \mathbb{Z}_\tau + \mathbb{Z}_\tau \xi^q = \Gamma_q \text{ such that } z^\star - \Phi \in P\}. \tag{54}$$

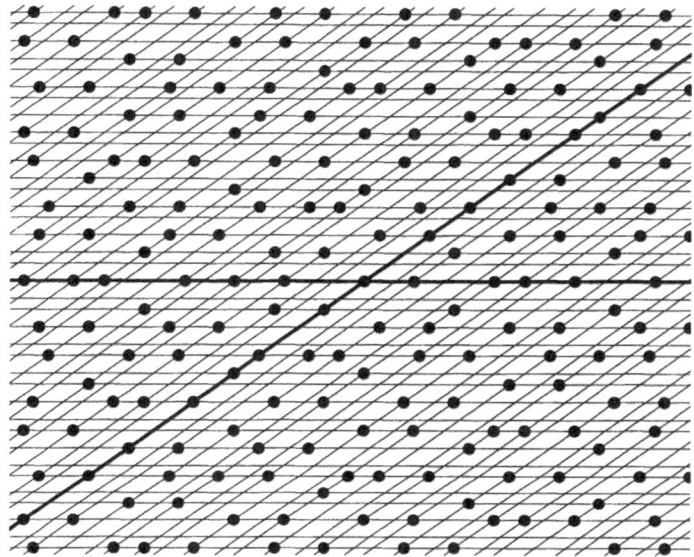

Fig. 5. — Penrose quasilattice as a subset of the τ-grid Γ_1.

A nice example of such an embedding/labelling is provided by Penrose tilings [37] and their diffraction patterns. We show in Figure 5 how the set of Penrose tiling vertices is actually a subset of the grid Γ_1. The Bragg peaks, beyond a given intensity, of the corresponding diffraction pattern are shown in Figure 6. For understanding Figure 5, we start from a τ-grid of the type $\Gamma_q = \mathbb{Z}_\tau + \mathbb{Z}_\tau \xi^q$, and proceed to precise algebraic filtering in order to get a Penrose quasilattice as a subset. The procedure is based on both algebraic colouring and conjugation [38]. The introduction of colour stems from the fact that the integral coordinates $(n_0, n_1, n_2, n_3, n_4)$ of a cyclotomic number $z = n_0 + n_1\xi + n_2\xi^2 + n_3\xi^3 + n_4\xi^4$ in $\mathbb{Z}[\xi]$ are not unique. The quantity $n_0 - n_1 + n_2 - n_3 + n_4$ is defined modulo 5, due to the identity $1 - \xi + \xi^2 - \xi^3 + \xi^4 = 0$. Therefore, there are five (algebraic) colours, each one corresponding to an equivalence class modulo 5. Hence there exists a colouring ring homomorphism $\phi : \mathbb{Z}[\xi] \rightarrow \mathbb{Z}_5 \equiv \mathbb{Z}/5\mathbb{Z}$ given by

$$z = n_0 + n_1\xi + n_2\xi^2 + n_3\xi^3 + n_4\xi^4 \longrightarrow \phi(z) = \overline{n_0 - n_1 + n_2 - n_3 + n_4}, \quad (55)$$

where the overbar designates the equivalence class modulo 5. Note that the colour of τ is $\bar{3}$, and the latter is the unique root of the equation $x^2 = x + 1$ in $\mathbb{Z}_5$. Also note that colour is left unchanged under algebraic conjugation, since $\phi(-\frac{1}{\tau}) = \phi(1 - \tau) = \bar{3}$, and $\phi(\xi^\star) = (\phi(\xi))^3 = \overline{-1}$. Let $\Phi = \sum_{i=0}^{4} \gamma_i \xi^{2i} \in \mathbb{Z}[\xi]$ such that $\sum_{i=0}^{4} \gamma_i = 0$. The set Π_Φ of vertices of the Penrose tiling with associated phason Φ is the union of four "coloured" discrete subsets $\Sigma(k) \subset$

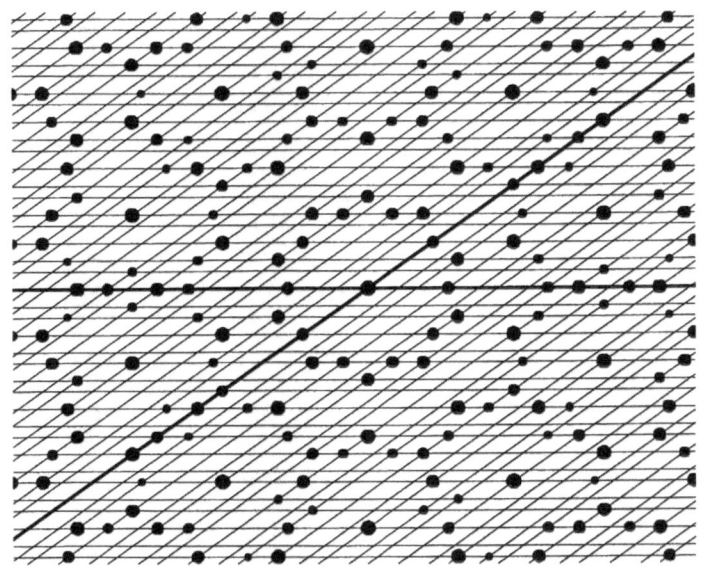

Fig. 6. — Diffraction pattern of Penrose quasilattice as a subset of the τ-grid Γ_1.

$\mathbb{Z}[\xi]$, one per colour $\bar{k} \in \{\bar{1}, \bar{2}, \bar{3}, \bar{4}, \}$, ($\bar{0}$ is excluded), and selected by imposing pentagonal windows in the Galois conjugate space. More precisely,

$$\Pi_\Phi = \bigcup_{k=1}^{4} \Sigma(k), \text{ where } \Sigma(k) = \{z \in \mathbb{Z}[\xi] \mid \phi(z) = \bar{k}, \, z^\star - \Phi \in P_{\bar{k}}\}. \qquad (56)$$

Here, $P_{\bar{1}}$ is the pentagonal convex hull of the five points $1, \xi^2, \xi^4, \xi^6, \xi^8$ (all with colour $\bar{1}$), $P_{\bar{4}} = -P_{\bar{1}}$, $P_{\bar{3}} = \tau P_{\bar{1}}$, and $P_{\bar{2}} = -\tau P_{\bar{1}}$. Note that the other pentagons have vertices with colour $\bar{4}$, $\bar{3}$, and $\bar{2}$ respectively. This is consistent with (56). Let us now denote by R_q the rhomboid convex hull of $\{1 + \xi^q, -1 + \xi^q, -1 - \xi^q, 1 - \xi^q\}$, for $q = 1, 2, 3, 4$. Those lozenges can also be written as $R_q = [-1, 1] + [-1, 1]\xi^q$, and it ensues from (16) and (48) that

$$\mathbb{Z}[\xi] \bigcap R_q = (\mathbb{Z}'_\tau + \mathbb{Z}'_\tau \xi^q) \bigcap R_q = \Gamma^\star_{q-3} \bigcap R_q. \qquad (57)$$

Next we notice that $P_{\bar{1}} \bigcup P_{\bar{4}} \bigcup \frac{1}{\tau} P_{\bar{2}} \bigcup \frac{1}{\tau} P_{\bar{3}} \subset R_q$ for $q = 2$ or 3, whereas

$$\frac{1}{\tau} P_{\bar{1}} \bigcup \frac{1}{\tau} P_{\bar{4}} \bigcup \frac{1}{\tau^2} P_{\bar{2}} \bigcup \frac{1}{\tau^2} P_{\bar{3}} \subset R_q$$

for $q = 1$ or 4. So a suitable scaling allows one to view Penrose vertices as elements of the labelling grid Γ_q. As a matter of fact, for $q = 1$ or 4, and with $\Phi = 0$, we have

$$\tau\Pi_0 = \bigcup_{k=1}^{4} \left\{ z \in \Gamma_q \mid \phi(z) = \overline{3k}, \ z^\star \in -\frac{1}{\tau} P_{\overline{k}} \right\}, \tag{58}$$

whereas for $q = 2$ or 3 we have

$$\tau^2\Pi_0 = \bigcup_{k=1}^{4} \left\{ z \in \Gamma_q \mid \phi(z) = \overline{-k}, \ z^\star \in \frac{1}{\tau^2} P_{\overline{k}} \right\}. \tag{59}$$

Now, an arbitrary Penrose tiling Π of the plane with two types of rhombic tiles, like it appears in familiar pictures, is as "democratic" as its one-dimensional counterpart, namely the Fibonacci chain of Section 3. This means that it is in some sense optimal in regard to the inclusion property

$$\Pi - \Pi \subset \Gamma_q, \tag{60}$$

for an appropriate choice of scale. For instance, if we put the length of the rhomb edges equal to τ^4, choose one of the vertices as the origin O, and one of the edges issued from 0 as the complex τ^4, then $\Pi \subset \Gamma_1$.

Finally, it is worthwhile here to mention a first exploration about the relevance of these τ-grids in experimental observations. In Figure 7 we show the coincidence between some of the nodes of Γ_1 and the most significant Bragg peaks of the diffraction pattern of a quasicrystal.

6. ARITHMETICS AND ALGEBRA OF THE β-INTEGERS

Most of the properties of the τ-numeration system which have been presented in the above are actually shared by a large class of other numbers, particularly by the quadratic unitary Pisot-Vijayaraghavan, among which we find the two other numbers relevant to quasicrystallography. First, let us indicate how a numeration system and its related basis expansion exist for any real number larger than one.

Let β be non-integer and $\beta > 1$. The β-expansion of an arbitrary positive real number x is the series $(\xi_l)_{-\infty \leq l \leq j}$ such that

$$x = \sum_{i=-\infty}^{j} \xi_i \beta^i \tag{61}$$

where j is the highest integer such that

$$\beta^j \leq x < \beta^{j+1}, \tag{62}$$

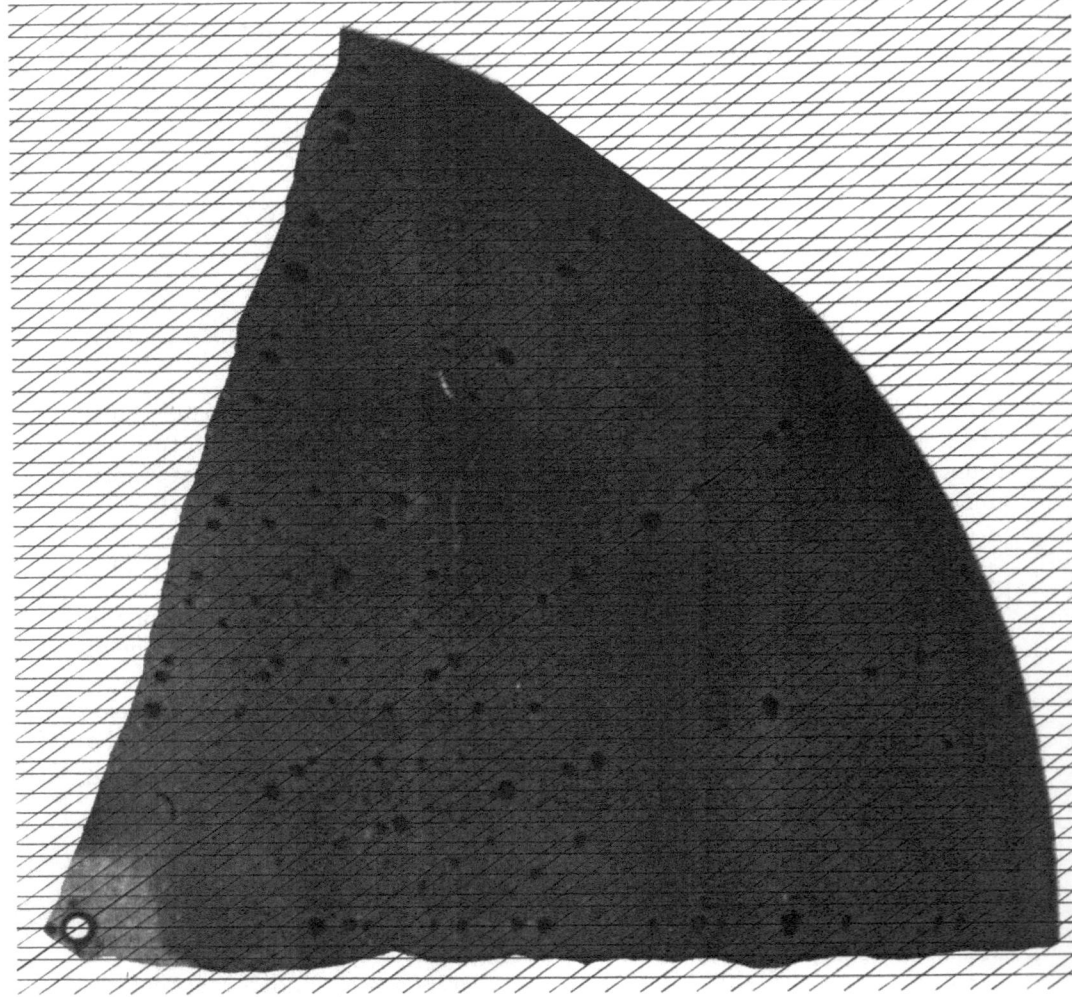

Fig. 7. — Monochromatic x-ray diffraction pattern ($Al_{63}Cu_{17.5}Co_{17.5}Si_2$, synchrotron LURE, Orsay) and τ-grid Γ_1, from Denoyer (in this volume).

the positive integers ξ_l assume their values in the alphabet

$$\{0, 1, 2, \ldots, [\beta]\} \tag{63}$$

and are computed by using the *greedy algorithm*. One recursively defines

$$\xi_j = [x/\beta^j], \quad r_j = \{x/\beta^j\} \text{ (the fractional part of } x/\beta^j),$$
$$\text{and for } l < j, \quad \xi_l = [\beta r_{l+1}], \quad r_l = \{\beta r_{l+1}\} \ldots,$$
$$\text{finally, if } j < 0, \text{ we put } \xi_0 = \xi_{-1} = \cdots = \xi_{j+1} = 0.$$

For short we also write

$$x = \xi_j \xi_{j-1} \xi_{j-2} \cdots \xi_0 . \xi_{-1} \xi_{-2} \xi_{-3} \cdots , \quad (e.g.\ 2 = 10.01 \text{ when } \beta = \tau). \tag{64}$$

The set of real numbers which have a zero fractional part in their β-expansion is named set of β-*integers* and is denoted by

$$\mathbb{Z}_\beta \overset{\text{def}}{=} \{\pm(\xi_j \beta^j + \xi_{j-1} \beta^{j-1} + \cdots + \xi_1 \beta + \xi_0)\} = \mathbb{Z}_\beta^+ \cup (-\mathbb{Z}_\beta^+), \tag{65}$$

where $\mathbb{Z}_\beta^+$ designates the set of non-negative β-integers.

Some configurations $\xi_j \xi_{j-1} \cdots \xi_l \cdots$ in the above definition are not possible. What is allowed and what is forbidden in the set of β-expansions is completely determined by what is called the *Rényi β-expansion of* 1:

$$d(1,\beta) = t_1 \beta^{-1} + t_2 \beta^{-2} + \cdots \tag{66}$$
$$= 0.t_1 t_2 \cdots t_l \cdots, \tag{67}$$

where $t_l \in \{0, 1, \ldots, [\beta]\}$. This expansion is reminiscent of the identity $1 = 0.99 \ldots 9 \ldots$ in the decimal system. It is obtained by the following process:

$$t_1 = [\beta], \quad r_1 = \{\beta\}, \cdots, t_l = [\beta r_{l-1}], \quad r_l = \{\beta r_{l-1}\} \cdots$$

or, equivalently,

$$t_l = [\beta T_\beta^{l-1}(1)],$$

where

$$T_\beta(x) = \beta x (\text{ mod } 1).$$

In this context, note that the greedy-algorithm coefficient ξ_l of a real number $x \in [0, 1)$ is also equal to

$$\xi_{-l} = [\beta T_\beta^{l-1}(x)].$$

We then have the β-expansion rule [30]:

Proposition 5. *No infinite sequence of positive integers is present in any β-expansion if itself and all its (one-sided) shifted are lexicographically larger or equal to:*

$$t_1 t_2 \ldots \quad \text{if the latter is infinite,}$$

and to:

$$(t_1 t_2 \ldots t_{m-1}(t_m - 1))^\omega \quad \text{if} \quad d(1,\beta) = 0.t_1 t_2 \ldots t_{m-1} t_m \text{ is finite.}$$

$(\ )^\omega$ *means that the word within $(\ )$ is indefinitely repeated.*

Therefore, once $d(1, \beta)$ is known, it becomes possible (in principle, but it may turn out to be unpracticable!) to build up $\mathbb{Z}_\beta$ by following the lexicographical order of the allowed sequences.

The countable set $\mathbb{Z}_\beta$ is naturally self-similar and symmetrical with respect to the origin:

$$\beta \mathbb{Z}_\beta \subset \mathbb{Z}_\beta \ , \ \mathbb{Z}_\beta = -\mathbb{Z}_\beta. \tag{68}$$

It tiles the line with a finite or infinite number of intervals separating two nearest neighbours $x_i < x_{i+1}$ with lengths of the tiles $l_i = x_{i+1} - x_i$. Now we are concerned by sets $\mathbb{Z}_\beta$ that be Delaunay and possibly Meyer. This is at a certain extent assured for the two following important results ([39] and [40]).

Proposition 6. *Suppose β is a Pisot number. Then the Rényi β-expansion of 1 is eventually periodic,*

$$d(1, \beta) = 0.t_1 t_2 \ldots t_m (t_{m+1} \ldots t_{m+p})^\omega.$$

When β is a Pisot number, then it follows that $\mathbb{Z}_\beta$ is a self-similar tiling of the line with a finite set of different tiles. The lengths of the tiles are $\{T_\beta^i(1), \ 0 \leq i \leq m + p - 1\}$ (see [41]). More precisely the lengths assume their values in the set

$$1, \beta - t_1, \beta^2 - t_1 \beta - t_2, \ldots,$$
$$\beta^{m+p-1} - t_1 \beta^{m+p-2} - \cdots - t_{m+p-1}.$$

Hence, if β is Pisot, then $\mathbb{Z}_\beta$ is Delaunay.

It can be also proven that, when β is a Pisot number, then the set $\mathbb{Z}_\beta$ of β-integers is a Meyer set, *i.e.* obeys (7) and (8). Actually, the following results will tell more about the finite sets X and Y appearing in (7) and (8). Of course, we are mainly concerned by the three quasicrystallographic cases (3, 4) and (5), but we also have (see [34]) results for general quadratic unit Pisot numbers, namely those β which are solution to

$$x^2 = ax + 1, \quad a \in \mathbb{Z}, \ a \geq 1 , \tag{69}$$

$$x^2 = ax - 1, \quad a \in \mathbb{Z}, \ a \geq 3. \tag{70}$$

In the sequel, the notation $\sum^P$ still designates the discrete subset obtained from the ring $\mathbb{Z}[\beta]$ through the algebraic sieving procedure:

$$\Sigma^P = \{x = m + n\beta \in \mathbb{Z}[\beta] \text{ such that } x' = m + n\beta' \in P\}. \tag{71}$$

Proposition 7. *Let β be the root > 1 of $x^2 - ax - 1$, $a \geq 1$. Then $\beta' = -\frac{1}{\beta}$ and we have*

$$\mathbb{Z}_\beta^+ = \Sigma^{(-1,\beta)} \cap \mathbb{R}^+ \text{ and } \mathbb{Z}_\beta^- = \Sigma^{(-\beta,1)} \cap \mathbb{R}^-, \tag{72}$$

$$\mathbb{Z}_\beta + \left\{0, \pm\frac{1}{\beta}, \ldots, \pm\frac{a}{\beta}\right\} \subset \frac{\mathbb{Z}_\beta}{\beta^2}, \tag{73}$$

$$\mathbb{Z}_\beta + \mathbb{Z}_\beta \subset \mathbb{Z}_\beta + \left\{0, \pm\frac{1}{\beta}\right\} \subset \frac{\mathbb{Z}_\beta}{\beta^2}, \tag{74}$$

$$\mathbb{Z}_\beta \times \mathbb{Z}_\beta \subset \mathbb{Z}_\beta + \left\{0, \pm\frac{1}{\beta}, \ldots, \pm\frac{a}{\beta}\right\} \subset \frac{\mathbb{Z}_\beta}{\beta^2}. \tag{75}$$

Proposition 8. *Let β be the root > 1 of $x^2 - ax + 1$, $a \geq 3$. Then $\beta' = \frac{1}{\beta}$ and we have*

$$\mathbb{Z}_\beta^+ = \Sigma^{[0,\beta)} \cap \mathbb{R}^+ \text{ and } \mathbb{Z}_\beta^- = \Sigma^{(-\beta,0]} \cap \mathbb{R}^-, \tag{76}$$

$$\mathbb{Z}_\beta \subset \Sigma^{(-\beta,\beta)} \subset \mathbb{Z}_\beta + \left\{0, \pm\frac{1}{\beta}\right\}, \tag{77}$$

$$\mathbb{Z}_\beta^+ + \mathbb{Z}_\beta^+ \subset \frac{\mathbb{Z}_\beta^+}{\beta}, \tag{78}$$

$$\mathbb{Z}_\beta^+ - \mathbb{Z}_\beta^+ \subset \mathbb{Z}_\beta + \left\{0, \pm\frac{1}{\beta}\right\}, \tag{79}$$

$$\mathbb{Z}_\beta \times \mathbb{Z}_\beta \subset \frac{\mathbb{Z}_\beta}{\beta}. \tag{80}$$

The above list (72–80) of properties is a departure point for developing a complete β-integer arithmetics. For instance, it can be proven [42] that $\mathbb{Z}_\beta$ is actually an abelian group isomorphic to $\mathbb{Z}$ through the correspondence

$$n \longrightarrow x_n \equiv n^{\text{th}} \ \beta\text{-integer} \tag{81}$$

already examplified in the $\beta = \tau$-case by (13). This allows us to conclude that this concept of β-integer is a natural tool for generalizing to quasilattices or more general aperiodic sets numerous features displayed by ordinary lattices.

REFERENCES

[1] Janot C., *Quasicrystals a Primer* (Clarendon Press, Oxford, 1992).
[2] Janot C. and Mosseri R., *Quasicrystals* Proceedings of the 5th International Conference, Avignon 1995 (World Scientific: Singapore, 1995).
[3] Steinhardt P.J. and Ostlund S., *The Physics of Quasicrystals* (World Scientific: Singapore, 1987).
[4] Senechal M., *Quasicrystals and Geometry* (Cambridge University Press, Cambridge, 1995).

[5] Moody R.V., Meyer Sets and their Duals, *The Mathematics of Aperiodic Long Range Order*, R.V. Moody, NATO-ASI Proceedings, Waterloo 1995 (Kluwer, Academic Publishers, 1996) 403.

[6] Levine D. and Steinhardt P.J., Quasicrystals I: Definitions and structure, *Phys. Rev. B* **34** (1986) 596-616.

[7] Socolar J. and Steinhardt P.J., Quasicrystals II: Unit-cell configurations, *Phys. Rev. B* **34** (1986) 617-647.

[8] Baake P., Kramer P., Schlottmann M. and Zeidler D.J., Planar patterns with fivefold symmetry as actions of periodic structures in 4-dim. space, *Int. J. Modern Phys. B* **4** (1990) 2217-2268.

[9] Baake P., Joseph D., Kramer P. and Schlottmann M., Root lattices and quasicrystals, *J. Phys. A: Math. Gen.* **23** (1990) L1037-L1041.

[10] Axel F. and Gratias D., *Beyond Quasicrystals, Les Houches 1994* (Les Éditions de Physique, Springer, 1995).

[11] Meyer Y., Algebraic numbers and harmonic analysis (North-Holland, 1972).

[12] Meyer Y., Quasicrystals, diophantine approximation and algebraic numbers, *Beyond Quasicrystals, Les Houches 1994*, edited by F. Axel and D. Gratias (Les Éditions de Physique, Springer, 1995).

[13] Lagarias J.C., Meyer's concept of quasicrystal and quasiregular sets, *Comm. Math. Phys.* **179** (1996) 365-376.

[14] Lagarias J.C., Geometric Models for quasicrystals I. Delone sets of finite type, *Discrete & Computational Geometry* (1998) to appear.

[15] Berman S. and Moody R.V., The algebraic theory of quasicrystals with five-fold symmetry, *J. Phys. A: Math. Gen.* **27** (1994) 115–129.

[16] Moody R.V. and Patera J., Quasicrystals and icosians, *J. Phys. A: Math. Gen.* **26** (1994) 2829–2853.

[17] Barache D., De Bièvre S. and Gazeau J.P., Affine Symmetry Semi-Groups for Quasicrystals, *Europhys. Lett.* **25** (1994) 435–440.

[18] Barache D., Champagne B. and Gazeau J.P., Pisot-Cyclotomic Quasilattices and their Symmetry Semi-groups, edited by J. Patera, *Fields Institute Monograph Series, Amer. Math. Soc.* **10** (1998).

[19] Pleasants P.A.B., The construction of quasicrystals with arbitrary symmetry group, *Proceedings of the 5th International Conference on Quasicrystals*, edited by Ch. Janot and R. Mosseri, Avignon (Word Scientific, Singapore, 1995).

[20] Gazeau J.P., Quasicrystals and their Symmetries, *Symmetries & structural properties of condensed matter* Zajaczkowo 1994, edited by T. Lulek (World Scientific: Singapore, 1995), 369.

[21] Gazeau J.P. and Patera J., Tau-wavelets of Haar *J. Phys. A: Math. Gen.* **29** (1996) 4549–4559.

[22] Gazeau J.P., Pisot-cyclotomic Integers for Quasicrystals, *The Mathematics of Aperiodic Long Range Order*, edited by R.V. Moody, NATO-ASI Proceedings, Waterloo 1995 (Kluwer, Academic Publishers, 1997).

[23] Gazeau J.P. and Spiridonov V., Toward discrete wavelets with irrational scaling factor, *J. Math. Phys.* **37** (1996) 3001–3013.

[24] Gazeau J.P., Canonical quasilattices for labelling quasicrystalline sites, *Group 21* Goslar 1996, edited by H.D. Doebner, W. Scherer and C. Schulte (World Scientific: Singapore, 1997).

[25] Elser V., Indexing problems in quasicrystal diffraction, *Phys. Rev. B* **32** (1985) 4892-4898.

[26] Gazeau J.P. and Lipinski D., Quasicrystals and their Symmetries, *Symmetries & structural properties of condensed matter* Zajaczkowo 1996, edited by T. Lulek (World Scientific: Singapore, 1997) 81.

[27] Bertin M.J., Decomps-Guilloux A., Grandet-Hugot M., Pathiaux-Delefosse M. and Schreiber J.P. *Pisot and Salem numbers* (Birkhäuser Verlag, 1992).

[28] Katz A., *A Short Introduction to Quasicrystallography*. From Number Theory to Physics, Les Houches 1989, edited by M. Waldschmidt, P. Moussa, J.M. Luck and C. Itzykson (Springer-Verlag, Berlin, 1992).

[29] Rényi A., Representations for real numbers and their ergodic properties, *Acta Math. Acad. Sci. Hungar.* **8** (1957) 477-493.

[30] Parry W., On the β-expansions of real numbers, *Acta Math. Acad. Sci. Hungar.* **11** (1960) 401-416.

[31] Frougny Ch. and Solomyak B., Finite beta-expansions, *Ergod. Th. & Dynam. Sys.* **12** (1992) 713-723.

[32] Frougny Ch., Fibonacci representations and finite automata, *I.E.E.E. Trans. Inform. Theory* **37** (1991) 393-399.

[33] Knuth D.E., *The art of computer programming* (Vols. 1, 2 and 3 Addison-Wesley, 1975).

[34] Burdik C., Frougny C., Gazeau J.P. and Krejcar R., Beta-Integers as Natural Counting Systems for Quasicrystals, *J. Phys. A: Math. Gen.* **31** (1998) 6449-6472.

[35] Hof A., *Diffraction by Aperiodic Structures* in *Mathematics of Aperiodic Long Range Order*, Proceedings NATO-ASI, Waterloo 1995, edited by R.V. Moody (Kluwer, Dordrecht, 1997).

[36] Gazeau J.P., Patera J. and Pelantova E., Tau-wavelets in the plane, *J. Math. Phys.* **39** (1998) 4201-4212.

[37] Penrose R., *Mathematical Intelligencer* **2** (1979) 32; Grünbaum G. and Sheppard G.C., *Tilings and Patterns* (Freeman, New York); de Bruijn N.G., *Indag. Math.* **84** (1981) 39; *Indag. Math.* **84** (1981) 53.

[38] Moody R.V. and Patera J., *Can. J. Phys.* **72** (1994) 442.

[39] Bertrand A., Développements en base de Pisot et répartition modulo 1, *C.R. Acad. Sc. Paris* **285** (1977) 419-421.

[40] Bertrand A., Comment écrire les nombres entiers dans une base qui n'est pas entière, *Acta Math. Acad. Sci. Hungar.* **54** (1989) 237-341.

[41] Thurston W., *Groups, tilings, and finite state automata* (AMS Colloquium Lecture Notes, Boulder, 1989).

[42] Frougny C., Gazeau J.P. and Krejcar R., β-integers as a group (1999) submitted.

Acoustic-Like Excitations
in Strongly Disordered Media

Given by E. Courtens

Written by E. Courtens and R. Vacher

Laboratoire des Verres, UMR 5587 du CNRS, Université Montpellier II, Place Eugène Bataillon, 34095 Montpellier Cedex 5, France

---

1. INTRODUCTION

The issue of the glassy state stands high up on the list of major unsolved problems in solid state physics. A central reason for the slow progress in this field is that it involves in an essential way relatively large length scales together with disorder. In fact, glasses look very much like crystals on the near neighbor scale, but they become isotropic at macroscopic scales. There must clearly exist an *intermediate length scale* where a structural crossover occurs [1]. Consider the example of silica, SiO_2. The macroscopic density of crystal quartz is 2.6 g/cm^3, whereas that of the glass is 2.2. Both systems are practically identical at the level of the SiO_4 tetrahedra. In the glass, on the near atomic scale, the arrangement is also quite similar to that of crystal quartz. Hence, the change of density from a local value close to 2.6 to the average value of 2.2, as well as the averaging over orientations, must result from fluctuations at some intermediate scale. In silica, one has reasons to believe that this intermediate scale might be of the order of 6 nm [2], which is nearly 40 times larger than the Si-O bond length. This is what makes the glass problem so difficult.

From this simple example it is clear that all the techniques at our disposal, whether experimental, theoretical, or numerical, are put to great strain in trying to gain a microscopic understanding of the glassy state. The difficulty for simulations and theory are evident. In the case of experiments, it turns out that this intermediate scale is too small for observations in real space, but also too large to apply the usual structural tools in reciprocal space.

It is for these reasons that the study of acoustic excitations in glasses, and of their behavior as the frequency increases, is of very much interest. At low acoustic frequencies glasses look homogeneous, and the excitations are plane waves propagating in a continuum. As the frequency is raised, the wave lengths approach the scale of the intermediate fluctuations and their scattering increases. One can anticipate that at sufficiently high frequencies the scattering by the inhomogeneities is so strong that propagating plane waves cease to exist and that the excitations become diffusive [3]. In fact, if the medium contained lots of voids, as it is the case for real mass fractals to be discussed in the first part of this lecture, one expects that soon above their crossover to strong scattering the excitations become truly localized [4]. This dynamical crossover provides a handle to investigate the structural one.

The crossover can also be identified in the low temperature thermal conductivity. In all solids, and above ≈ 1 K, the thermal conductivity κ increases with a power of the temperature T. A specificity of glasses is that they exhibit a plateau in κ in a region around 10 K [5]. The dominant phonons at this temperature have frequencies of the order of 1 THz[1]. The standard explanation for the plateau is that these phonons cease to propagate [7], which corresponds to the crossover. Another universal feature of glasses is observed in their density of vibrational states (DOS) at these frequencies. It exceeds the prediction of the Debye model, and the excess produces a hump in the curve C/T^3 *vs.* T, where C is the specific heat [5]. In both neutron and Raman spectroscopies there is a corresponding additional spectral weight, often called the "Boson peak" [8], as it obeys Bose-Einstein statistics. In contrast, relaxational contributions, which correspond to temperature induced structural changes in the glass, may appear in the same frequency range as T is increased, and therefore their spectral intensity does not simply vary with Bose-Einstein statistics.

A central feature of the Boson peak is that it is observed over a large range of scattering vectors $\mathbf{Q}$. One can invoke at least three causes for this peak. The first is that acoustic excitations at these frequencies are so strongly scattered by structural inhomogeneities that they contribute to light or neutron scattering at all $\mathbf{Q}$-values. There can be several reasons for a sizeable increase of the apparent acoustic DOS, including contributions from flat regions in dispersion curves [9]. A second cause for the Boson peak, that may dominate in polymeric or other organic glasses, is the presence of many other low frequency modes arising from molecular motions. These should in fact interact with acoustic

---

[1] Energies, and also frequencies, will be expressed in GHz (10^9 Hz) or in THz (10^{12} Hz), but also in cm^{-1} or in meV as the case may be; 1 meV equals 242 GHz or 8.07 cm^{-1}. From $k_B T = \hbar\omega$, 1 K equals 20.84 GHz or 0.0862 meV. However, owing to the powers of ω entering into the calculation of thermal properties [6], the dominant acoustic phonons in these problems occur at $\hbar\omega$ equals 3 to 5 times $k_B T$, from which 10 K roughly corresponds to 1 THz.

excitations and contribute to their strong scattering. A final cause is very specific to the structural crossover discussed above. As pointed out by the regreted Prof. Alexander at this meeting, one may expect in glasses regions that are buckled[2] and regions that are under tension [10]. In other words glasses are frozen in a state that is relatively close to a mechanical instability. As a result there are additional low frequency modes, as nicely shown in a recent simulation [11]. The inhomogeneities at the intermediate scale might be responsible for the soft potentials that have been invoked with considerable success to describe the spectral and thermal properties of glasses [12].

As a first example of disordered media, results on laboratory fractals will be presented. In this case considerable progress has been achieved on all fronts, theory, simulations, and experiments. The fractal scaling laws [13, 14] turned out to be very powerful to produce theoretical predictions as well as to evaluate data. Many successful simulations could also be made on systems of significant sizes [15]. The important features of the structures and of the vibrations turned out to be well accessible to experiments on *silica aerogels* [16]. These are materials that can form rather ideal fractals that can be handled in the laboratory, as described below. A clear structural crossover from the homogeneous regime at large sizes to the fractal regime at small sizes was identified in these aerogels [17]. This structural crossover could unambiguously be related to a dynamical one where strong scattering sets in [18]. In this case, the crossover separates phonons at low frequencies from *fractons* at high frequencies. The name fracton designates an eigenvibration of a fractal structure. Hence, aerogels turned out to be an excellent testing ground to learn how to handle the collective vibrations of disordered systems. As explained below, important dynamical aspects of disorder do remain whether the system is fractal or not. In the final part of these lectures, we turn to the high frequency acoustic excitations of glasses. This is an extremely active field currently. Many results have been presented in the literature, and some are still heavily debated. Examples will be given, and recent progress that was achieved, owing in particular to recent developments in scattering experiments, will be discussed.

2. THE CASE OF MASS-FRACTAL MEDIA

In this section, we deal with real objects (or with their models), made of particles or grains of typical size (radius) a, of volume v_a, and of mass m_a. The material within the grains has a density $\rho_a = m_a/v_a$. One can define mass fractals as objects assembled from such connected grains, whose mass within a

---

[2] Buckling is the mechanical instability of a column which is compressed along its axis from both ends. The "buckled" regions spoken of here, are regions where the axial pressure is greater than the threshold for this instability and where the original structure collapsed.

sphere of radius $L > a$, *centered on a grain*[3], scales on the average like [19]

$$M(L) \approx m_a \times (L/a)^D. \tag{1}$$

Here, $D < 3$ is the fractal dimension of the object. As the volume V of a sphere scales like L^3 (we do not bother about factors like $4\pi/3$), the average density of the fractal within a sphere of radius L centered on a grain is given by

$$\rho_f(L) = M(L)/V(L) \approx \rho_a \times (L/a)^{D-3}. \tag{2}$$

Here, we used $\rho_a \approx m_a/a^3$. By definition, since these equations require that there is a grain at the center of the sphere, $\rho_f(L)$ relates to the correlation function as it is defined for fractal objects in Appendix A, equations (A.7–A.9).

As D is smaller than 3, one sees from (2) that $\rho_f(L)$ decreases with increasing L. If L could tend to infinity the object would have zero density! In practice, for a given sample, the upper value of L will result from the preparation recipe. One should note that there will always be a biggest possible L imposed by the mechanical stability of the structure. It is known that the thermal stability sets that limit if the grains are rather small in terms of number of atomic diameters, whereas it is gravitational collapse that sets the stability limit for large grains [20]. We call ξ the upper value of L imposed by the preparation recipe of a given sample. It is called the *fractal persistence length* (or the fractal correlation length) of that object. A real material is assembled from fractal blobs of upper size ξ. The macroscopic density ρ of such a material just equals $\rho_f(L)$ at ξ,

$$\rho = \rho_a \times (\xi/a)^{D-3}. \tag{3}$$

2.1. The structure of mass fractals

2.1.1. *Mass fractals in simulations*

Simulations are interesting for themselves, they allow to check analytical predictions, and they are extremely useful in providing a mental picture. It is the latter aspect which is discussed now, with the help of the simplest possible example, that of *percolation* networks.

Site percolation – Take a piece of squared paper. Let the surface of each square represent the equivalent of a *site*. Each site can be connected by *bonds* to its four neighbors (we assume that no bonds are possible along diagonals). First consider that all sites are empty, and with a pencil start filling them (blackening them) randomly with probability p. As neighboring sites are filled, let the bond between them automatically exist. As p is increased towards a critical value p_c, a *percolation threshold* will be reached where a continuous

---

[3] This is an essential point, as the average mass within an *arbitrary* volume is always proportional to the volume, and thus scales like L^3 rather than like L^D.

Fig. 1. — Examples of site (left) and bond (right) percolation in 2 dimensions for 68 × 68 realisations. The "infinite" cluster is shown in dark, the finite ones in grey. Empty sites on the left are white, while the unoccupied bonds on the right are not shown. (This illustration was kindly provided by Stoll.)

path is first established between two opposite sides of the piece of paper. For a small sheet, the average value of p_c which is found by repeating this process will be slightly larger than the thermodynamic value. Critical percolation is illustrated on the left of Figure 1. At that p there exists one large cluster, which contains the connecting path, and which is called the "*infinite*" cluster, this by reference to the thermodynamic limit where this cluster really contains an infinite number of occupied sites. There are also many finite disconnected clusters, of different sizes. If one continued filling in the sites, the space occupied by the infinite cluster would grow much faster than just from the new filled sites, this at the expense of the finite clusters that progressively attach to it, starting primarily with the largest ones. If we imagine *removing the finite clusters*, the mass distributed in the remaining infinite cluster (if the simulation were large enough) is a fractal. This process is called *site percolation* [21].

Bond percolation – One could proceed in another manner. Imagine that all sites are initially occupied, but without any bond between them. With a pencil start filling randomly the bonds with probability p. At a certain point, $p \approx p_c$ (this gives another value for p_c), the first continuous path is established. Again an infinite cluster is formed, and there are many finite clusters not attached to it by any bond. *Removing these finite clusters*, the remaining mass shows a fractal distribution. This is called *bond percolation* [21], and an example is shown on the right of Figure 1.

Comparing the two pictures, we note that site percolation clusters appear to have many small voids. Their origin lies in the initial random filling of sites. This leaves holes having *nothing* to do with fractality, just with *white noise*. The argument can be made fully quantitative, and the reader is refered to [22] for details. The mass fractality is due to something else: it is produced by the removal of finite clusters *at all scales* ("all" if $p = p_c$). This leaves *fractal blobs* at all scales, *i.e.* from the size ξ of the paper down to the size a of a single square. Blobs of size L contain smaller blobs of all sizes down to a. It is this embedding of blobs which is typical of mass fractality.

2.1.2. Silica aerogels

Silica occurs in a large variety of forms [23]. In particular many porous varieties of silica are known [16]. Some of these are prepared from solutions that are gelled, and the wet gels are then dried in some fashion. If this drying is done above the critical point of the solvent, no meniscus can form at the liquid-gas-solid interfaces. This preserves the mechanically fragile internal gel structure [24]. The result is a monolithic, highly porous solid, called an *aerogel*. There are many varieties of aerogels depending on the preparation of the wet gel. In several preparation methods, one grows particles in the solution before gelation. Typically, one starts from an alkoxysilane, $Si(OR)_4$, where R$-$ is CH_3-, C_2H_5-, ..., which is mixed with water, alcohol, and a pH-modifier that serves as catalyst. The hydrolysis of $Si(OR)_4$ forms silicon hydroxide in several steps, and the hydroxide groups polycondense into siloxane bonds, $\equiv Si - O - Si \equiv$. The pH modifies the various reaction rates. Under "neutral" condition (no catalyst intentionally added) the hydrolysis is slow and incomplete, and the condensation rapid. This produces so-called N-aerogels which grow like branched polymers, with small, polydisperse, fluffy particles. The addition of base catalyst increases the hydrolysis rate and the solubility of SiO_2. Larger grains grow, while smaller ones tend to redissolve. The aggregation of grains is initially prevented by their charge. Once the particles are sufficiently large they finally form clusters. These aggregate until the solution gels. This leads to B-aerogels. For either N- or B-aerogels, the final size of clusters, and thus the macroscopic density, can be adjusted by changing the relative amount of alcohol in the initial mixture. There exist many other variants to the preparation, such as acid catalysis, two-step reaction processes, drying with solvent exchange, etc. In each case the structure is influenced by the preparation, so that a rather careful structural study generally is a prerequisite to other physical investigations. It turns out that suitably prepared N- and B-aerogels form, over a certain range of lengths, $a < L < \xi$, remarkably good mass fractals.

Fractal structure of silica aerogels – As explained in Appendix A, D and ξ can be determined from $S(Q)$ measured by small-angle scattering. Data are usually obtained with cold neutrons (SANS), or also with X-rays, giving a Q-range typically from $\sim 2 \times 10^{-3}$ to ~ 0.3 Å^{-1}. This allows to observe fractal behavior from ~ 1000 Å, the upper limit on the measurable ξ, down to ~ 10 Å, the lower limit on the measurable grain size a. The results shown in Figure 2 were obtained on N-aerogels [17]. In Figure 2a, the samples are labelled by their macroscopic densities, ρ, in kg/m^3. The intensities are normalized to the same volume of water. In the homogeneous ($Q < \xi^{-1}$) and fractal regimes,

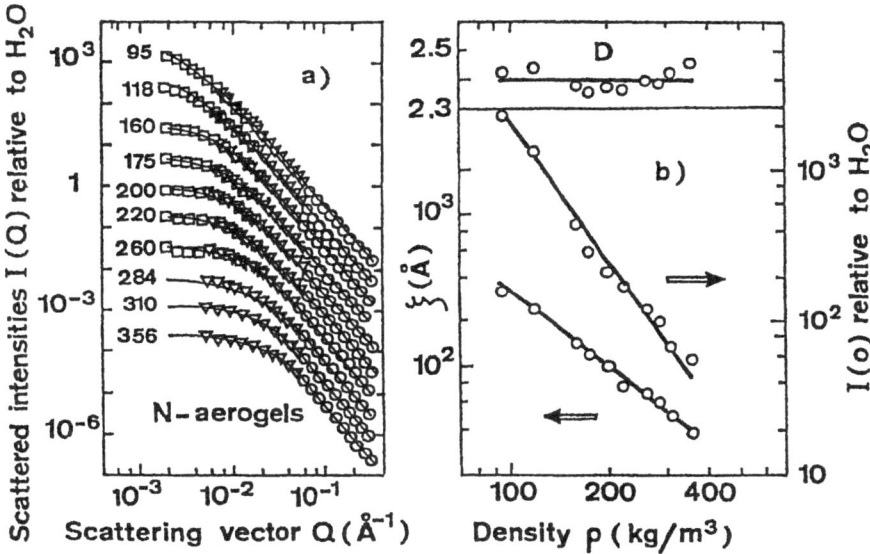

Fig. 2. — a) The structure factor of N-aerogels. The curves are labelled with the densities in kg/m³. Successive intensities, from top to bottom, were divided by increasing powers of 4 to separate the curves. Different symbols represent different detector positions, and the lines are the fits explained in the text. b) The parameters extracted from the fits in a); on top, the fractal dimension D, and below, the scaling of ξ and of $I(0)$.

excellent fits are obtained using simply a smooth exponential crossover [25],

$$g(r) - 1 \propto r^{D-3} \exp(-r/\xi), \tag{4}$$

where $g(r)$ is the pair correlation function as defined in (A.8). The Fourier transform of (4) gives the lines in Figure 2a. In the fractal range the intensity is simply proportional to Q^{-D}. The fits have three parameters, D, ξ, and $I(0)$, the scattered intensity at $Q = 0$. The resulting values of these parameters are shown in Figure 2b. The first observation is that each sample has approximately the same $D = 2.40 \pm 0.03$. Further, the power law for ξ corresponds to $\xi \propto \rho^{-1/(3-D)}$, with *the same* average D. The power law for $I(0)$, $I(0) \propto \rho^{-(2D-3)/(3-D)}$, implies that the intensities in the fractal region superpose when normalized to the sample masses. We return to the deep significance of these observations in the following section.

Similar experiments were performed with B-aerogels [26]. In that case the scattered intensities are not monotonous functions of Q. In the region $Q \approx \xi^{-1}$ they show a hump which cannot be fitted with the Fourier transform of (4).

The physical origin for this maximum is the existence of a prefered size for the fractal blobs that result from the growth. This produces a pseudo-periodicity at the most probable size of the blobs [17, 27]. Excellent fits to the scattering curves are obtained using diffusion-limited cluster-cluster aggregation (DLCA) simulations [28]. The maximum in $S(Q)$ is found to occur at a $Q_{\max} = 2.75/\xi$. For these materials one finds $D = 1.72$, a value that agrees with DLCA. One also obtains $\xi \propto \rho^{-1/(3-D)}$, while the scattered intensities again superpose in the fractal region when normalized to the sample masses.

2.2. The concept of mutually self-similar series or MSSS

The scaling of physical properties in the fractal range, $a < L < \xi$, implies that these properties obey power-law dependences on L. This is the case for the density itself, by its very definition (2). We saw in the previous section how this can actually be measured in a scattering experiment. The power law Q^{-D} directly results from the mass scaling (1). However, many other properties should also scale. Take for example the bulk modulus K. If one had only a single sample, a measurement of this scaling would require a microscopic method of determining the bulk modulus in function of L, $K(L)$. Although, depending on the property of interest, this might sometimes be possible, clearly it would not generally be an easy experimental task.

A way around is to use what we have called a *mutually self-similar series* of samples, or MSSS [17]. These are materials that are identical in the fractal regime, and differ only by their persistence length ξ. To take the above example, the *macroscopic* bulk modulus for one such sample is $K(\xi)$. So it suffices to study the ξ-dependence of $K(\xi)$ on many samples to determine the scaling $K(L)$.

This assumes that the samples can *really* be made identical, except for ξ. It is not sufficient that different samples have the same fractal dimension D, as illustrated in Figure 3a which shows schematically the fractal density $\rho_f(L)$ for two materials. The two samples (1) and (2) cannot be expected to have scalable properties at ξ_1 and ξ_2. Figure 3b shows a necessary condition for a MSSS: the points A and B must line up with the fractal slope $\rho_f \propto L^{D-3}$. Hence, if we have an independent way of determining ξ on the different samples, one must find $\xi \propto \rho^{-1/(3-D)}$. This is the deep significance of the scaling of ξ found in the previous section. Similarly, the curves for $S(Q)$ must also superpose, as found there. Depending on the actual property of interest, it might further be important, contrary to what is sketched in Figure 3b, to have identical particle sizes a, and densities ρ_a. Finally, if the property depended on the connection between the grains (as would the bulk modulus to return to the previous example), it were also important to have similar connections between grains, which cannot be checked simply by an elastic scattering experiment. It turns out that suitably prepared N- and B-aerogels can form MSSS, not only for what the structure is concerned, but also for the dynamics, as explained below. The above implies that the properties to be measured are not affected by

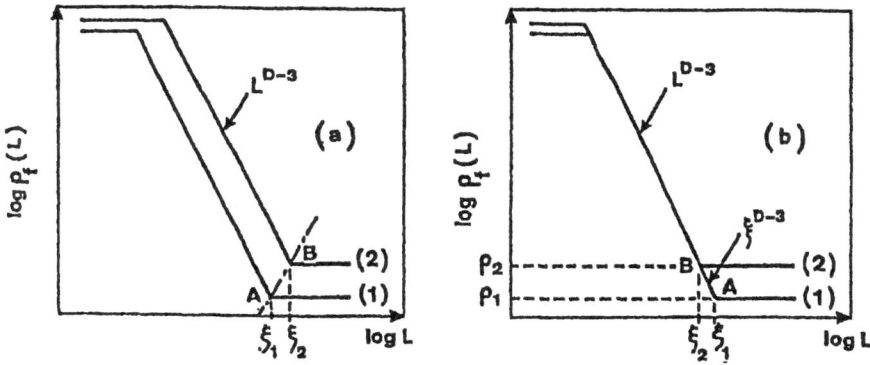

Fig. 3. — (a) Schematic presentation of the density $\rho_f(L)$ of two samples having the same D, but not forming a MSSS. (b) The same for a MSSS.

extraneous macropores. Macropores are always present owing to statistics [29], and these do not have such an adverse effect.

2.3.   The dynamics of mass fractals

2.3.1.   *Phonons and fractons: A simple example*

Consider the very simple model of N masses ($i = 1$ to N) connected by springs. Assume that the masses can only displace in one direction, by quantities u_i. Assume further that the restoring forces are simply proportional to the strains with direct neighbors, $u_j - u_i$. This is called *scalar* elasticity [10, 30]. Usual elasticity is tensorial of course, displacements are vectors, and nearest neighbor forces include not only stretching but also at least bending. In spite of its simplicity, the scalar elastic model captures some of the essential ingredients of the dynamics of fractal systems. To fix the ideas, the masses could be the occupied sites of a large percolation system of the type shown in Figure 1, with periodic boundary conditions. The support is then fractal from the grain size up to a persistence length of the order of either ξ or the box size. To find the *harmonic* vibrations of such a model one diagonalizes its dynamical matrix. For a fully connected system, one finds N modes, of which only one is the uniform translation mode with $\omega = 0$, and $N - 1$ are at frequencies $\omega_\alpha \neq 0$, with eigenvectors u_i^α, orthogonal to the uniform mode, so that $\sum_i u_i^\alpha = 0$, for all α.

It helps to inspect some of these modes to find out what happens. At very small ω one has acoustic phonons. However, as soon as the wavelength becomes of the order of the persistence length, the modes localize. Hence, above the corresponding frequency, the modes have non-negligible intensity only over a

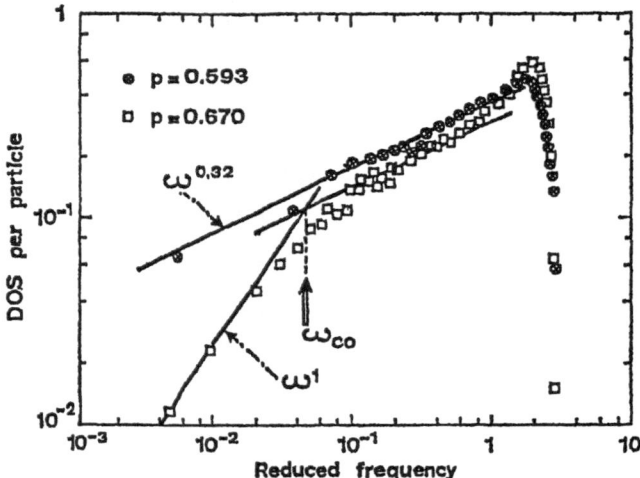

Fig. 4. — The DOS calculated for a 700×700, $d = 2$, site percolation model, for two values of p. The DOS is normalized to 1 per site, and the frequency to that of the highest phonon $\omega_{\max} = \sqrt{4d}$. The dots correspond to $p \approx p_c$. For $p > p_c$, a crossover from the phonon regime to the fracton regime is seen at ω_{co}. The humps near the upper end of the fracton regime correspond to the more tightly connected regions attaching fractal blobs.

restricted region of space, of size λ_α for mode α. Images of eigenvectors are found *e.g.* in [31]. These modes cease to be acoustic phonons and are called *fractons* [13, 14]. The change of regime can be seen as a crossover in the density of vibrational states (DOS), $\mathcal{N}(\omega)$. Figure 4 illustrates the situation for a two-dimensional ($d = 2$) percolation system at two values of p [32]. For p well above p_c, there is a phonon regime at large length scales (small ω),

$$\mathcal{N}(\omega) \propto \omega^{d-1}, \tag{5}$$

where $d - 1 = 1$ in Figure 4. It crosses over to a fracton regime,

$$\mathcal{N}(\omega) \propto \omega^{\tilde{d}-1}, \tag{6}$$

as the scale of the excitations decreases. This occurs at a crossover frequency ω_{co}. The new index $\tilde{d}$ is called the *fracton* or *spectral* dimension. For $p \approx p_c$, one sees only the fracton regime in Figure 4. The slope $\tilde{d} - 1 \cong 0.32$ in that case is a property of percolation systems in scalar elasticity [13]. The humps at the high frequency end of the fracton regime are produced by the tightly connected regions. When ξ decreases, the abundance of these regions increases, and therefore the overall density of fractal states decreases [32].

The size of a fracton is a concept that can be precisely defined by an averaging over sites in which the square amplitudes $(u_i^\alpha)^2$ are taken as weights [31]. Similarly, one can define local averages, for example $\overline{\omega_i^2}$, by averaging with the same weights over all fractons. Finally, if the system is sufficiently large, or by taking a sufficient number of realizations, one can average over fractons in a small frequency interval $\Delta\omega$ about ω to define an ω-dependent average, which for λ_α is $\lambda(\omega) = \langle \lambda_\alpha \rangle_{\Delta\omega}$. From the simulations one finds that: (1) fractons of *all* frequencies are felt *everywhere*; (2) more connected regions have higher $\overline{\omega_i^2}$; (3) the average $\lambda(\omega)$ follows a scaling law, which can be written [4, 13, 14]

$$\lambda(\omega) \propto \omega^{-\tilde{d}/D}. \tag{7}$$

It is important to understand that (6) actually follows from (7). Many proofs exist [4, 13], and the relation is vindicated by simulations. Maybe the simplest way is to remark that fracton states are on a fractal support. To λ^{-1} corresponds a dominant wave vector $Q_{\max}$ in a Fourier analysis. Counting the states in the usual way, their density must obey $\mathcal{N}(\omega)\, d\omega \propto dQ^D \propto d\lambda^{-D} \propto d\omega^{\tilde{d}}$, where (7) was used for the last relation. This proves (6). If the support was tightly connected at all scales, one should expect $\tilde{d} = D$. However, if the number of connections increases as λ decreases, then ω increases faster than Q, and $\tilde{d} < D$. Hence, in general, $\tilde{d} \leq D$. This is why the effective $\tilde{d}$ depends so strongly on the range of forces, even if these remain always short range [33].

2.3.2. The single length scale postulate, or SLSP

This postulate states that for fractons all dynamically relevant lengths, whether $Q_{\max}^{-1}$, the scattering length, the localization length, or else, must simply be proportional to the fracton size $\lambda(\omega)$ [4]. This powerful statement allows to predict scaling relations. Its validity was demonstrated by computer simulations for the bond-percolation model [22], as illustrated for $d = 2$ in Figure 5. This figure compares the scaling region for several relevant lengths:

  i) the size $\lambda(\omega)$ itself, calculated as explained above;
  ii) the Thouless localization length $\mathcal{L}$, defined by [34]
     $[\mathcal{L}(\omega)]^D \equiv \sum_i (\mid u_i^\alpha \mid^2)^2 / \sum_i (\mid u_i^\alpha \mid^4)$;
  iii) its root-mean-square fluctuation $\Delta\mathcal{L}(\omega)$.

All are seen to scale with a common slope, $-\tilde{d}/D = -64/91$, an exponent which happens to be known for the $2d$ bond-percolation model. Similarly, scaling must also imply that the dynamical structure factor, $S(Q,\omega)$, follows a master curve. This quantity is defined in Appendix B. Its scaling is nicely confirmed by simulations [22]. It seems thus that the SLSP is a valid working hypothesis.

2.3.3. Inelastic scattering and the vibrations of aerogels

Aerogels are *excellent* systems to investigate fracton dynamics. Their acoustic persistence length can be made of the order of the wavelength of visible light.

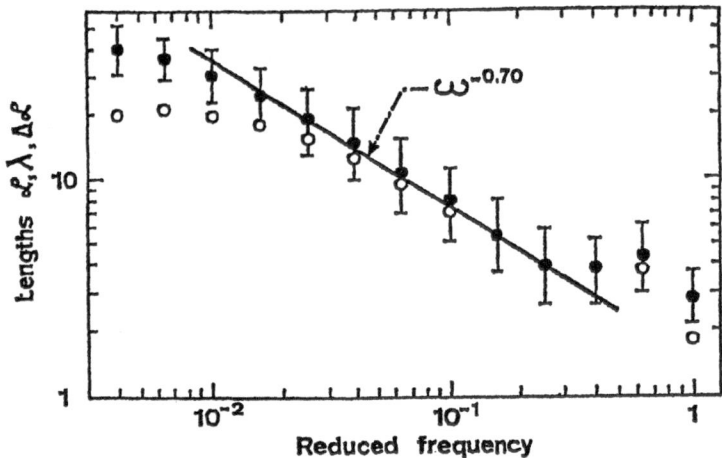

Fig. 5. — The fracton size $\lambda(\omega)$ (open symbols) and the Thouless localization length $\mathcal{L}(\omega)$ (solid symbols) for a $d = 2$ bond-percolation model. The results are averages over 40 68 × 68 realizations. The vertical bars are not error bars (the actual errors are smaller than the size of the symbols) but they represent the root-mean-square fluctuations $\Delta\mathcal{L}(\omega)$. In the scaling region, their constant length implies that $\Delta\mathcal{L}/\mathcal{L}$ is constant, and thus that $\Delta\mathcal{L}$ also scales with the common slope $-\tilde{d}/D$.

Hence, the phonon-fracton crossover at ω_{co} can be investigated by Brillouin light scattering [18, 26, 35]. With incoherent neutron scattering, one can separately measure $\mathcal{N}(\omega)$ [36–38]. Light scattering at frequencies well above ω_{co}, which one might call somewhat arbitrarily Raman scattering, provides another possibility to investigate fractons [39, 40]. Finally, thermal properties give additional information on fracton dynamics [41, 42]. The size of this course does not allow a full discussion of all these aspects. The interested reader is encouraged to refer to recent reviews [15, 16]. We concentrate here on Brillouin scattering results, as they provide a useful introduction to the section on glasses.

Brillouin scattering near the phonon-fracton crossover – Brillouin scattering was measured on MSSS of N- and B-aerogels [18, 26, 35], in function of the densities ρ and of the scattering angle. The latter fixes the momentum exchanged in the experiments, Q. For small Q values, and for dense samples, the spectra consit of narrow lines. These broaden rapidly as the crossover is approached, either upon increasing Q or decreasing ρ. This is due to the strong scattering of acoustic phonons by structural disorder, which smears the definition of the phonon wave vector q, and thereby produces a Rayleigh scattering contribution to the phonon linewidth, $\Gamma \propto \omega^{d+1} = \omega^4$ [4, 43]. It is important

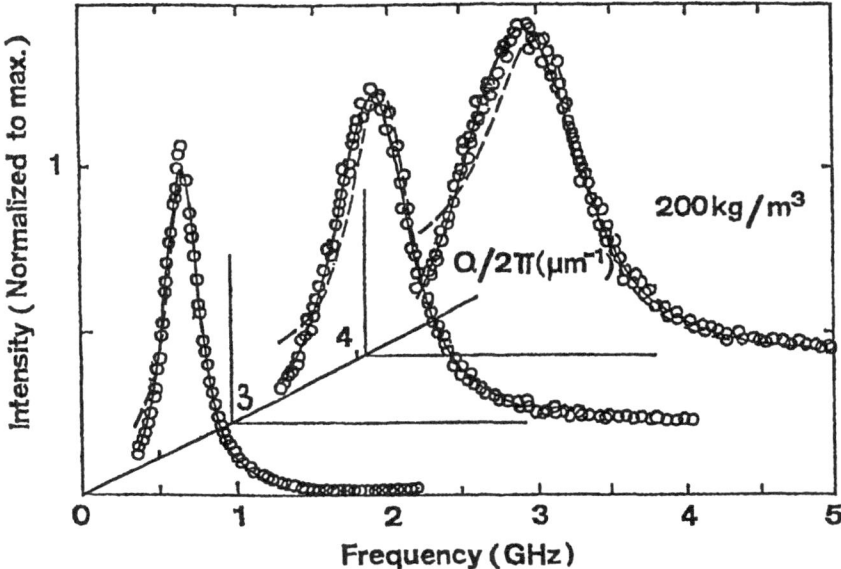

Fig. 6. — Typical Brillouin light scattering spectra obtained at various scattering angles on a N-aerogel. The scattering angle defines the Q-values indicated on the inclined axis. The crossover is found at $q_{co} \cong 3.4 \ \mu m^{-1}$ on that scale. Hence, the left (front) spectrum is in the phonon regime, the central one is near crossover, and the right (back) one is in the fracton regime. The dots are normalized photocounts, the dashed lines illustrate best fits with a DHO, while the solid lines are fits using the crossover expressions, equations (8) to (10).

to understand that this is not lifetime broadening, but rather broadening by structural disorder, totally unrelated to the lifetime of the excitations. The experiment probes excitations having a Fourier component at the scattering vector Q. As the wave vector q of the phonons looses its sharpness, the line effectively become *inhomogeneously* broadened, *i.e.* what is observed at a given Q is a contribution from many vibrational eigenmodes. For fractons, in particular if their characteristic length $\lambda < Q^{-1}$, one always has a Fourier component at Q. This is a regime akin to Raman scattering from localized molecular vibrations, with the difference that the fractal structure generates a broad range of eigenfrequencies. The fracton linewidths are thus essentially independent of the temperature T [18].

A simple-minded interpretation of the spectra – In a first attempt, we seeked to extract the peak position ω_p and the half widths at half height Γ of the Brillouin lines [18, 44]. It is evident, from the above discussion, that a damped harmonic oscillator response (DHO), which is the lineshape appropriate to

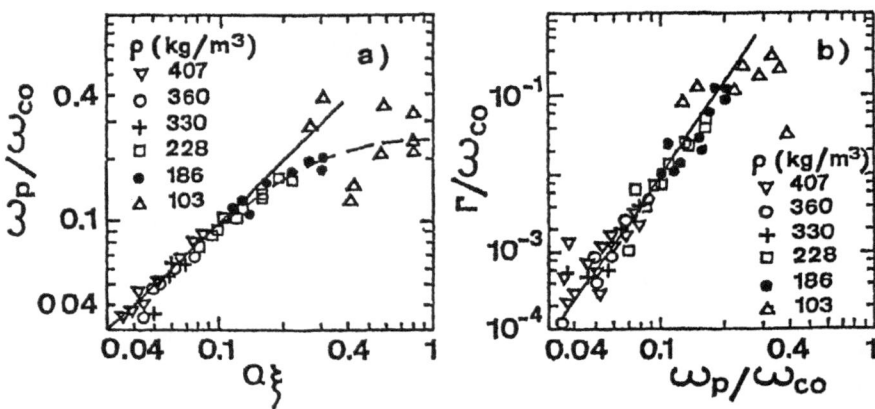

Fig. 7. — a) Scaling of the phonon frequencies, ω_p, extracted from the peak positions, using the value of $\xi = 1/q_{co}$ taken from neutron scattering, and with a velocity exponent $x = 1.4$. The straight line is the asymptotic limit and the dashed one a guide to the eye. b) Scaling of the half width Γ extracted from the spectra. The straight line is the Rayleigh scattering law.

describe lifetime broadening, does not necessarily provide the correct line profile in the present situation. This point is clearly illustrated by the dashed lines in Figure 6 which represent the best DHO fits to some observed Brillouin spectra [45]. One sees that the DHO-profile is much more asymmetric than the experiments. Therefore, in a somewhat naive approach, the line profiles of N-aerogels were adjusted to either lorentzians or powers of lorentzians [18]. From such fits, we extracted values for ω_p and Γ [44], as shown in Figure 7. It is seen that ω_p saturates like an acoustic dispersion branch, but at very small Q-values, and this more rapidly the lighter the gel. One also notes that $\Gamma \propto \omega^4$.

From this analysis, one can extract several scaling exponents:

  i) the exponent x for the velocity in the phonon regime, $c_0 \propto \rho^x$, where c_0 is defined by $\omega_p = c_0 Q$ in the phonon regime;

 ii) the exponent y for the crossover frequency, $\omega_{co} \propto \rho^y$, where ω_{co} is defined here as the frequency where $\Gamma \propto \omega_p^4$ first reaches $\Gamma = \omega_p$;

iii) and finally, the exponent ς for a crossover length, $L \propto \rho^{-\varsigma}$, where L is derived from the high-Q saturation of the apparent dispersion relation, $\omega_p(Q)$.

One defines a wave vector at crossover, q_{co}, by the point (q_{co}, ω_{co}) in the (Q, ω) plane where the extrapolated phonon regime, $\omega_p = c_0 Q$, intercepts the fracton regime, $\omega \propto Q^{D/\bar{d}}$. It is clear that $L \propto q_{co}^{-1} \propto \xi_{ac}$, where ξ_{ac} is a persistence length appropriate to the elastic waves [29]. Thus, $\varsigma = 1/(3-D)$. For a MSSS, c_0 scales like $q_{co}^{D/\tilde{d}-1}$, from which $x = (D - \tilde{d})/\tilde{d}(3 - D)$. Finally, by similar arguments, one finds $y = x + \varsigma$. From these scalings of the data in Figure 7, we estimated $D = 2.36$, and $\tilde{d} = 1.25$ [18, 44]. This approach does not use the richness of the spectral shapes. It also makes believe that the dispersion that appears in Figure 7a is similar to the dispersion at the edge of a Brillouin zone, although it is a completely different phenomenon having to do with the scattering spectrum of localized excitations.

Accounting for spectral shapes – Theories of the spectral shapes were developed in parallel to our scattering measurements [46, 47]. The smearing in q of the phonons leads to a dependence in frequency (rather than in q which is not a well defined quantity for the vibrations) of the parameters entering the spectral function $F(Q, \omega)$. These parameters are the linewidth, $\Gamma(\omega)$, and the velocity, $c(\omega)$. From a Green function analysis coupled with an effective medium approximation the authors of [47] find

$$F(Q, \omega) = \frac{c^2 Q^2}{\omega} \frac{\Gamma}{(\omega^2 + \Gamma^2 - c^2 Q^2)^2 + 4\Gamma^2 c^2 Q^2}. \tag{8}$$

At low $\omega \ll \omega_{co}$, c is equal to c_0 and $\Gamma \propto \omega^4$, the Rayleigh scattering law. At high frequency, $\omega \gg \omega_{co}$, one has reached the Ioffe-Regel limit $\Gamma \cong \omega$ (for plane waves) [48], while $c = \omega/Q \propto \omega^z$, with $z = 1 - \tilde{d}/D$ from the fracton dispersion (7). Complications arise in the intermediate regime. The effective medium approximation on a percolation model leads to discontinuities at ω_{co} [47]. Such discontinuities seem to arise from the approximation, and they are not apparent in the real spectra. Therefore, we invented empirical crossover expressions [35],

$$c = c_0[1 + (\omega/\omega_{co})^m]^{z/m}, \tag{9}$$

$$\Gamma = W\omega \frac{(\omega/\omega_{co})^3}{[1 + (\omega/\omega_{co})^m]^{3/m}}. \tag{10}$$

They have the proper asymptotic behavior. There is only one additional crossover parameter, m. This is not a mass (!), but a crossover exponent that fixes how rapidly the phonon regime changes into the fracton regime. Although spectral shapes for $Q \approx q_{co}$ depend significantly on m, the values obtained for c_0 and ω_{co} are not strongly correlated with m. The other parameter entering (10), $W \approx 1$, sets the proportionality between Γ and ω in the Ioffe-Regel limit, $\Gamma = W\omega$. Empirically, we seem to find that more "homogeneous" systems tend to exhibit smaller values of W.

The scattered intensity is related to the spectral shape by

$$I(Q, \omega) = AQ^2 n_B F(Q, \omega). \tag{11}$$

A is a constant amplitude coefficient that should be independent of ω and Q. The factor Q^2 in (11) arises from the expansion of the density correlation function in the displacements [49], and n_B is the Bose factor, equal to $k_B T/\hbar\omega$

for the experiments shown here. This coupling to light is the so-called "direct coupling" mechanism, in which the modulation of the dielectric constant directly results from the time dependence of the particle positions. It is the same dependence which is generally active in neutron or X-ray scattering. In light scattering there are other possible couplings, as discussed in [49]. These appear to become active at higher frequencies, in a region generally called "Raman scattering" although this term might be misleading for fractons (see, *e.g.* [26,40]). Around the Brillouin peak the scattering is dominated by direct coupling, as can be seen from its power-law dependence at frequencies just above ω_p [50]. In practice, the constancy of A is difficult to check in light scattering, as changes of Q also modify the scattering volume. Further, one must note that equations (8–11) are in a sense *ad hoc*. Instead, what should really be done is to sum over all the vibrational modes [46]. The approximation that leads to (8) is not fully controlled. This may well lead to a dependence of A on Q, especially above q_{co}, where the modes are definitely not plane waves.

Measurements at 8 values of Q were performed on a high quality MSSS of N-aerogels with 10 different densities [35]. The entire data are adjusted with the same values of z and m, but to each spectrum there are different q_{co}, ω_{co}. The parameter A also varies. The fits are performed iteratively, searching for consistency between the selected z and the fracton dispersion curve (7) derived from a plot such as that in Figure 8 [50]. This relies on the MSSS-property. Excellent fits are obtained with $z = 0.47$, $m = 2$, and $W = 1$. Examples are shown in Figure 6. An acoustic value for D, written D_{ac}, is then extracted from the scaling of q_{co} with ρ. We find $D_{ac} = 2.46 \pm 0.03$, in good agreement with the structural determination $D = 2.40 \pm 0.03$. Finally, from D_{ac} and z, one finds $\tilde{d} = 1.3 \pm 0.1$. This determination depends on dynamical MSSS. It is nicely confirmed by measurements of $\mathcal{N}(\omega)$ on single samples by neutron scattering [38,51]. With this scaling, Brillouin spectra from different samples can be superposed. In fact, the spectra depend only on Q/q_{co} and on ω/ω_{co} [52].

Similar measurements on a MSSS of B-aerogels with $D = 1.72$, give $m = 4$ and $z = 0.35$, from which $\tilde{d} = 1.1 \pm 0.1$ [26]. In this case the fits are improved by setting $W = 0.77$. One finds $D_{ac} \cong D$. The higher value of m indicates a sharper crossover to the fracton regime, consistent with the existence of fractal blobs that produce the hump at $\sim 1/\xi$.

3.   THE CASE OF GLASSES

As already discussed in Section 1, our knowledge of the structure of glasses is very limited, even in the simplest case of single component glasses. Returning to the example of vitreous silica, v-SiO_2, which is by far the most investigated glass to date, X-ray and neutron diffraction experiments show that the O-Si-O angle inside the SiO_4 tetrahedron is practically the same as it is in the crystal [1]. The position and the width of the peak characteristic of the Si-Si distance in the correlation function $g(r)$ allows to evaluate the average angle

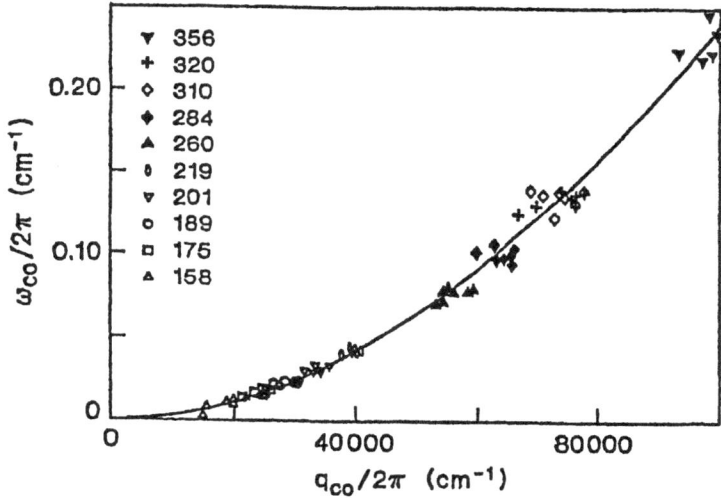

Fig. 8. — The fracton dispersion curve, equation (7), derived from the Brillouin determination of ω_{co} and q_{co} on a MSSS of N-aerogels. The solid line corresponds to $D/\tilde{d} = 1.88$, or $\tilde{d} \approx 1.3$.

between adjacent tetrahedra as well as its spread. The latter is already quite large in v-SiO_2 [1]. At larger distances, the features in $g(r)$ become less and less marked, indicating an increased amount of disorder [1]. In some particular cases, complementary probes such as Raman spectroscopy or NMR have allowed to identify superstructural units. This is the case for boroxol rings and other building blocks in vitreous boron oxide [53,54], or for rings in v-SiO_2, as discussed in [55]. The latter studies are of considerable interest, as little else is actually known about the structural organization of the glass at intermediate length scales that was discussed in the Introduction.

As a consequence of disorder, one expects that harmonic plane waves will not be the normal modes of vibration of glasses, except for sufficiently long wavelength acoustic phonons. Ideally, one would like to observe the normal modes, to measure their frequencies, and also their wave vectors as far as they have one. It would be of primary interest to determine their spatial extent if they are non-propagative. Further information can be obtained from the density of vibrational states. Owing to anharmonicities, the normal modes interact, leading to finite lifetimes, or, for propagating waves, to limited mean free paths.

Specific heat and inelastic neutron scattering give information on the DOS, which can also be determined to some extent and more indirectly by Raman scattering. Thermal conductivity allows to evaluate the mean free path of propagating acoustic phonons. Information on the wave vector $\mathbf{q}$ and the mean free path of plane acoustic waves can also be obtained by ultrasonic and thermal pulse propagation experiments, and by Brillouin scattering of light, X-rays, and neutrons. However, at high acoustic frequencies, scattering experiments measure a projection of the eigenmodes on the scattering vector $\mathbf{Q}$, while the excitations might not be plane waves. In this section, we will summarize the essential information and show recent results concerning the frequency regime below $\sim$20 meV, *i.e.* in the domain where acoustic phonons usually dominate in crystals.

3.1. What is already established

3.1.1. Propagation of acoustic phonons

Ultrasonic measurements performed at frequencies of the order of 1 MHz to 10 GHz, *i.e.* for v-SiO$_2$ at wavelengths ranging from $\sim$6 mm down to $\sim$0.6 μm, have demonstrated the existence of harmonic plane waves whose mean free path is much longer than the wavelength [56]. Brillouin scattering of light allowed to observe both longitudinal and transverse phonons up to 35 and 16 GHz, respectively [59]. At these frequencies, the linewidth Γ is still about 1000 times smaller than the frequency. Tunneling experiments extended this observation up to 400 GHz [60]. The acoustic phonons have nearly linear dispersion branches over the whole frequency range. Thermal pulse propagation experiments demonstrated that Γ is approximately proportional to ω^2 from 1 to 400 GHz over a broad temperature interval, from $\sim$ 80 K to above 300 K [61,62]. The law $\Gamma \propto \omega^2$ being produced by viscous damping, and since the proportionality constant is a strong function of T [59], this indicates that the main origin of broadening in this regime is a finite lifetime of the excitations.

In the temperature range where usually acoustic phonons dominate the thermal properties, *i.e.* below $\sim$ 10 K, the measured specific heat of glasses is much higher than the prediction of the Debye model, showing the existence of a large *excess* of modes. This is illustrated in Figure 9, which compares the situation in crystal quartz, q-SiO$_2$, where the Debye prediction is excellent, to that in v-SiO$_2$, where it fails. In particular, measurements at very low temperatures, indicate an almost frequency independent excess density of states in the relevant frequency range. This is demonstrated in Figure 10a, where this constant density of excess states leads to a specific heat nearly proportional to T, and much larger than the calculated Debye value which is $\propto T^3$. As shown in Figure 10b, in the same temperature range the thermal conductivity is much lower than that of the corresponding crystal, and it varies approximately like T^2, instead of the T^3 law found in the crystal and predicted by the Debye density of states. Hence, the excess modes do not conduct heat and also scatter the acoustic phonons. The whole set of results –thermal measurements, acoustic damping, pulse-echo experiments, among others– has been

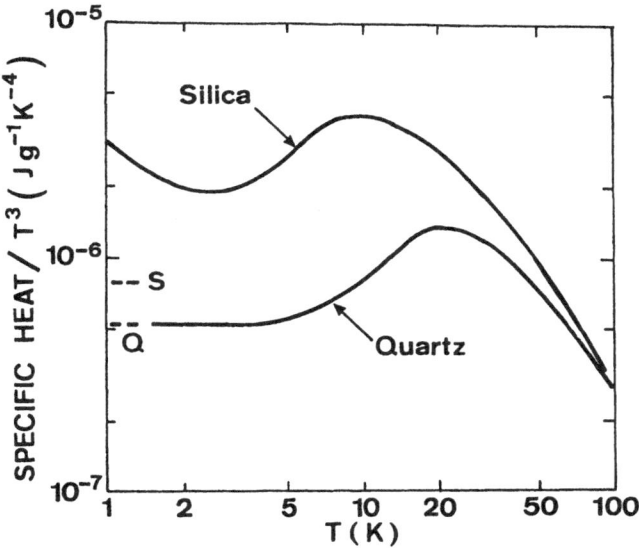

Fig. 9. — The specific heat of silicon dioxide above 1 K, plotted as C_P/T^3. The dashed lines labelled S and Q are Debye specific heats calculated from measured sound velocities of silica and quartz, respectively (from [57]).

convincingly explained by assuming the existence of two-level systems which couple to phonons *via* resonant and relaxational processes [5].

3.1.2. Density of states

Above 400 GHz, no direct measurement of phonon propagation could be performed until recently. There is however much information available on the DOS. A first series of results is derived from the specific heat, C_P. For most glasses, C_P/T^3 plotted *vs.* T, shows a hump around 10 K [7] as illustrated in Figure 9 for vitreous silica.

The presence of a considerable excess in the DOS has been confirmed by neutron-scattering experiments [63]. In materials where incoherent scattering dominates, the dynamical structure factor $S(Q, \omega)$ is directly proportional to the DOS [64]. However, in most inorganic glasses, and in particular in v-SiO$_2$, the intensity comes from *coherent* scattering. In that case, one remarks that acoustic phonons at ~ 1 THz, if they were propagating with the sound velocity, would have a rather small wave vector q of about 0.1 Å^{-1}. Neutron scattering measurements, performed at much larger scattering vectors Q in the range from about 1 Å^{-1} to 5 Å^{-1}, and averaged over this broad Q-range, allow the determination of the DOS in an *incoherent approximation* [65]. In v-SiO$_2$ and at room temperature, $S(Q, \omega)$ shows a broad peak around 1 THz. From that data one can extract the DOS as illustrated in Figure 11. When plotted as

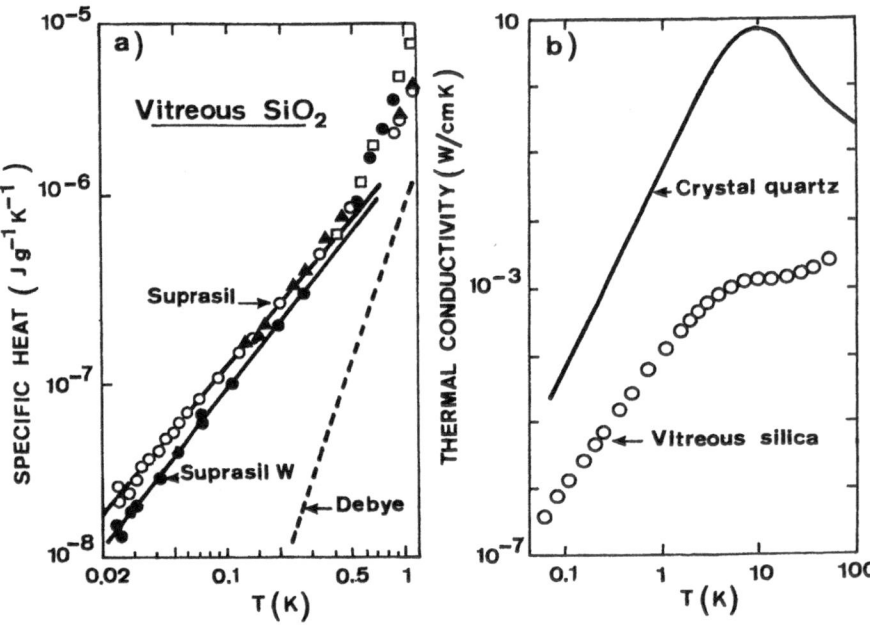

Fig. 10. — a) The specific heat of silica at very low temperatures. Samples of different quality have slightly different densities of excess states (from [57]); b) the thermal conductivity of vitreous silica compared to that of crystal quartz (from [58]).

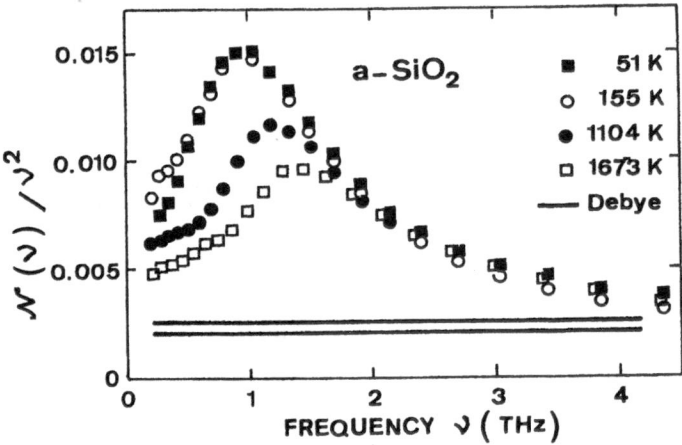

Fig. 11. — Density of states of vitreous silica obtained in neutron scattering at four temperatures. The solid lines represent the Debye density of states for the lowest (upper line) and highest (lower line) temperatures (from [66]).

$\mathcal{N}(\nu)/\nu^2$ (where $\nu = \omega/2\pi$), it reveals a maximum in its excess over the Debye value, maximum also located around 1 THz at room T. The corresponding $\mathcal{N}(\nu)$ shows a maximum around 3 THz, in good agreement with the specific heat. In the region of the peak in $S(Q,\omega)$, this DOS is about 5 times larger than the Debye prediction for acoustic modes [63]. The peak in Figure 11 is what is called the Boson peak.

Boson peaks are also observed in Raman scattering. In this case the scattering vector $\mathbf{Q}$ is very small, about 3 orders of magnitude smaller than in neutron scattering. The coupling of these very high frequency excitations to light of relatively long wavelength is clearly a more complex mechanism [67]. The intensity is often expressed as,

$$I(\omega) = \frac{\mathcal{C}(\omega)}{\omega} n(\omega) \mathcal{N}(\omega), \qquad (12)$$

where $\mathcal{C}(\omega)$ is a coupling coefficient and $n(\omega)$ is the Bose factor [68]. Taking $\mathcal{C} \propto \omega$, one finds that the Boson peak in v-SiO$_2$, and in many other glasses, is very similar in Raman and in inelastic neutron scattering [69].

3.1.3. Thermal conductivity

In a large number of glasses, the thermal conductivity κ increases with T going up to a few K, then remains nearly constant in temperature up to $\sim$15 K, and increases again with increasing temperature [7,70]. The flat region is called the *plateau*. The most commonly accepted explanation for its origin is that acoustic phonons cease to propagate above a crossover frequency, ω_{co}, and thus that they cannot carry heat [7]. A simple calculation in the Debye approximation shows that an extrapolation to high frequencies of the law $\Gamma \propto \omega^2$ observed up to 400 GHz is unable to produce a plateau [71]. Further, Brillouin light scattering has demonstrated that the law $\Gamma \propto \omega^2$ is of dynamical origin since Γ decreases strongly when the temperature decreases below 50 K [59]. Hence, this phonon damping process cannot be the cause for the low κ at low T. The most likely origin for the plateau is an elastic (Rayleigh) scattering of acoustic phonons by the fluctuations of density, connectivity, and local anisotropy associated with the structural disorder in glasses. This produces a temperature-independent $\Gamma \propto \omega^4$ [43]. Plane waves cease to exist when the Ioffe-Regel condition $\Gamma \approx \omega$ is fulfilled [48]. This first happens at the crossover frequency ω_{co}, and remains thereafter. A fit of κ for v-SiO$_2$ gives $\omega_{co}/2\pi \simeq 1$ THz [71]. The "soft-potential" model [12] provides a specific mechanism to account for the strong scattering of phonons. It assumes a resonant interaction of phonons with local oscillators characterized by a potential of the form

$$V(x) \propto \mathcal{E}_0 \left[\eta \left(x/a \right)^2 + \xi \left(x/a \right)^3 + \left(x/a \right)^4 \right]. \qquad (13)$$

Here, x is a generalized coordinate, $\mathcal{E}_0$ is a characteristic energy of the order of the binding energy, and a is an elementary size. Depending on the values of η and ξ, this model predicts tunneling systems, thermally activated defects, or nearly harmonic local modes. For large positive values of η, resonant absorption of the phonons by the quasi-harmonic oscillators lead to a mean free path $\propto \omega^{-4}$, independently of T. In this sense, the soft potentials may be viewed as one of the particular mechanisms contributing to Rayleigh scattering. In summary, thermal conductivity provides strong indications for a crossover from propagating phonons to non-propagating modes, and this crossover occurs around 1 THz in v-SiO$_2$.

3.2. Spectroscopy of acoustic excitations in the terahertz regime – three remarks

3.2.1. Coherent and incoherent interactions

The only manner to gain additional information on very high frequency vibrations, which are not expected to be propagating plane waves, appears to be measuring the dynamical structure factor, $S(\mathbf{Q}, \omega)$. Here $\mathbf{Q}$ is the scattering vector set by the conditions of an experiment, not to be confused with a wave vector $\mathbf{q}$ which probably could not be defined for sufficiently high frequency excitations. We consider here exclusively coherent scattering *mechanisms*, such as the scattering of neutrons by the average scattering lengths of atoms, or Thomson scattering of X-rays, or also the scattering of light by fluctuations of the dielectric tensor. The *interaction* between the radiation and the vibrations can however be either coherent or incoherent in the sense that it can either conserve energy and quasi-momentum or not conserve momentum. The former situation is the usual one in Brillouin scattering, for which

$$\hbar\mathbf{Q} = \hbar\mathbf{k}_s - \hbar\mathbf{k}_i, \tag{14}$$

and

$$\hbar\omega = \hbar\omega_s - \hbar\omega_i, \tag{15}$$

where $(\mathbf{k}_i, \omega_i)$ and $(\mathbf{k}_s, \omega_s)$ are the wavevectors and frequencies of the incident and scattered quanta, respectively. Then $(\mathbf{Q}, \omega)$ is the Fourier component investigated, and for sufficiently well-defined plane waves $(\mathbf{q}, \omega)$, one observes sharp peaks at $\mathbf{Q} = \mathbf{q}$.

In crystals, there is an alternate manner to conserve quasi-momentum, which is to use the periodicity of the lattice which gives a set of Bragg vectors $\mathbf{G}$ in reciprocal space. Then, (14) can be replaced by

$$\mathbf{Q} + \mathbf{G} = \mathbf{k}_s - \mathbf{k}_i. \tag{16}$$

In neutron scattering, the value of ω that can be reached depends of course on the size of k_i, k_s, since the energies are $E_{i,s} \equiv \hbar\omega_{i,s} = \hbar^2 k_{i,s}^2/2m$, where m is here the neutron mass. To reach high ω, one must have sufficiently large k values, and (16) offers a way to investigate small Q's at large k. It also gives a way to have $\mathbf{Q}$ not parallel to $\mathbf{k}_s - \mathbf{k}_i$, which is crucial to observe transverse modes in one-phonon scattering. Indeed, neutron scattering by fluctuations of the density of scattering lengths only senses motions (polarisation vectors of vibrations) that have a component along $\mathbf{k}_s - \mathbf{k}_i$.

In glasses one does not have well-defined values of $\mathbf{G}$, but rather broad Debye-Scherrer rings of the structure factor $S(\mathbf{K})$. This nevertheless allows similar "Umklapp" scattering processes

$$\mathbf{Q} + \mathbf{K} = \mathbf{k}_s - \mathbf{k}_i, \tag{17}$$

but now there is a distribution of $\mathbf{K}$ given by $S(\mathbf{K})$. This is also coherent scattering, and it allows to observe contributions from transverse modes as long as these have sufficiently well-defined wave vectors $\mathbf{q}$ [65, 72].

On the other hand, some interactions do not conserve quasi-momentum. In light scattering, if one looks at modes of small extension ℓ, one will have $Q\ell \ll 1$. The situation is somewhat similar to Raman scattering from molecules in a solution. These modes contribute to the scattering spectrum whatever the size of the scattering vector Q. The result reflects the whole density of states [69], and this also happens for the scattering of neutrons from local modes. At the other extreme, one can also consider situations where Q is very large compared to the typical ℓ^{-1}, $Q\ell \gg 1$. This can easily be done in neutron scattering. In that case the contributions from different scatterers add up quasi incoherently.

3.2.2. Experimental possibilities

Neutron scattering – In trying to perform Brillouin scattering with neutrons, one is confronted with the difficulty of verifying the kinematic conditions (14-15). If θ is the scattering angle (the angle between $\mathbf{k}_s$ and $\mathbf{k}_i$), the triangle condition (14) leads immediately to, [73]

$$\hbar^2 Q^2/2m = E_s + E_i - 2\sqrt{E_s E_i} \cos\theta. \tag{18}$$

For a given θ, and for a given E_i (or E_s), (18) and (15) establish a relation between Q and ω. At very small θ, and for small Q, this relation amounts to $\omega \approx vQ$, where v is the neutron velocity. At larger θ, one generally finds that $\omega < vQ$. This means that in practice the kinematic condition cannot be satisfied unless the neutron velocity v is sufficiently larger than the "sound" velocity c to be investigated. Unless c were rather small (~ 2000 to 3000 m/s), it implies that one would need neutrons of rather high energy (above thermal) and using these one just cannot achieve the energy resolution of interest (better than 1 meV). In that case neutron scattering can still be very useful to investigate the excitations that have become non-propagative, whose size is smaller than q_{co}^{-1}, and for which a broad spectrum should be obtained. However it precludes a study of the crossover itself.

For glasses of low c, and provided q_{co} is not too small, neutrons offer two major advantages:

1) very sharp resolution functions, nearly Gaussian for three axis spectrometry (TAS), and that can be sharply triangular for time-of-flight spectrometry (TOF);

2) easy absolute calibration of the intensities.

Inelastic X-ray scattering – An inelastic X-ray spectrometer with a practical resolution slightly better than 2 meV FWHM is now in operation at the European Synchrotron Radiation Facility (ESRF) [74]. It uses strongly pre-filtered radiation from an undulator with near backscattering monochromator and analyzers. The temperature difference between monochromator and analysers is scanned to obtain the spectra. Using the Si (11 11 11) reflection, the X-ray energy is 21.7 keV. With electromagnetic radiation the kinematic conditions (14-15) can always be satisfied. The main issue here is the instrument *contrast*. Glasses are disordered materials that give a strong quasi-elastic peak in the static structure factor $S(Q)$ even at small Q-values. The instrumental function being lorentzian-like, the wings of the elastic peak reach into the ω-regions of interest. They add to the generally weak Brillouin scattering signal, making it more noisy and more difficult to analyze. The instrument is thus not suited to investigate glasses with a relatively strong $S(Q)$. Owing to X-ray absorption, it is also restricted to samples with sufficiently light atoms. The Q-values that can be achieved start from ~ 1 nm^{-1}, with a $\Delta Q \approx \pm 0.1$ to ± 0.2 nm^{-1} depending on the angular acceptance of the analyzer. In most cases, this lower limit in Q is of the order of the expected q_{co}. However we have hope to find materials for which q_{co} is sufficiently large so that significant information could be acquired at Q and ω values below the crossover.

3.2.3. Choice of a spectral function

Homogeneously broadened spectra – In dealing with plane waves that are damped by anharmonic effects, the spectral function is well approximated by a damped harmonic oscillator (DHO), as easily shown,

$$F(Q,\omega) = \frac{2\Gamma\omega}{(\Omega^2 - \omega^2)^2 + 4\omega^2\Gamma^2}. \tag{19}$$

Here Γ is the HWHM, and $\Omega = cQ$ where c is the wave velocity. The corresponding spectral shape, as in (11), assuming "direct" coupling to density fluctuations (which is the case for both neutron and X-ray scattering), can be written

$$I(Q,\omega) = AQ^2 e^{-2W} n_B F(Q,\omega), \tag{20}$$

where at these high scattering vectors it may be necessary to account for the Debye-Waller factor e^{-2W}. This description is expected to apply to homogeneously damped longitudinal waves (the only ones that couple). For $k_B T > \hbar\omega$, and neglecting the Debye-Waller factor, the integral of I over ω is a *constant* $A' = \pi A k_B T/\hbar c^2$. In the case of scattering near a Bragg point $\mathbf{G}$ in a crystal, these equations are modified as G^2 replaces Q^2 in (20), and the integral of I becomes proportional to $1/\Omega^2$, a well known property in neutron spectroscopy of acoustic modes.

Considering the shape of $I(Q,\omega)$, and taking again the limit $k_B T > \hbar\omega$, one should remark that $I(\omega = 0) = A'2\Gamma/\pi\Omega^2$, $I(\omega = \Omega) = A'/2\pi\Gamma$, and

$I(\omega \gg \Omega) = \Omega^2 2\Gamma/\pi\omega^4$. When Γ becomes appreciable, I is relatively large at the center of the spectrum since $I(0)/I(\Omega) = (2\Gamma/\Omega)^2$, while it always decays rapidly in its wings ($\propto \omega^{-4}$).

The case of Rayleigh scattered vibrations – As seen above for fractons, Rayleigh scattering leads to $\Gamma \propto \omega^4$ in 3 dimensions (in d-dimensions the law is $\Gamma \propto \omega^{d+1}$, a fact not to be forgotten when considering simulation results). The rapid increase of Γ with ω terminates with a Ioffe-Regel crossover at $\Gamma = W\omega$, where $W \simeq 1$. As already pointed out in [75], in that situation the parameters entering the spectral response can only depend on ω, since there is no wave vector q. The spectral response (8), together with the crossover expression (9-10), is *not restricted to fractals*. It rather represents, admitedly in a somewhat *ad hoc* manner, the crossover to strong scattering. We thus propose that it could also be used as a first approximation near the end of the acoustic branches in glasses. We call it the crossover spectral response (CSR).

The meaning of the exponent z entering (9) is then not related to a fracton dimension $\tilde{d}$, but it represents in an approximative way the change of the density of states from a law $\mathcal{N}(\omega) \propto \omega^2$ to a law $\mathcal{N}(\omega) \propto \omega^{2-3z}$. Figure 12 illustrates the form of $I(\omega, Q)$ obtained from (8-10, 20) across the crossover. For $Q \ll q_{co}$, and thus $\Gamma \ll cQ$ in the region of interest, the shape of the peak is practically identical to that of the DHO. To see this one just remarks that $(\omega^2 + \Gamma^2 - c^2Q^2)^2 + 4\Gamma^2 c^2 Q^2$ in the denominator of (8) can also be written $(\omega^2 - \Gamma^2 - c^2Q^2)^2 + 4\Gamma^2\omega^2$, which is equivalent to the denominator in (19) if one takes $\Omega^2 = c^2Q^2 + \Gamma^2$. However, strictly speaking the CSR is always 0 for $\omega = 0$ which distinguishes it markedly from the DHO when cQ becomes large. For $Q > q_{co}$, the CSR remains small near $\omega = 0$, and it develops a slowly varying high frequency wing beyond a peak near $\omega = \omega_{co}$ [78], as exhibited in Figure 12. This should clearly distinguish it from the DHO.

3.3. Some studies near the end of acoustic branches in glasses

3.3.1. Neutron scattering from glassy selenium, g-Se

When quenched from its liquid state, Se easily forms a polymer-like glass at $T_g \approx 317$ K. Se nuclei are good coherent neutron scatterers, and the system offers many advantages for a neutron study:

(i) most important, the longitudinal sound velocity in the glass is small, $c_L \approx$ 1800 m/s. This allows to observe a Brillouin signal at small scattering angles in triple-axis spectrometry (TAS).

(ii) Se being an elemental solid, the absolute calibration of the signals is particularly straightforward.

(iii) Se has a substantial absorption cross-section, already somewhat larger than the scattering cross-section for thermal neutrons. This is very favorable as it practically eliminates the need for multiple scattering

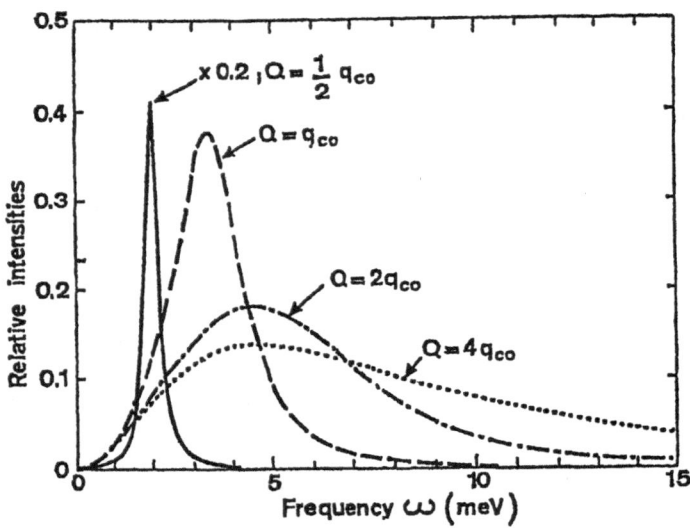

Fig. 12. — Example of the CSR calculated with equations (8–10, 20) using parameters that are appropriate to describe the spectra in v-SiO$_2$: $W = 1$, $m = 2$, $z = 0.37$, and $c = 3.9$ meV/nm^{-1}. There is a single crossover at $\omega_{co} = 3.9$ meV so that $q_{co} = 1.0$ nm^{-1}. The temperature is 295 K, and Debye-Waller effects are neglected.

corrections, which otherwise can be tedious and introduce quite some uncertainty.

Figure 13 illustrates the (Q, ω)-regions that are accessible with the two spectrometers that have been used for these experiments. The expected (q, ω)-dispersion is also sketched. The TAS, utilizing the higher neutron energies, and smaller scattering angles, allows to explore the region of the coherent Brillouin signals. The time-of-flight (TOF) spectrometer, at lower neutron energies, and larger angles, provides data at higher Q-values and at low ω, with an excellent resolution in ω. Figure 14 shows typical neutron results obtained with these two instruments [76]. With the TAS, the Brillouin signal at small (Q, ω) is immediately recognizable prior to any data analysis. The actual analysis reveals that the neutron scattering signal consists in fact of three components [76, 77],

$$S(Q, \omega) = S_{\mathrm{B}} + S_{\mathrm{LM}} + S_{\mathrm{U}}. \tag{21}$$

S_{B} is the Brillouin signal that can be described by equations (8–10, 20). S_{LM} is a signal produced by low frequency local modes (LM). These scatter in an incoherent approximation. Hence, $S_{\mathrm{LM}} \propto Q^2 Z_{\mathrm{LM}}(\omega)$, where Z_{LM} is the density of states of these modes. Owing to its Q^2-dependence, this signal

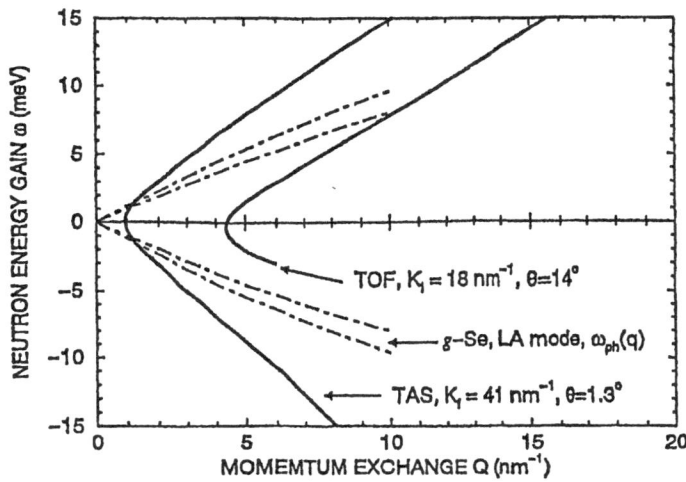

Fig. 13. — Regions of (Q, ω)-space that were explored with the instruments IN8 (TAS) and IN5 (TOF), both at the Institute Laue-Langevin in Grenoble, France. The curves that are shown correspond to the smallest scattering angles, and the regions that can be explored lie to the right of these curves. The expected LA-dispersion of g-Se, $\omega_{ph}(q)$, is also shown, from [76].

rapidly overwhelms S_B as Q is increased, as seen already for $Q = 7.5$ nm^{-1} in Figure 14a. Finally, S_U is "Umklapp" scattering as explained in Section 3.2.1. Under approximations that are valid here, one can show that $S_U \stackrel{\sim}{\propto} Q^2 S(Q)$ [65]. For this reason, that part of the signal increases faster than Q^2 in going from the curve at 12 nm^{-1} to the one at 20 nm^{-1} in Figure 14b. This increase results from the large change in $S(Q)$. The same figure shows that the ω-dependence of S_U exhibits a rapid decrease as ω increases beyond ~ 3 meV. The analysis of S_B with $(8 - 10, 20)$ gives $\omega_{co} \approx 3.4$ meV. This is the value appropriate for longitudinal excitations. The ω-dependence of S_U shows that also transverse modes, which dominate S_U to more than 80%, also cease to have a definite q-vector above ω_{co}. On the other hand the peak in Z_{LM}/ω^2 is around $\omega_B \approx 2$ meV, as seen from the dashed line in Figure 14b. This corresponds to the Boson peak in this glass. Thus, $\omega_{co} \approx \omega_B$. In this particular case, the LM's being so strong, it is conceivable that the crossover to strong acoustic scattering is much influenced by translational-vibrational coupling with local modes.

3.3.2. Neutron and X-ray scattering from v-SiO_2

It is obvious from Figure 13 that in the case of v-SiO_2, where $v_L \approx 5900$ m/s, a value three times larger than in g-Se, there is no chance to observe Brillouin

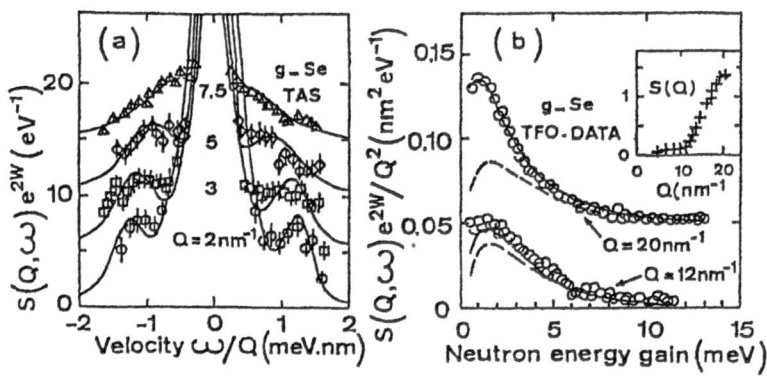

Fig. 14. — Neutron scattering results on g-Se. (a) $S(Q,\omega)$, corrected for the Debye-Waller factor, from TAS measurements. The curves are displaced vertically in steps of 5. The solid lines are fits with $S_B + S_{LM}$. One should note that the abcissa is normalized by the Q-value for each curve. (b) The results from TOF measurements at higher Q-values. As S_{LM} increases with Q^2, here it is $S(Q,\omega)/Q^2$ which is presented, again corrected for Debye-Waller effects. Two slices of the data, of thickness 2 nm^{-1} are shown, the upper one displaced vertically by 0.05. The dashed lines are the LM-component alone. The solid line is the overall adjustment with $S_B + S_{LM} + S_U$, whereby the relative weight of S_B becomes negligible for $Q > 12$ nm^{-1}. The difference between the dashed and solid lines is essentially S_U/Q^2, which grows like the static structure factor $S(Q)$ obtained from the same measurement and shown in the inset.

scattering with neutrons within the present capabilities of the instruments. This is a typical case where one wishes to use X-ray scattering. The signals are weak [78], and thus a better S/N ratio can be obtained at higher temperatures [79]. These latter results are shown in Figure 15, together with an analysis in terms of the CSR [80]. It should be noted that owing to the strong elastic scattering one cannot conclude from the spectra alone what is the appropriate spectral response. In fact, the same data was originally fitted with a DHO up to $Q = 6$ nm^{-1} [79]. With the CSR we find at this elevated temperature that $q_{co} \approx 1.5$ nm^{-1}, much below $Q = 6$ nm^{-1}, and that $\omega_{co} \approx 6$ meV [80]. Results obtained with extensive data accumulation at room temperature [81], when similarly analyzed give $\omega_{co} \approx 4$ meV and $q_{co} \approx 1$ nm^{-1}, like in [78]. This agrees with the thermal conductivity plateau as explained in Section 3.1.3 and in [71]. The observed increase of q_{co} with the strong elevation of temperature suggests that v-SiO$_2$ becomes more "homogeneous" on the medium range scale as T_g is approached. All these results are consistent with the view that acoustic phonons cease to propagate near and above q_{co}. The temperature variation of ω_{co} also follows the displacement of the Boson peak with T, as shown in

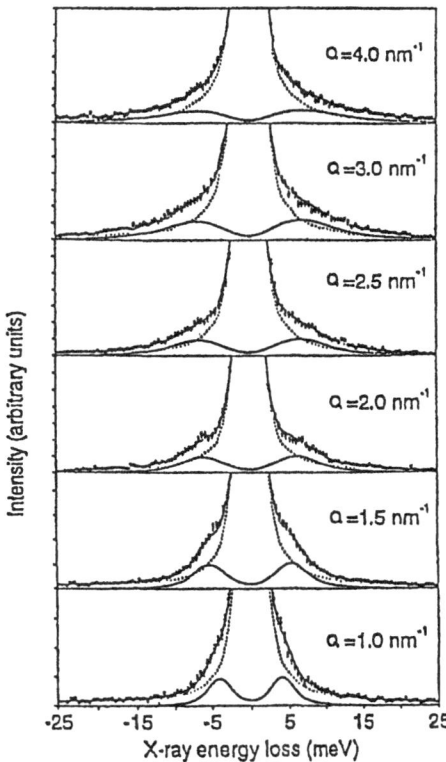

Fig. 15. — Inelastic X-ray scattering spectra of v-SiO_2 at elevated temperature (data points from [79]) adjusted to the CSR with the same parameters as in Figure 12, except that $\omega_{co} = 6.1$ meV, or $q_{co} = 1.5$ nm^{-1} (line). The inelastic signal convoluted with the instrument response is also shown separately (line), as well as the elastic peak plus background (dashes), from [80].

[66], confirming once more that ω_B and ω_{co} have about the same value. It is also interesting to consider TOF neutron-scattering data (Fig. 16) as these provide accurate frequency information on strongly scattered modes as well as localized ones at Q-values above q_{co}. As seen in [66] (Fig. 1b), the "local" mode contribution, growing like Q^2, is negligible below 10 nm^{-1}. At these low Q-values, the only contribution to the true signal comes from S_B in equation (21). Above 10 nm^{-1}, there is both a contribution in Q^2, like S_{LM}, and one in $Q^2S(Q)$, like S_U. It thus makes sense to attempt fitting the data of Figure 16 with the CSR, and these are the solid lines shown through the data points. The amplitude coefficient that is found (see inset) is actually constant up to

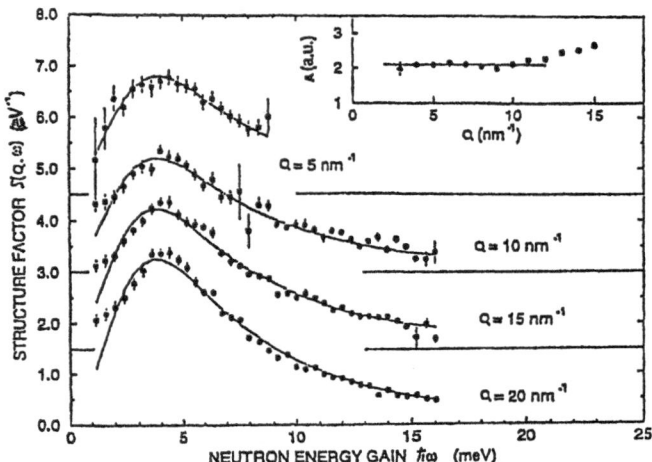

Fig. 16. — TOF neutron scattering results on v-SiO$_2$ at room temperature. Four slices of thickness 5 nm^{-1} are shown, adjusted to the CSR as explained in [78]. Fits to thinner slices, of thickness 1 nm^{-1}, give a constant prefactor A up to $Q = 10$ nm^{-1}, as shown in the inset. (From [78].)

10 nm^{-1}, while it starts growing thereafter. Thus, above 10 nm^{-1}, the signal probably contains a superposition of S_B and S_{LM}. The latter could be related to the tetrahedra rotation modes as proposed by Buchenau *et al.* [63]. Also seen in Figure 16, at low $\omega < \sim 3$ meV, there is an additional scattering contribution that could very well relate to S_U.

3.3.3. Outlook

The results on glasses that have just been exposed are rather encouraging. However, they are still far from showing the amount of detail and the level of understanding that have been obtained on aerogels. One reason for the difficulty is that the spectroscopy of the modes is at the limit of current instrumental capabilities. Another deep reason is that we have at our disposal no structural tool to investigate the sizes that are of interest for the dynamics. This is of course why the dynamics in itself is so interesting, as it might ultimately inform on aspects of the structure. At this stage, it is essential in performing experiments to carefully select the materials so as to match the capabilities of the spectroscopy. So, for example the choice of Se, with its exceptionally low sound velocity – Section 3.3.1 –, was made to match neutron spectrometers. In this sense it appears that there remain interesting possibilities in making use of the capabilities of current inelastic X-ray scattering. The materials that have been investigated so far with that technique all have thermal conductivity plateaus around 10 K, and the estimate for their crossover is typically at 4 meV or below, with $q_{co} \approx 1$ nm^{-1}. The latter figure is the lower limit for the measurable Q, so that practically all inelastic X-ray spectra so far were obtained for $Q \geq q_{co}$. This is clearly not a good situation to explore crossover phenomena.

It were thus of interest to investigate with this technique materials that are likely to be more "homogeneous" on intermediate length scales, so that they would have a higher value of q_{co}. As discovered by Bridgman [82], v-SiO$_2$ can be permanently densified from $\rho = 2.2$ g/cm^3 to $\rho \approx 2.6$ g/cm^3 when submitted to elevated pressures, especially at somewhat elevated temperatures. The latter value of ρ is about that of crystal quartz. The thermal conductivity plateau of densified silica, d-SiO$_2$, is indeed found at a significantly higher temperature than in normal silica [83]. Since the plateau is around 20 K rather than at 10 K, it implies that ω_{co} should be around 8 meV rather than at the value of 4 meV found in v-SiO$_2$. Hence, the value of q_{co} should be around 2 nm^{-1} rather than at 1 nm^{-1}. This seems to be confirmed by Raman observations of the Boson peak. Its maximum reaches near 90 cm^{-1} in d-SiO$_2$, while it is around 50 cm^{-1} in v-SiO$_2$, as shown $e.g.$ in [84]. Thus, inelastic X-ray experiments on d-SiO$_2$ should allow, for the first time with that instrument, to explore the crossover from propagating acoustic waves to strongly scattered phonons, and this in a material that should be rather homogeneous at the corresponding length scale. In that respect one also notes that $S(Q)$ at small Q is much smaller in d-SiO$_2$ than in v-SiO$_2$ [85]. Therefore, the disturbance caused by the wings of the elastic signal should also be significantly smaller in d-SiO$_2$, which will be very favorable for the observation of the inelastic signal. We hope being able to perform that crucial experiment fairly soon[4].

4. CONCLUSIONS

The first part of these two lectures on acoustic excitations of strongly disordered media was devoted to a brief review of the structure and vibrations of fractal materials, with an application to silica aerogels. In that case, a very satisfactory degree of understanding and self-consistency could be obtained. On the one hand, important information on the structure could be gained by small-angle scattering. On the other hand, the materials being disordered at a scale that can be investigated with light scattering, very detailed vibrational spectra could be measured, revealing the crossover from propagating phonons to localized fractons. The connection between the structural and vibrational results can be made owing to the scaling properties of fractals. A high degree of internal consistency was in fact achieved in measurements over the entire fractal range, including changes of regime from bending-dominated to stretching dominated

---

[4] *Footnote added in proof*: this experiment has now proven very succesful. The results will be published by E. Rat *et al.*, *Phys. Rev. Lett.* **83** (1999) 7.

vibrations [38,86], as well as other results that were not described here for lack of space.

However, the reader should not gain the impression that all issues concerning the vibrations of fractals are now fully settled. There remain a number of open questions, such as the nature of the Raman scattering mechanisms and their scaling [49,87], the origin of the depolarized scattering and of the slow crossover into the asymptotic Raman regime [39], the exact nature of the bending- and stretching-dominated fracton modes, the origin of the observed ω-dependence of the Debye-Waller factor [37], among others. New formal developments, very large realistic simulations, as well as more experiments on well characterized materials, would be very useful in these contexts.

The second part of these lectures was devoted to the situation near the end of acoustic branches in glasses. This is a less mature, challenging, and very active field of current investigations. Part of the difficulty is that we have very little information on the structure of glasses at the scales that seem to be of interest, *i.e.* at lengths of several nanometers. There exists a theoretical prediction that glasses are likely to be quite inhomogeneous at such scales, at least *dynamically*, with regions under tension and others that would have collapsed by buckling [10]. There is unfortunately very little in terms of direct experimental confirmation. The common structural tools in Fourier space usually do not reach to these sizes, and if they do they can only check fluctuations about the average density, which are likely to be minute. Interestingly, in atomic force microscopy of fractured surfaces of glasses, at mesoscopic scales, rather well defined "entities" with a height of typically 5 to 10 nm have been observed [88]. The authors report that this is quite a universal property, independent from the more macroscopic features of the different fracture zones, and also from the chemical nature of the glass. They have also checked that water is not involved. Hence, this might just be a signature of the intrinsic inhomogeneity of these materials, possibly related to the surface relaxation that occurs when regions under different tensions are exposed by brittle fracture. These questions about the medium-range structure make it all the more important to investigate the corresponding dynamics.

On glassy selenium, we were able using neutron scattering to separate quite clearly the contribution of phonons from that of local modes. Neutron Brillouin scattering was in fact observed, as well as Umklapp scattering of transverse phonons *via* the static structure factor. These observations reveal that phonons cease to propagate as plane waves at frequencies above ~ 1 THz, corresponding in this case to $q_{co} \sim 3$ nm^{-1}, or to a length of ~ 20 Å, considerably larger than the nearest neighbor distance. The phonon crossover also corresponds approximately to the position of the Boson peak, suggesting that rotational-vibrational coupling might, in this case, be an important factor contributing to the end of the acoustic branches. In the case of vitreous silica, q_{co} is even smaller, ~ 1 nm^{-1} at room temperature. At small scattering vectors ($Q \approx q_{co}$), the contribution from local modes to the dynamical structure factor is negligible, and the Boson peak is likely to contain a proportionally larger "acoustic" component at these small wave-vectors. One finds then more precisely that $\omega_B \cong \omega_{co}$.

Particular caution is required in the interpretation of inelastic X-ray scattering data, as *e.g.* in [78–80, 89]. For $Q > q_{co}$, Q is only the scattering vector probed in experiments, but it has lost its meaning of quantum number, or wave-vector, for the excitations. The latter have a mean size that still depend on ω [90]. This leads to an apparent dispersion when one plots the peak position, ω_p, in $S(Q, \omega)$, *vs.* Q. One should however refrain from interpreting $\omega_p(Q)$ as a *dispersion curve*. For example, the derivative, $\partial \omega_p / \partial Q$, is unrelated to a group velocity. Similar words of caution of course apply to simulation results, *e.g.* to these in [91].

To summarize the current situation in glasses:

- Brillouin spectra obtained with X-rays and neutrons can be interpreted with a model that assumes that high frequency phonons crossover into a regime of strong scattering, at $\omega_{co} \sim 1$ THz.

- This interpretation agrees with the abundant data on thermal properties of glasses.

- The Boson peak at ω_B generally results from a combination of several types of vibrations.

- One has $\omega_B \approx \omega_{co}$, either that local modes participate to the strong scattering of acoustic excitations, or, if there are few local modes, that acoustic excitations themselves produce the Boson peak.

Acknowledgments

The recent work on glasses benefited greatly from collaborations with M. Foret and B. Hehlen at the Laboratoire des Verres, with H. Casalta and B. Dorner at the ILL, and with J.-B. Suck in Chemnitz. Our thinking concerning these problems was considerably influenced by our collaboration with the late Prof. Shlomo Alexander, as well as by conversations with Profs. Jacques Pelous and Ray Orbach.

Appendix A

Density correlation functions and static structure factors

The structure factor is a quantity generally accessible to either neutron or X-ray scattering. It characterizes the relative positions of atom pairs, and it is the first level of information about the structure beyond its mere density. For simplicity, we take the case of neutron scattering as it is somewhat easier. Neutrons are mainly scattered by the nuclei and thus, at the scale of interest for structures, they are scattered by points. Point scattering is isotropic and it can be characterized by a scattering amplitude, called in this case the *scattering length* b_i associated with each atom i. On the opposite, since X-rays are scattered by the more extended electron clouds, their interaction with atoms involves *angular dependent* atomic-form factors.

The density of neutron scattering lengths can be written

$$\rho_b(\mathbf{r}, t) = \sum_i b_i \, \delta[\mathbf{r} - \mathbf{R}_i(t)], \qquad (A.1)$$

where $\mathbf{R}_i$ is the position of atom i at time t, and δ is the Dirac delta function. The scattering produced by the variations of b within the same atomic species, owing for example to isotopes or to different nuclear spin orientations, provides no information about the relative atomic positions. This *incoherent scattering* must often be subtracted somehow from the signal to just keep the coherent contribution. In what follows we assume that this has been done, and that the b_i's are simply the *coherent* scattering lengths $\bar{b}_i$, *i.e.* the average b_i's for each atomic species of interest. We also take them real, neglecting absorption.

The quantity that controls coherent neutron scattering is the correlation function of the densities of scattering lengths [64],

$$G_{\text{coh}}(\mathbf{r}) = \frac{1}{N} \int \langle \rho_b(\mathbf{r}') \rho_b(\mathbf{r} + \mathbf{r}') \rangle \mathrm{d}\mathbf{r}', \qquad (A.2)$$

where N is the number of atoms in the scattering volume and the $\langle \, \rangle$ designates an average over states and realizations. In a quantum mechanical description, which is actually needed to study the dynamics, the positions $\mathbf{R}_i(t)$ are operators, but at equal times these operators commute. From (A.1) and (A.2), it then follows that $G_{\text{coh}}(\mathbf{r})$ can also be written

$$G_{\text{coh}}(\mathbf{r}) \equiv \frac{1}{N} \sum_{i,j} b_i b_j \langle \delta(\mathbf{r} - \mathbf{R}_{ij}) \rangle, \qquad (A.3)$$

In this expression $\mathbf{R}_{ij} = \mathbf{R}_i - \mathbf{R}_j$, both taken at the same time t, for example $t = 0$. It is assumed that the sample is large enough not to worry about the surface in (A.3). This is generally the case in practice, except for very fine grained powders. In simulations, this is usually achieved by using periodic boundary conditions.

In actual experiments, the spectrometer fixes the momentum exchange $\mathbf{Q}$,

$$\mathbf{Q} = \mathbf{k}_s - \mathbf{k}_i, \qquad (A.4)$$

where $\mathbf{k}_i$ and $\mathbf{k}_s$ are the incident and scattered neutron wave vectors, respectively. If the experiment is not resolved in energy, it measures the coherent differential cross-section of the sample $\mathrm{d}\sigma$ per unit solid angle $\mathrm{d}\Omega$,

$$\left(\frac{\mathrm{d}\sigma}{\mathrm{d}\Omega}\right)_{\text{coh}} \equiv N S_{\text{coh}}(\mathbf{Q}) = \sum_{i,j} b_i b_j \langle \mathrm{e}^{-i\mathbf{Q}\cdot\mathbf{R}_i} \mathrm{e}^{i\mathbf{Q}\cdot\mathbf{R}_j} \rangle, \qquad (A.5)$$

The first identity defines the *static structure factor*, $S_{\mathrm{coh}}(\mathbf{Q})$. It is obvious from (A.2) and (A.5) that

$$S_{\mathrm{coh}}(\mathbf{Q}) = \int G_{\mathrm{coh}}(\mathbf{r})\, e^{-i\mathbf{Q}\cdot\mathbf{r}}\, d\mathbf{r}. \qquad (A.6)$$

Owing to the normalization by N in (A.5), the expression (A.6) amounts to an effective differential cross-section per atom.

Isotropic media and pair distribution function – The materials of interest here are isotropic, so that $G_{\mathrm{coh}}(\mathbf{r})$ only depends on the length r, and can be written $G_{\mathrm{coh}}(r)$. The correlation function also becomes uniform at large distances, meaning that $G_{\mathrm{coh}}(r \to \infty) = \bar{b}\, \rho_N$, where $\bar{b}$ is the sample averaged scattering length, and ρ_N is just the macroscopic number density. Thus $S_{\mathrm{coh}}(\mathbf{Q})$ in (A.6) includes a delta function at the origin. To analyze this in more details, and to simplify the discussion, we assume that instead of n distinct atomic species, there were only one. The b_i's then factor out from the above definitions of ρ, G, and S. These are then equally applicable to the "grains" used in simulations. In the following, b is just replaced by 1, so that (A.3) becomes

$$G(r) \equiv \frac{1}{N}\sum_{i,j}\langle \delta(\mathbf{r} - \mathbf{R}_{ij})\rangle, \qquad (A.7)$$

where N is the number of grains in the sample, and $\mathbf{R}_{ij}$ refers to their relative positions. This correlation function can be expressed in terms of the pair distribution function $g(r)$, which is proportional to the average probability of finding a grain in a unit volume at distance r from a central grain. It is customary to normalize this probability by the number density $\rho_N = N/V$, where V is the sample Euclidean volume. With this normalization, $g(r) \to 1$ for $r \to \infty$. In the case of fractals, it will also be assumed that V is much larger than the cube of the persistence length defined in Section 2, $V \gg \xi^3$. The distribution function, taken around the grain at $\mathbf{R}_i$, is then

$$g(r) \equiv \frac{V}{N}\sum_{j}{}'\langle \delta(\mathbf{r} - \mathbf{R}_{ij})\rangle = \frac{V}{N}[G(r) - \delta(r)]. \qquad (A.8)$$

The prime on the summation sign means that $j = i$ is excluded from the sum. For large r, $G(r)$ tends to ρ_N, and indeed $g(r)$ tends to 1. As r approaches the grain size a, $g(r)$ feels the *excluded volume* about the central grain. The function depends there on the grain size and shape. If the shape can be approximated by a sphere, and if the polydispersity is moderate, then the problem is easily handled in Fourier space. If there were n different species, one would of course have to define $n(n-1)/2$ distinct pair correlation functions.

The case of fractals – To perform the sum in (A.8), we take grain i at the center of a fractal blob in the sense of (1). Then $\sum_{j}{}'\langle \delta(\mathbf{r}-\mathbf{R}_{ij})\rangle$ at fixed r represents the mean number density of grains at distance r from a central grain. In terms of (2) it equals $\mathrm{d}(M/m_a)/\mathrm{d}V$. Introducing this in (A.8), and since $\rho_N = \rho/m_a$, one finds

$$g(r) = \frac{m_a}{\rho} \frac{\mathrm{d}(M/m_a)}{\mathrm{d}V} \approx \rho_f(r)/\rho, \tag{A.9}$$

where a factor $D/3 \approx 1$ was neglected. Writing $g(r) = \rho_f(r)/\rho$, it has the proper asymptotic limit, $g(r) = 1$ at $r \gg \xi$.

With ρ_f given by (2), (A.9) is only valid in the fractal region. Many models have been made to describe the crossover to the homogeneous regime taking place as r approaches ξ. Some are reviewed in [28]. This aspect lacks universality and often requires simulations. For example, in the case of diffusion-limited cluster-cluster aggregation [92,93], only simulations were able so far to provide the actual form of $g(r)$ [28].

The structure factor of isotropic media – The static structure factor assuming one atomic species can be written

$$S(\mathbf{Q}) = \int G(r)\,\mathrm{e}^{-i\mathbf{Q}\cdot\mathbf{r}}\,\mathrm{d}\mathbf{r}. \tag{A.10}$$

Since $G(r)$ tends to the constant ρ_N at large r, (A.10) contains a term $\delta(\mathbf{Q})$. Subtracting this term and using (A.8), one finds for $Q \neq 0$

$$S(\mathbf{Q}) = 1 + \rho_N \int [g(r) - 1]\,\mathrm{e}^{-i\mathbf{Q}\cdot\mathbf{r}}\,\mathrm{d}\mathbf{r}. \tag{A.11}$$

Performing the angular integration, (A.11) gives

$$S(Q) = 1 + \rho_N \int_0^\infty \frac{\sin Qr}{Qr}\,[g(r) - 1]\,4\pi r^2 \mathrm{d}r \quad \text{(for } Q \neq 0\text{)}. \tag{A.12}$$

Appendix B

Dynamical structure factor

As in Appendix A, only the coherent scattering of neutrons is considered for this discussion. (A.2) is now replaced by a time-dependent correlation function

$$G_{\mathrm{coh}}(\mathbf{r}, t) = \frac{1}{N} \int \langle \rho_b(\mathbf{r}', 0)\rho_b(\mathbf{r} + \mathbf{r}', t)\rangle \mathrm{d}\mathbf{r}', \tag{B.1}$$

where ρ_b is defined by (A.1). The order of the time-dependent operators in (B.1) is important, as the positions generally do not commute at different times. This is responsible for the Stokes-Antistokes asymmetry in the scattering.

The coherent inelastic differential cross-section of the sample, $d\sigma$ per unit solid angle $d\Omega$, and unit scattered energy dE_s, is given by [64]

$$\left(\frac{d^2\sigma}{d\Omega\,dE_s}\right)_{\text{coh}} = N\frac{k_s}{k_i}S(\mathbf{Q},\omega),\tag{B.2}$$

where the *dynamic structure factor*, $S(\mathbf{Q},\omega)$ is given by

$$S(\mathbf{Q},\omega) = \frac{1}{2\pi\hbar}\int_{-\infty}^{+\infty} G_{\text{coh}}(\mathbf{r},t)\,e^{i\omega t - i\mathbf{Q}\cdot\mathbf{r}}\,d\mathbf{r}\,dt.\tag{B.3}$$

In (B.2), the factor k_s/k_i enters because the velocities of the scattered and incoming neutrons intervene in the definition of the corresponding neutron fluxes, and the ratio of the latter defines the scattering cross-section σ. The Planck constant $\hbar$ enters the expression (B.3), because the differential cross-section is traditionally written per unit energy, while the Fourier transform in (B.3) is in terms of frequency. It must be emphasized that in the above definitions $S(\mathbf{Q},\omega)$ includes the scattering lengths.

Local modes – For a solid, one can replace $\mathbf{R}_i(t)$ in (A.1) by

$$\mathbf{R}_i(t) = \mathbf{R}_i + \mathbf{u}_i(t),\tag{B.4}$$

where $\mathbf{R}_i$ is the average position of atom i, and $\mathbf{u}_i(t)$ is the fluctuation about that position. The spatial Fourier transformation in (B.3) can be written in terms of the spatial transforms of ρ_b,

$$\rho_b(\mathbf{Q},t) = \sum_i b_i\,e^{-i\mathbf{Q}\cdot(\mathbf{R}_i+\mathbf{u}_i)}.\tag{B.5}$$

For harmonic local modes, the latter can be analyzed as a sum over orthogonal eigenvibrations α of frequency ω_α and vibrational amplitudes

$$\mathbf{u}_{i\alpha}(t) = \mathbf{u}_{i\alpha}(0)\,e^{-i\omega_\alpha t}.\tag{B.6}$$

This gives,

$$\rho_b(\mathbf{Q},t) = \rho_{b0}(\mathbf{Q}) + \sum_\alpha \rho_\alpha(\mathbf{Q},t) + \mathcal{O}(u^2).\tag{B.7}$$

Here, $\rho_{b0}(\mathbf{Q})$ is the quantity entering the static structure factor, $\rho_{b0}(\mathbf{Q}) = \sum_i b_i e^{-i\mathbf{Q}\cdot\mathbf{R}_i}$, while

$$\rho_\alpha(\mathbf{Q},t) = -i\,e^{-i\omega_\alpha t}\sum_i [\mathbf{Q}\cdot\mathbf{u}_{i\alpha}(0)]\,e^{-i\mathbf{Q}\cdot\mathbf{R}_i} = e^{-i\omega_\alpha t}\,\rho_\alpha(\mathbf{Q}).\tag{B.8}$$

This defines the quantity $\rho_\alpha(\mathbf{Q})$. Introducing (B.7) and (B.8) in (B.3) using (B.1), gives [49]

$$S(\mathbf{Q},\omega) = \frac{1}{\hbar}\sum_\alpha \delta(\omega - \omega_\alpha)\,\rho_\alpha^*(\mathbf{Q})\,\rho_\alpha(\mathbf{Q}).\tag{B.9}$$

This neglected higher order terms, *i.e.* mostly the Debye-Waller factors.

It is clear from (B.8) and (B.9) that $S(\mathbf{Q}, \omega)$ contains the square modulus of $\mathbf{Q} \cdot \mathbf{u}_{i\alpha}(0)$. In other words, in neutron scattering, as well as in X-ray scattering, one senses the projection of the displacements along the direction of $\mathbf{Q}$. The same is true for propagating acoustic modes [64]. There is however a possibility to explore TA-modes through Umklapp scattering via the static structure factor, as explained in Section 3.2.1. Owing to the factor u^2, the result (B.9) is proportional to n/ω, where n is the Bose factor, and $1/\omega$ is a harmonic oscillator factor that results from mode quantization, but whose existence can also be derived from purely classical concepts [4].

REFERENCES

[1] A.C. Wright, in Amorphous Insulators and Semiconductors, edited by M.F. Thorpe and M.I. Mitkova, *NATO ASI Series 3. High Technology* **23** (Kluwer, Dordrecht, 1997) 83-132, and references therein.

[2] M. Foret, E. Courtens, R. Vacher and J.-B. Suck, *Phys. Rev. Lett.* **77** (1996) 3831.

[3] See, *e.g.*, J. Fabian and P.B. Allen, *Phys. Rev. Lett.* **77** (1996) 3839.

[4] S. Alexander, *Phys. Rev. B* **40** (1989) 7953.

[5] See, *e.g.*, edited by W.A. Phillips, Amorphous Solids: Low Temperature Properties (Springer Verlag, Berlin, 1981).

[6] C. Kittel, Introduction to Solid State Physics (Wiley, New York, 1967).

[7] R.C. Zeller and R.O. Pohl, *Phys. Rev. B* **4** (1971) 2029.

[8] V.K. Malinovski and A.P. Sokolov, *Solid State Commun.* **57** (1986) 757.

[9] N. Bilir and W.A. Phillips, *Philos. Mag.* **32** (1975) 113.

[10] S. Alexander, *Physics Reports* **296** (1998) 65.

[11] W. Schirmacher, G. Diezemann and C. Ganter, *Phys. Rev. Lett.* **81** (1998) 136.

[12] D.A. Parshin, *Fiz. Tverd. Tela* **36** (1994) 1809; *Phys. Solid State* **36** (1994) 991, and references therein.

[13] S. Alexander and R. Orbach, *J. Phys. Lett.* **43** (1982) L625.

[14] R. Rammal and G. Toulouse, *J. Phys. Lett.* **44** (1983) L13.

[15] For a review, see T. Nakayama, K. Yakubo and R. Orbach, *Rev. Mod. Phys.* **66** (1994) 381.

[16] For a recent review, see E. Courtens and R. Vacher, in Amorphous Insulators and Semiconductors, edited by M.F. Thorpe and M.I. Mitkova, *NATO ASI Series 3. High Technology* **23** (Kluwer, Dordrecht, 1997) 255-288.

[17] R. Vacher, T. Woignier, J. Pelous and E. Courtens, *Phys. Rev. B* **37** (1988) 6500.

[18] E. Courtens, J. Pelous, J. Phalippou, R. Vacher and T. Woignier, *Phys. Rev. Lett.* **58** (1987) 128.

[19] B. Mandelbrot, Fractals: Form, Chance and Dimension (Freeman, San Francisco, 1977).

[20] Y. Kantor and T.A. Witten, *J. Phys. Lett.* **45** (1984) L675.
[21] D. Stauffer, Introduction to Percolation Theory (Taylor & Francis, London, 1985).
[22] E. Stoll, M. Kolb and E. Courtens, *Phys. Rev. Lett.* **68** (1992) 2472.
[23] R.K. Iler, The Chemistry of Silica (Wiley, New York, 1979).
[24] S.S. Kistler, *J. Phys. Chem.* **36** (1932) 52.
[25] T. Frelthoft, K.J. Kjems and S.K. Sinha, *Phys. Rev. B* **33** (1986) 269.
[26] E. Anglaret, A. Hasmy, E. Courtens, J. Pelous and R. Vacher, *Europhys. Lett.* **28** (1994) 591.
[27] M. Foret, J. Pelous and R. Vacher, *J. Phys. I France* **2** (1992) 791.
[28] A. Hasmy, M. Foret, J. Pelous and R. Jullien, *Phys. Rev. B* **50** (1994) 6006.
[29] R. Vacher, E. Courtens, E. Stoll, M. Böffgen and H. Rothuizen, *J. Phys. Condens. Matter* **3** (1991) 6531.
[30] S. Alexander, *J. Phys. France* **45** (1984) 1939.
[31] E. Courtens, R. Vacher and E. Stoll, *Physica D* **38** (1989) 41.
[32] K. Yakubo, E. Courtens and T. Nakayama, *Phys. Rev. B* **42** (1990) 1078.
[33] E. Stoll and E. Courtens, *Z. Phys. B* **81** (1990) 1.
[34] D.J. Thouless, *Phys. Rep.* **C13** (1974) 94.
[35] E. Courtens, R. Vacher, J. Pelous and T. Woignier, *Europhys. Lett.* **6** (1988) 245.
[36] R. Vacher, T. Woignier, J. Pelous, G. Coddens and E. Courtens, *Europhys. Lett.* **8** (1989) 161.
[37] R. Vacher, E. Courtens, G. Coddens, J. Pelous and T. Woignier, *Phys. Rev. B* **39** (1989) 7384.
[38] R. Vacher, E. Courtens, G. Coddens, A. Heidemann, Y. Tsujimi, J. Pelous and M. Foret, *Phys. Rev. Lett.* **65** (1990) 1008.
[39] Y. Tsujimi, E. Courtens, J. Pelous and R. Vacher, *Phys. Rev. Lett.* **60** (1988) 2757.
[40] E. Courtens and R. Vacher, *Philos. Mag. B* **65** (1992) 347.
[41] R. Calemczuk, A.M. de Goer, B. Salce, R. Maynard and A. Zarembowitch, *Europhys. Lett.* **3** (1987), 1205.
[42] A. Bernasconi, T. Sleator, D. Posselt, J.K. Kjems and H.R. Ott, *Phys. Rev. B* **45** (1992) 10363.
[43] P.G. Klemens, in Solid State Physics, edited by F. Seitz and D. Turnbull (Academic Press, New York, 1958) 1-98.
[44] E. Courtens and R. Vacher, *Z. Phys. B* **68** (1987) 355.
[45] R. Vacher, M. Foret, E. Courtens, J. Pelous and J.-B. Suck, *Philos. Mag. B* **77** (1998) 373.
[46] A. Aharony, O. Entin-Wohlman, S. Alexander and R. Orbach, *Philos. Mag. B* **56** (1987) 949.
[47] G. Polatsek and O. Entin-Wohlman, *Phys. Rev. B* **37** (1988) 7726.
[48] A.F. Ioffe and A.R. Regel, *Prog. Semicond.* **4** (1960) 237.
[49] S. Alexander, E. Courtens and R. Vacher, *Physica A* **195** (1993) 286.
[50] E. Courtens and R. Vacher, *Proc. R. Soc. Lond.* **A423** (1989) 55.

[51] E. Courtens, C. Lartigue, F. Mezei, R. Vacher, G. Coddens, M. Foret, J. Pelous and T. Woignier, *Z. Phys. B* **79** (1990) 1.

[52] E. Courtens and R. Vacher, in Random Fluctuations and Pattern Growth: Experiments and Models, edited by H.E. Stanley and N. Ostrowsky, *NATO ASI Series E.* **157** (Kluwer, Utrecht, 1988) 20.

[53] F.L. Galeener, *Solid State Commun.* **44** (1982) 1037.

[54] J.W. Zwanziger, K.K. Olsen, S.L. Tagg and R.E. Youngman, in Amorphous Insulators and Semiconductors, edited by Thorpe M.F. and Mitkova M.I., *NATO ASI Series 3. High Technology* **23** (Kluwer, Dordrecht, 1997) 245-254.

[55] A. Pasquarello and R. Car, *Phys. Rev. Lett.* **80** (1998) 5145.

[56] S. Hunklinger and W. Arnold, in Physical Acoustics, edited by W.P. Mason and R.N. Thurston (Academic Press, New York, 1976), Vol. XII, 155.

[57] R.O. Pohl, in [5], 29.

[58] A.C. Anderson, in [5], 65.

[59] R. Vacher, J. Pelous and E. Courtens, *Phys. Rev. B* **56** (1997) R481, and references therein.

[60] M. Rothenfusser, W. Dietsche and H. Kinder, in Phonon Scattering in Condensed Matter, edited by W. Eisenmenger, K. Lassmann and S. Döttinger (Springer, Berlin, 1984) 419.

[61] T.C. Zhu, H.J. Maris and J. Tauc, *Phys. Rev. B* **44** (1991) 4281.

[62] C.J. Morath and H.J. Maris, *Phys. Rev. B* **54** (1996) 203.

[63] U. Buchenau, M. Prager, N. Nücker, A.J. Dianoux, N. Ahmad and Phillips W.A., *Phys. Rev. B* **34** (1986) 5665.

[64] S.W. Lovesey, Theory of Neutron Scattering from Condensed Matter (Clarendon, Oxford, 1984).

[65] U. Buchenau, *Z. Phys. B* **58** (1985) 181.

[66] A. Wischnewski, U. Buchenau, A.J. Dianoux, W.A. Kamitakahara and J.L. Zaretsky, *Phys. Rev. B* **57** (1998) 2663.

[67] J. Jäckle, in The Physics of Non-Crystalline Solids, edited by G.H. Frischat (Trans Tech Publications, Aedermannsdorf, Switzerland, 1977) 568.

[68] R. Shuker and R.W. Gammon, *Phys. Rev. Lett.* **25** (1970) 222.

[69] A.P. Sokolov, U. Buchenau, W. Steffen, B. Frick and A. Wischnewski, *Phys. Rev. B* **52** (1995) R9815.

[70] J.E. Graebner, B. Golding and L.C. Allen, *Phys. Rev. B* **34** (1986) 5696.

[71] A.K. Raychaudhuri, *Phys. Rev. B* **39** (1989) 1927.

[72] J.M. Carpenter and C.A. Pelizzari, *Phys. Rev. B* **12** (1975) 2391.

[73] J.-B. Suck, P.A. Egelstaff, R.A. Robinson, D.S. Sivia and A.D. Taylor, *J. Non-Cryst. Sol.* **150** (1992) 245.

[74] F. Sette, G. Ruocco, M. Krisch, C. Masciovecchio and R. Verbeni, *Phys. Scripta* **T66** (1996) 48.

[75] J. Jäckle and K. Fröböse, *J. Phys. F: Metal Physics* **9** (1979) 967.

[76] M. Foret, B. Hehlen, G. Taillades, E. Courtens, R. Vacher, H. Casalta and B. Dorner, *Phys. Rev. Lett.* **81** (1998) 2100.

[77] M. Foret, B. Hehlen, E. Courtens, R. Vacher, H. Casalta and B. Dorner, *Physica B*, Proceedings of Phonons 98 (to be published).
[78] M. Foret, E. Courtens, R. Vacher and J.-B. Suck, *Phys. Rev. Lett.* **77** (1996) 3831.
[79] P. Benassi, M. Krisch, C. Masciovecchio, V. Mazzacurati, G. Monaco, G. Ruocco, F. Sette and R. Verbeni, *Phys. Rev. Lett.* **77** (1996) 3835.
[80] M. Foret, E. Courtens, R. Vacher and J.-B. Suck, *Phys. Rev. Lett.* **78** (1997) 4669.
[81] C. Masciovecchio, G. Ruocco, F. Sette, P. Benassi, A. Cunsulo, M. Krisch, V. Mazzacurati, A. Mermet, G. Monaco and R. Verbeni, *Phys. Rev. B* **55** (1997) 8049.
[82] P.W. Bridgman and I. Šimon, *J. Appl. Phys.* **24** (1953) 405.
[83] Zhu Da-Ming, *Phys. Rev. B* **50** (1994) 6053.
[84] S. Sugai, H. Sotokawa, D. Kyokana and A. Onodera, *Physica B* **219-220** (1996) 293.
[85] M. Arai (private communication).
[86] S. Feng, *Phys. Rev. B* **32** (1985) 5793.
[87] G. Viliani, R. Dell'Anna, O. Pilla, M. Montagna, G. Ruocco, G. Signorelli and V. Mazzacurati, *Phys. Rev. B* **52** (1995) 3346.
[88] E. Guilloteau, H. Arribart, F. Creuzet, *Mat. Res. Soc. Symp. Proc.* **409** (1996) 365.
[89] P. Benassi, M. Krisch, C. Masciovecchio, V. Mazzacurati, G. Monaco, G. Ruocco, F. Sette and R. Verbeni, *Phys. Rev. Lett.* **78** (1997) 4670.
[90] E. Duval and A. Mermet, *Phys. Rev. B* **58** (1998) 8159.
[91] R. Dell'Anna, G. Ruocco, M. Sampoli and G. Viliani, *Phys. Rev. Lett.* **80** (1998) 1236.
[92] P. Meakin, *Phys. Rev. Lett.* **51** (1983) 1119.
[93] M. Kolb, R. Botet and R. Jullien, *Phys. Rev. Lett.* **51** (1989) 1123.

Intermittent Dynamics and Ageing in Glassy Systems

J.-Ph. Bouchaud

*Service de Physique de l'État Condensé, Centre d'Études de Saclay,
Orme des Merisiers, 91191 Gif-sur-Yvette Cedex, France*

1. INTRODUCTION

A great variety of systems exhibit slow "glassy" dynamics: glasses of all kinds [1], of course, but also spin-glasses [2], pinned "defects" such as Bloch walls, vortices in superconductors, charge density waves, dislocations, etc. interacting with randomly placed impurities [3]. Another class of systems where surprisingly slow dynamics can occur are soft glassy materials, such as foams, dense emulsions or granular materials, which attracted a lot of interest recently [4–6]. Common features in the dynamics of these systems include:

- a fast increase of the relaxation time $\tau(T)$ as the temperature T is decreased (or as the density is increased). This increase is either a power law divergence near the critical temperature (*e.g.* in spin glasses), or a Vogel-Fulcher like divergence, in any case faster than a simple activated divergence.

- A decoupling between fast and slow modes. The dynamical structure factor of glasses for example, exhibit a two steps relaxation process, where the correlation function first decays quite rapidly to a plateau value (β relaxation), and then on much longer time scales $\simeq \tau(T)$ to zero (α relaxation). The first stage can be loosely associated to the motion of each molecule in its "cage" (or to the fast spin oscillations around a frozen magnetisation, or else to the small wavelength oscillations of – say – a dislocation in the potential created by the surrounding impurities). The second, slow stage is associated to collective rearrangments of the system on larger length scales, which usually proceeds via thermal activation. Correspondingly, excess low frequency noise and dissipation appear. For example, the elastic moduli of soft glassy material reveal extremely large low frequency contributions.

- History dependence and ageing: one of the most striking effects in glassy systems is the ageing phenomenon, which means that the correlation and response of the system become time dependent, as if the system "stifened" with time. For example, the frequency dependent susceptibility of spin-glasses decreases with the waiting time t_w during which the system is left at the working temperature (below the glass temperature). A good fit of this decay is given by [2, 7]:

$$\chi(\omega, t_w) = \chi_{eq}(\omega) + A(\omega t_w)^{x-1} \qquad x < 1 \qquad (1)$$

where $\chi_{eq}(\omega)$ is the equilibrium contribution, and x a certain (temperature dependent) exponent. Similarly, the correlation function of the system ceases to be invariant under time translation. For example, the spin-spin correlation function in spin-glasses, or the structure factor in glasses, which measure how far the system moves away from itself between times t_w and $t_w + t$, cannot be written as a function of t only. Rather, the correlation function can be decomposed into an equilibrium part (corresponding to $\chi_{eq}(\omega)$ above) and an ageing part, which is often a function of the *ratio* t/t_w rather than of the difference t. Qualitatively, this means that the time needed for the system to relax is fixed by the time during which the system was prepared. Other history dependent effects have been reported in disordered systems [8]; one of the most striking are the small temperature cycles in spin-glasses which reveal strong *memory* effects [9].

- Response to an external field. One of the most important properties of pinned defects is their response to an external driving field; for example, the response of a Bloch wall to a magnetic field, a dislocation to a stress field, or a vortex to a current. Often, the technological properties of the material under consideration is governed by the response of its defects to external sollicitations. Typically, a finite threshold force must be reached (at least at zero temperature) before the object moves. The velocity is thus zero below a certain critical force (the depinning transition [10]), and grows in a non trivial manner above. Similarly, the rheological behaviour of soft glassy materials often presents strong non linearities: in some cases a finite yield stress is needed for the material to flow and for the shear rate (the analogue of the velocity) to be non zero [4].

2. A SIMPLE MODEL: TRAPS AND INTERMITTENT DYNAMICS

Let us introduce the following simplified picture of a glass: the motion of a given particle can be thought of as taking place in a random potential created by its neighbours [11]. Since the motion of the other molecules is extremely slow at low temperatures, one can assume that this random potential has a static component. The random potential has minima ("traps"), where the particle performs fast oscillations and from which it escapes through thermal activation and "hops" to the neighbouring sites. Similarly, the motion of a linear object (say a vortex) pinned by impurities is a succession of hops between locally favourable conformations [12]. Here, however, an interesting idea appears (which might actually also be important for glasses), which is the fact that changes of conformations requiring large scale moves are much slower than small scale moves. This separation of time scales suggests a hierarchical picture for the dynamics, where (on a given time window) small scale hops between "traps" take place while the large scale degrees of freedom are frozen [13]. The separation between the two types of dynamics (equilibrium and ageing) takes place for length scales such that the typical pinning energy is of the same order of magnitude as the thermal energy. Although much more difficult to visualize, this idea may also pertain to the phase-space dynamics of spin-glasses, and motivate a tree-like organisation of the "traps" [13, 14], as suggested by mean-field models [15]. Such a picture is actually very helpful to understand the remarkable memory effects alluded to above [9].

Coming back to the simple real space model of diffusion among regularly placed traps, the important quantity is the distribution of barrier heights $\rho(\Delta E)$, which in turn determines the distribution of local *trapping times* τ through the relation

$$\tau = \tau_0 \exp \left[\frac{\Delta E}{T} \right],$$

where τ_0 is a microscopic time scale (typically the oscillation period within the trap) and $\rho(\Delta E)$ describes the strength of the "cages" created by the neighbouring particles. Let us consider the case where $\rho(\Delta E)$ decreases for large ΔE as

$$\exp - \left(\frac{\Delta E}{T_0} \right)^{1+\nu}.$$

In the following, $\vec{r}(t)$ will denote the position of the particle at time t. When $\nu = 0$ (exponential distribution of barrier heights), one can show [4, 11, 14, 16] that there is a true dynamical phase transition for a critical value $x = 1$, where $x \equiv T/T_0$. This phase transition separates:

- a high temperature *liquid* phase where the correlation function

$$C_q(t_w + t, t_w) = \langle \exp[i\vec{q} \cdot (\vec{r}(t + t_w) - \vec{r}(t_w))] \rangle$$

is time translation invariant

$$C_q(t_w + t, t_w) \equiv C_q(t)$$

and decays to zero for large times as t^{1-x}, and

- a low temperature *glass* phase where ergodicity is "weakly broken", in the sense that [2]:

$$\lim_{t\to\infty} \lim_{t_w\to\infty} C_q(t_w + t, t_w)0 \tag{2}$$

(note the order of limits). More physically, this means that $C_q(t_w + t, t_w)$ has an "ageing" component which is a function of t/t_w, decaying on a time scale proportional to the waiting time t_w itself. For $T < T_0$, the asymptotic diffusion constant of the particle is zero [16, 17]. Translated into a frequency dependent susceptibility, one finds a formula close to equation (1) above, with $x = T/T_0$.

The basic mechanism leading to this "phase transition" is the fact that the distribution of trapping times decays very slowly for large τ's as τ^{-1-x}. This has actually been observed directly on numerical (Molecular Dynamics) simulations [16]. When $T < T_0$, the average trapping time diverges; the longest trapping time encountered is then always of the same order of magnitude as the observation time t_w. The dynamics becomes extremely intermittent in the sense that most of the time *nothing* happens; hopping events occur on a "fractal" set with vanishing density as $t_w \to \infty$: this is deeply related to non stationarity and ageing in this model [14, 18].

For $\nu > 0$ the "glass" phase transition only takes place, strictly speaking, at $T = 0$. However, the relaxation time $\tau(T)$ beyond which $C_q(t_w + t, t_w)$ decays to zero diverges strongly as the temperature is decreased:

$$\tau(T) \propto \tau_0 \exp\left(\frac{T_0}{T}\right)^{1+\frac{1}{\nu}}. \tag{3}$$

For $t_w \gg \tau(T)$, the relaxation function only depends on the time difference t and is well approximated by a stretched exponential with an exponent

$$\beta \simeq \left(\frac{T}{T_0}\right)^{\frac{1}{2}+\frac{1}{2\nu}}.$$

Furthermore, $C_q(t)$ obeys an approximate "time temperature" superposition principle, *i.e.*

$$C_q(t) \simeq f_q\left(\frac{t}{\tau(T)}\right)$$

in the α regime [11]. Finally, when $t_w, t \ll \tau(T)$, the correlation function is ageing very much like what happens in the case $\nu = 0$, $T < T_0$ [11]. In particular, the α-peak in the susceptibility spectrum shifts to low frequencies as t_w is increased.

This simple trap model can be extended in several directions. First, one can consider a hierarchical organisation of "traps within traps within traps...", to account for the hierarchy of time scales (or energy scales) corresponding to different length scales. This leads to a much richer model, with the possibility of different "glass" transition temperatures, corresponding to the freezing of the motion on smaller and smaller length scales [2, 13, 14]. This model should be directly relevant to describe the dynamics of pinned domain walls (or other elastic defects), where one knows that the energy barriers grow as a function of lengthscales.

Another very interesting generalization proposed in the context of soft glassy materials is to endow the above simple trap model with mechanical properties [4]. This allows one to obtain stress-strain (or shear rate) relations, which exhibit many interesting similarities with experimental data. For example, for $2 > x > 1$, one finds that the elastic moduli vary as power-laws of the frequency; for $x < 1$, a finite yield stress does appear, characteristic of the glass (solid) phase [4].

3. RELATION WITH MODE-COUPLING DESCRIPTIONS

The diffusion of a particle in a quenched disordered potential can be studied in a completely different way, by analyzing the problem in large dimensions of space, where one can establish *exact*[1] equations [19] relating the two-time correlation function $C(t + t_w, t_w) = \langle \vec{r}(t + t_w) \cdot \vec{r}(t_w) \rangle$ and the two-time response to an external force $R(t + t_w, t_w)$. For temperatures higher than a certain T_c, one finds that C and R are actually *time translation invariant*, and furthermore that the fluctuation-dissipation theorem

$$R(t) = -\frac{1}{T} \Theta(t) \frac{\partial C(t)}{\partial t}$$

is obeyed. In this case, one can then eliminate $R(t)$ and find an equation for $C(t)$ which, interestingly, *is precisely the schematic Mode-Coupling equation* proposed to describe supercooled liquids, with a kernel which is related to the correlation of the random potential. Hence, the physical content of the (schematic) Mode Coupling Theory (MCT) is clear: it is a mean-field description of a single particle in a static random potential. It is very important to identify a well defined Hamiltonian which corresponds to the MCT equations: from this point of view, one should note that the same equations can be obtained starting with some mean-field models of *spin-glasses* [20, 21]. The important point is thus that MCT implicitly assumes the presence of some *quenched disorder* which should rather, in reality, be "self-induced" by the dynamics itself – but see next section.

Coming back to the equations relating C and R, one can now investigate the glass phase $T < T_c$ [22]. In this case, the correlation and response function

---

[1] Assuming the random potential is Gaussian.

cease to be functions of t only. More precisely, $C(t_w + t, t_w)$ can be written as the sum of an "equilibrium" contribution $C_{eq}(t)$ which only depends on t, and an "ageing" part which depends on the ratio[2] $u = \frac{t}{t_w}$, $\mathcal{C}(u)$. The same decomposition holds for the response function; however, the ageing parts of C and R are related by an "anomalous" fluctuation dissipation theorem, where T is replaced by an effective temperature $\frac{T}{X}$, with $X \leq 1$ [23, 24]. The violation of the fluctuation dissipation theorem in non-equilibrium situations, first suggested by mean field models, has now been observed numerically in many different systems (binary Lennard-Jones, spin-glasses in finite dimensions, etc.) [25]. Thus, again, the system exhibits ageing: for a finite waiting time t_w after the quench below T_c, one expects that the susceptibility $\chi(\omega, t_w)$ still exhibits two peaks: a high frequency β-peak very similar to the high temperature $(T > T_c)$ one, and a low frequency α-peak which reaches a maximum at a frequency $\omega_\alpha \simeq \frac{1}{t_w}$, and thus progressively disappears as $t_w \to \infty$. An interesting prediction of this low temperature extension of MCT is that the high frequency part of the "ageing" α-peak behaves as $(\omega t_w)^{-b}$, while the low frequency "foot" of the β-peak behaves as ω^a, with the following relation between a, b, and the "anomaly" X [19, 22]:

$$X \frac{\Gamma^2[1+b]}{\Gamma[1+2b]} = \frac{\Gamma^2[1-a]}{\Gamma[1-2a]}. \tag{4}$$

This equation generalizes the famous MCT relation [1] between a and b for $T > T_c$, for which $X \equiv 1$. Some recent numerical investigation of ageing in Lennard-Jones systems, in the spirit of MCT, can be found in [29].

Hence, the simple minded "trap" models lead to predictions which are, overall, very similar to those of the MCT. It is thus natural to wonder whether the above two descriptions are in fact equivalent. This is not so: the motion of a particle in infinite space dimensions (which leads to the schematic MCT equations) is peculiar because the particle actually never reaches the bottom of an energy well: there is always a direction along which it can escape and lower its energy [23, 26, 27][3]. Hence, ageing within the low temperature MCT is not related at all to barrier crossing and activated processes, but rather to a slow, endless descent in phase space. On the other hand, in any finite dimension, the particle reaches a local minimum of the potential in a finite time, and activation is then crucial to leave these minima. From a theoretical point of view, it would be extremely interesting to understand how finite dimensional effects and the residual motion of the surrounding particles (which makes the random potential time dependent) can be included in a systematic way within the Mode-Coupling framework.

---

[2] Actually the ratio could be $u = \frac{h(t_w + t)}{h(t_w)}$, with a more complicated function h: see references [2, 7, 19, 23] for a more precise discussion of this point.

[3] This does not mean, however, that the particle reaches the lowest energy states: it is localized in regions where the energy is still very high [23, 28].

4. SELF-INDUCED QUENCHED DISORDER AND OPEN QUESTIONS

An important difference between spin-glasses or other disordered systems and structural glasses or soft glassy materials is the absence of "quenched" disorder in the latter case. However, as mentionned above, the phenomenology of both types of systems is remarkably close. Furthermore, from a theoretical point of view, the description of these problems often takes very similar paths. For example, the most popular theory of glasses (MCT) is actually an exact mean-field theory for certain disordered systems. In a series of recent works, several authors [30–34] have shown the existence of spin-glass like behaviour in systems with frustration but *without quenched disorder*. These systems thus provide natural spin analogues of glass formers. Although their microscopic description does remain remote from that of structural glasses (in particular because they involve infinite range interactions), they provide at least an existence proof to the phenomenon of "self-induced disorder". This is quite important from a conceptual point of view.

From an experimental point of view, Charge Density Wave systems (see among others [35]) have recently been shown to behave very much like disordered systems [36], with however a very small density of impurities, suggesting that incommensurability effects alone (inducing some frustration) might be sufficient to generate "self-induced disorder" [37].

We shall concentrate here on a simple spin system with frustration but without disorder, which is the problem of "low autocorrelation binary sequences" (LABS). This is an important problem from communication theory [38], which bears some relationship with the problem of random number generators. It was restated in physical terms by Bernasconi [39] as follows: take a one dimensional chain of Ising spins $S_i = \pm 1$, $i = 1...N$. Compute the correlation function at distance k

$$C_k = \sum_{i,j=1}^{N} S_i S_j \delta_{j,i+k} \tag{5}$$

and define the Hamiltonian of the system as

$$E(\{S\}) = \frac{1}{2(N-1)} \sum_{k=1}^{N-1} C_k^2 \tag{6}$$

from which the partition function can be computed. Two versions of this problem have actually been studied, differing in the choice of boundary conditions. Due to the long-range nature of the interactions implied by the above form of the Hamiltonian, they present significative differences. A first version studied in references [30,31,40] has free boundary conditions, where the correlation function C_k is defined with the sum over the spin indices i going from 1 to $N - k$. Another choice, studied in [31] is that of periodic boundary conditions, where one can define C_k through a sum over the spin indices i going from 1 to

$N-1$. We shall restrict here to the case with free boundary conditions. It was shown by Monte Carlo simulations that there exists a finite temperature freezing region in a temperature range around $T \simeq 0.1$, with a weak cooling rate dependance of the final low temperature energy. Computations spanning very long time scales have been perfomed at low temperature by using an efficient Monte Carlo algorithm [40, 41], and reveal a clear ageing effect, characterized by a t/t_w scaling. In this case, the presence of "traps" in phase space with a broad distribution of trapping times [40] is rather convincingly observed for finite N.

The analytical study of the LABS model is in itself very interesting. Despite its simplicity, we know of no direct solution. A rather indirect way of proceeding is to replace the non-disordered LABS model by a "fiduciary" model, in the sense of being as faithful as possible to the original model, but now with quenched disorder. The basic idea consists in considering the model at hand as one special sample of an ensemble of systems containing quenched disorder. In the case with free boundary conditions this is achieved as follows [30, 31]: one defines a "disordered" correlation function

$$C_k^d = \sum_{ij} M^{(k)}{}_{ij} S_i S_j , \qquad (7)$$

where $M^{(k)}$ is a matrix with random elements, equal to 0 or 1, with the only constraint that

$$\sum_{ij} M_{ij}^{(k)} = N - k.$$

The original problem is a particular choice of $M^{(k)}$, where the only nonzero elements are on the kth diagonal. The hope is that this particular case is a generic case, and this is actually not at all obvious. (For example, it would be nonsense to claim that a ferromagnet is a special instance of a spin-glass with $J_{ij} = \pm J$ couplings, where all J_{ij} happen to be equal to $+J$: the ferromagnet is just a very atypical sample.) There is in fact quite a bit of educated guesswork involved in the choice of the ensemble of disordered system, of which the original model is argued to be a generic member – see below. In the present case, the original model is extremely frustrated due to the long-range and conflicting nature of the interactions, two features which are indeed retained by the Hamiltonian defined using equation (7).

Now, the crucial remark is that if the model is indeed generic, its static properties can be obtained by means of the replica method, where the averaging is performed over the fictitious disorder.

In the case at hand, the resulting free energy indeed turns out to be a good approximation of the original model in the high temperature, replica symmetric phase. This approximation actually corresponds to the one proposed by Golay [38] using different arguments; as seen from a high temperature expansion, this approximation is however not exact. Its main virtue is to predict the existence of a static phase transition at a temperature $T_s = 0.0476$, below which a breaking of replica symmetry is needed, as in spin-glasses. The low temperature phase is characterized by a residual entropy density which is linear in T, but small (less than 10^{-5} per spin at T_s). From a glass point of view this phase transition can be seen as the resolution of an entropy crisis appearing at an extrapolated Kauzman temperature which is very close to T_s. The prediction for the ground state energy density, $E_0/N \simeq 0.0202$, is compatible with a large N extrapolation of the ground state energies found by exact enumeration on small samples. On the other hand, it does substantially differ from the apparent ground state energy extracted from Monte Carlo simulations, even after extrapolating to very small cooling rate. One can actually also compute the dynamical temperature corresponding to the MCT T_c discussed above, where the system ceases to equilibrate, and finds $T_c = 0.103$, not far from where simulations suggest.

We have thus seen how some frustrated spin systems without disorder can be solved (approximately, or even exactly at least in the high T phase for the LABS with periodic boundary conditions [31]), following a rather interesting strategy. This strategy consists in substituting the original problem by a "fiduciary" one with quenched disorder, and solving the disordered system using, e.g., the replica method to obtain the static properties and information about the transitions. There is unfortunately no systematic method for choosing the fiduciary model so far. The examples studied so far show the importance of symmetry considerations in the choice of the fiduciary disordered problem, and suggest as a criterion that this disordered model should be as "close" as possible to the original one in the high temperature (liquid) phase. This strategy is reminiscent of the very fruitful approach to energy levels in complex nuclei through the study of fiduciary random Hamiltonians with the proper symmetries. In the present case one does not yet understand when such an approach is justified, if it is only restricted to finite time dynamics or if it does apply to thermodynamical properties. In some cases (see [21, 31, 33]), the "spin-glasses without disorder" explicitly involve pseudo random numbers in the sense that the spin couplings are deterministic, but very rapidly oscillating. Finally, as discussed in the previous subsection, self-consistent (Mode-Coupling) approximations of non disordered models often lead to equations which are exact for some adequately chosen disordered systems [22]. In a loose sense, one expects that the slow dynamics at low temperatures originate from some degrees of freedom which freeze and play the role of an effectively quenched disorder field for the other degrees of freedom.

Acknowledgements

The ideas presented here have been developed with many collaborators and friends, in particular L. Cugliandolo, D. Dean, J. Kurchan, M. Mézard, C. Monthus and E. Vincent. Discussions on related topics with F. Alberici, A. Baldassarri, K. Biljakovic, M. Cates, A. Compte, P. Doussineau, P. Hébraud, A. Levelut, F. Lequeux, E. Pitard and P. Sollich have been most enjoyable.

REFERENCES

[1] For reviews, see W. Götze, in *Liquids, freezing and glass transition*, Les Houches, edited by J.P. Hansen, D. Levesque and J. Zinn-Justin (North Holland, 1989); see also W. Götze and L. Sjögren, *Rep. Prog. Phys.* **55** (1992) 241.

[2] For a review, see J.P. Bouchaud, L. Cugliandolo, J. Kurchan and M. Mézard, *Out of Equilibrium dynamics in spin-glasses and other glassy systems*, in "Spin-glasses and Random Fields", edited by A.P. Young (World Scientific, 1998), and references therein.

[3] For a recent review, see T. Giamarchi, P. Le Doussal, in "Spin-glasses and Random Fields", edited by A.P. Young (World Scientific, 1998), and references therein.

[4] P. Sollich, F. Lequeux, P. Hebraud and M. Cates, *Phys. Rev. Lett.* **70** (1997) 2020; P. Sollich, *Phys. Rev. E* **58** (1998) 738.

[5] E.R. Nowak, J.B. Knight, E. Ben-Naim, H.M. Jaeger and S.R. Nagel, *Phys. Rev. E* **57** (1998) 1971.

[6] M. Nicodemi and A. Coniglio, *Phys. Rev. Lett.* **82** (1999) 916.

[7] E. Vincent, J. Hammann, M. Ocio, J.P. Bouchaud and L. Cugliandolo, cond-mat/9607224, *Proceedings of the 1996 Sitges Conference of Glassy Systems*, edited by I. Rubi (Springer, 1997).

[8] see *e.g.* F. Alberici, P. Doussineau and A. Levelut, *J. Phys. I France* **7** (1997) 329; F. Alberici, J.P. Bouchaud, L. Cugliandolo, J. Doussineau and A. Levelut, *Phys. Rev. Lett.* **81** (1998) 4987.

[9] see E. Vincent, J. Hammann and M. Ocio, in Recent Progress in Random Magnets, edited by D.H. Ryan (World Scientific, Singapore, 1992), and also: K. Jonasson, E. Vincent, J. Hammann, J.P, Bouchaud and P. Nordblad, *Phys. Rev. Lett.* **81** (1998) 3243.

[10] For reviews, see H. Leschhorn, T. Nattermann, S. Stepanow and L.H. Tang, *Annalen der Physik* **6** (1997) 1; D.S. Fisher, in "Fondamental Problems in Statistical Mechanics IX" (Springer, 1998).

[11] C. Monthus and J.P. Bouchaud, *J. Phys. A* **29** (1996) 3847.

[12] H. Yoshino, *J. Phys. A* **29** (1996) 1421; *Phys. Rev. Lett.* **81** (1998) 1483.

[13] L. Balents, J.-P. Bouchaud and M. Mézard, *J. Phys. I France* **6** (1996) 1007.

[14] J.P. Bouchaud and D.S. Dean, *J. Phys. I France* **5** (1995) 265.

[15] M. Mézard, G. Parisi and M.A. Virasoro, *Spin Glass Theory and Beyond* (World Scientific, Singapore, 1987).

[16] T. Odagaki, *Phys. Rev. Lett.* **75** (1995) 3701.

[17] J.P. Bouchaud and A. Georges, *Phys. Rep.* **195** (1990) 127.

[18] F. Bardou, J.P. Bouchaud, O. Emile, C. Cohen-Tannouji and A. Aspect, *Phys. Rev. Lett.* **72** (1994) 203; and A. Aspect, F. Bardou, J.P. Bouchaud and C. Cohen-Tannouji, in preparation.

[19] S. Franz and M. Mézard, *Europhys. Lett.* **26** (1994) 209; *Physica A* **209** (1994) 1; L. Cugliandolo and P. Le Doussal, *Phys. Rev. E* **53** (1996) 1525.

[20] T.R. Kirkpatrick and D. Thirumalai, *Phys. Rev. B* **36** (1987) 5388.
[21] S. Franz and J. Hertz, *Phys. Rev. Lett.* **74** (1995) 2114.
[22] J.P. Bouchaud, L. Cugliandolo, J. Kurchan and M. Mézard, *Physica A* **226** (1996) 243.
[23] L. Cugliandolo and J. Kurchan, *Phys. Rev. Lett.* **71** (1993) 173; *J. Phys. A* **27** (1994) 5749.
[24] L. Cugliandolo, J. Kurchan and L. Peliti, *Phys. Rev. E* **55** (1997) 3898.
[25] see *e.g.* E. Marinari, G. Parisi, F. Ricci-Tersenghi and J. Ruiz-Lorenzo, *J. Phys. A* **31** (1998) 2611.
[26] J. Kurchan and L. Laloux, *J. Phys. A* **29** (1996) 1929.
[27] A. Barrat and M. Mézard, *J. Phys. I France* **5** (1995) 941.
[28] A. Barrat, R. Burioni and M. Mézard, *J. Phys. A* **29** (1996) 1311.
[29] W. Kob and J.L. Barrat, "Fluctuations, response and ageing dynamics in simple glass forming liquids out of equilibrium", cond-mat/9905248.
[30] J.P. Bouchaud and M. Mézard, *J. Phys. I France* **4** (1994) 1109.
[31] E. Marinari, G. Parisi and F. Ritort, *J. Phys. A* **27** (1994) 7615; E. Marinari, G. Parisi and F. Ritort, *J. Phys. A* **27** (1994) 7647.
[32] L.F. Cugliandolo, J. Kurchan, G. Parisi and F. Ritort, *Phys. Rev. Lett.* **74** (1995) 1012; L.F. Cugliandolo, J. Kurchan, R. Monasson and G. Parisi, *J. Phys. A* **29** (1996) 1347.
[33] P. Chandra, L.B. Ioffe and D. Sherrington, *Phys. Rev. Lett.* **75** (1995) 713; P. Chandra, M.V. Feigel'man and L.B. Ioffe, *Phys. Rev. Lett.* **76** (1996) 4805; P. Chandra, M.V. Feigel'man, M.E. Gershenson and L.B. Ioffe, in *Complex Behaviour of Glassy Systems*, edited by M. Rubi and C. Perez-Vicente (Springer-Verlag, 1997); P. Chandra, M.V. Feigel'man, L.B. Ioffe and I. Kagan, *Phys. Rev. B* **58** (1997) 11553.
[34] See also the interesting work by S. Obuhkov, D. Kobzev, D. Perchak, M. Rubinstein, and C. Reuner, H. Löwen and J.L. Barrat, On rotating hard rods (preprints, 1995).
[35] G. Aeppli and P. Chandra, *Science* (1997).
[36] K. Biljakovic, D. Staresinic, K. Hosseini, W. Brütting, H. Berger and F. Lévy, *Physica B* **244** (1998) 167.
[37] G.P. Tsironis and S. Aubry, *Slow relaxation induced by breathers in nonlinear lattices*, Saclay-preprint (1996).
[38] M.J.E. Golay, *IEEE IT* **23** (1977) 43; *IEEE IT* **28** (1982) 543.
[39] J. Bernasconi, *J. Phys. France* **48** (1987) 559.
[40] W. Krauth and M. Mézard, *Z. Phys. B* **97** (1995) 127; W. Krauth and O. Pluchery, *J. Phys. A* **27** (1994) L715.
[41] A.B. Bortz, M.H. Kalos and J.L. Lebowitz, *J. Comput. Phys.* **17** (1975) 10.

A Short Introduction to Ergodic Theory and Its Applications

F.M. Dekking

Stieltjes Institute for Mathematics, Delft University of Technology, Faculty ITS, Mekelweg 4, 2628 CD Delft, The Netherlands

---

1. DYNAMICAL SYSTEMS

In general, a dynamical system is a space X with a family of maps from X to itself. Often there is a notion of time. We speak of a *continuous time dynamical system* if the family of maps has the form $(T_t)_{t \in \mathbb{R}}$ where $T_t : X \to X$, and the T_t satisfy $T_{s+t} = T_s \circ T_t$ for all real s and t. In these notes we shall only consider *discrete time dynamical systems* which are determined by a single map $T : X \to X$ and its iterates T^n, defined by $T^n = T \circ T \circ \cdots \circ T$, n times. In ergodic theory the space X, or rather a sigma algebra[1] of subsets of X, is moreover equipped with a measure μ which is *preserved* by T:

$$\mu(T^{-1}(B)) = \mu(B)$$

for all subsets B of X in the sigma algebra. Note that we should require that T is measurable, *i.e.* that T has the property that $T^{-1}(B)$ is also in this sigma algebra for all such B. In the sequel we shall only consider measures μ such that $\mu(X) = 1$, so that we can make probabilistic statements on the behaviour of the dynamical system $(X, T, \mathcal{B}, \mu)$.

1.1. Examples

Example 1. Coin tossing and beyond

---

[1] This is a collection of subsets of X which contains X, and is closed for the operations of taking complements and countable unions.

Repeated independent tossing of a coin, which has probability p to turn up heads (coded by 1), and $1 - p$ to yield tails (coded by 0) can be described as a dynamical system by taking $X = \{0,1\}^{\mathbb{Z}}$ and $T : X \to X$ the shift map defined by

$$(Tx)_n = x_{n+1} \quad n \in \mathbb{Z}, \quad \text{if } x = \cdots x_{-1} x_0 x_1 \cdots$$

The probability measure μ should be defined on so-called *cylinder sets*

$$[C] = \{x \in X : x_1 = c_1, \ldots, x_m = c_m\} \quad \text{for} \quad C = c_1 c_2 \ldots c_m$$

where $c_i \in \{0, 1\}$ for $i = 1, \ldots, m$ by

$$\mu([C]) = p^{\sum_{i=1}^m c_i}(1 - p)^{m - \sum_{i=1}^m c_i}.$$

(So for example $\mu([1]) = p$, and $\mu([100]) = p(1 - p)^2$.)

Interestingly, *any* real valued stationary random process $(Y_n)_{n \in \mathbb{Z}}$ can be represented as a dynamical system in a similar way. Suppose the Y_n are defined on a probability space $(\Omega, \mathcal{F}, \mathbb{P})$. Stationarity of the process is defined by the requirement

$$\mathbb{P}(Y_{n+1} \leq y_1, \ldots, Y_{n+m} \leq y_m) = \mathbb{P}(Y_1 \leq y_1, \ldots, Y_m \leq y_m)$$

for all integers n, m. Now take $X = \mathbb{R}^{\mathbb{Z}}$ and let $T : X \to X$ be the shift map as above, and define $\varphi : \Omega \to X$ by

$$(\varphi(\omega))_n = Y_n(\omega) \qquad n \in \mathbb{Z},$$

and μ on the sigma algebra of Borel sets[2] of X by

$$\mu(B) = \mathbb{P}(\varphi^{-1}(B)).$$

Then the stationarity of (Y_n) implies[3] that μ is preserved by T:

$\mu(T^{-1}((-\infty, c_1] \times \cdots \times (-\infty, c_m])) =$
$\mathbb{P}\{\omega \in \Omega : \varphi(\omega) \in T^{-1}((-\infty, c_1] \times \cdots \times (-\infty, c_m]\} =$
$\mathbb{P}\{\omega \in \Omega : T(\varphi(\omega)) \in (-\infty, c_1] \times \cdots \times (-\infty, c_m])\} =$
$\mathbb{P}\{\omega \in \Omega : Y_2(\omega) \in (-\infty, c_1], \ldots, Y_{m+1}(\omega) \in (-\infty, c_m])\} =$
$\mathbb{P}\{\omega \in \Omega : Y_1(\omega) \in (-\infty, c_1], \ldots, Y_m(\omega) \in (-\infty, c_m])\} =$
$\mathbb{P}\{\omega \in \Omega : \varphi(\omega) \in (-\infty, c_1], \ldots, (\varphi(\omega))_m \in (-\infty, c_m]\} =$
$\mu((-\infty, c_1] \times \cdots \times (-\infty, c_m]).$

Note that in particular, if we had already $\Omega = \mathbb{R}^{\mathbb{Z}} = X$, then $Y_n = Y_0 \circ T^n$, *i.e.*, the stationary random process is determined by a dynamical system plus a function (the function $Y_0 : X \to \mathbb{R}$).

---

[2] This is the smallest sigma algebra that contains all "cylinder sets" of the form $\{x \in X : x_1 \leq c_1, \ldots, x_m \leq c_m\}, c_i \in \mathbb{R}$ for $i = 1, \ldots, m$.
[3] Standard results in measure theory tell us that it is enough to check the invariance for all "cylinder sets".

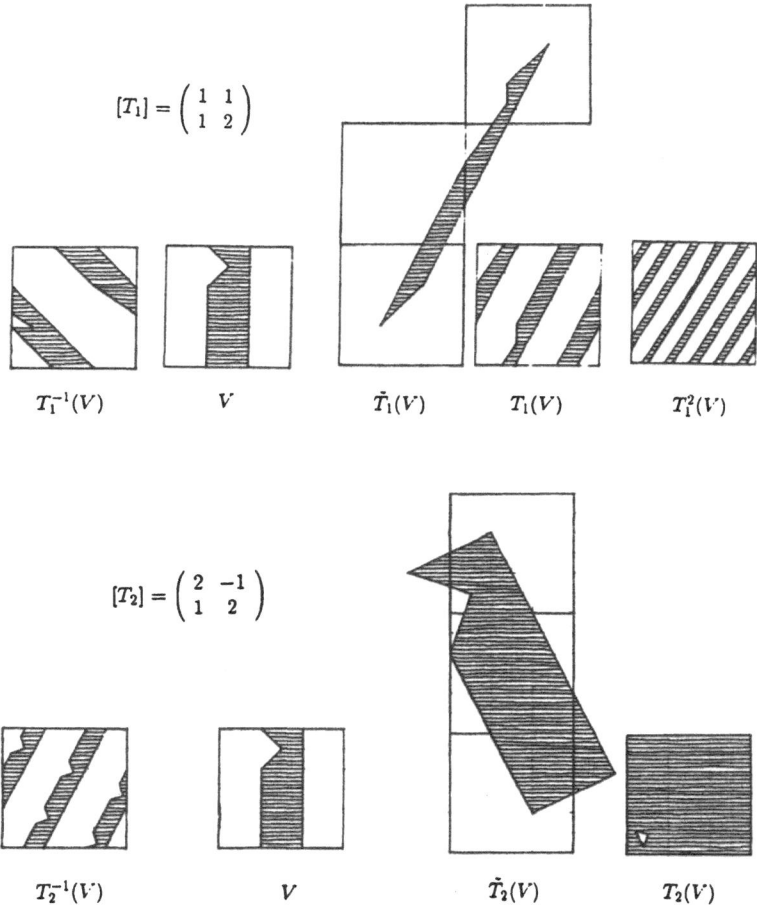

Fig. 1. — The iterated action of two maps on the torus (represented as a square with opposite edges identified); $\tilde{T}_1$ and $\tilde{T}_2$ denote the corresponding maps in the plane.

Example 2. Linear map on the torus

Here the space is $X = \mathbb{T}^d$ the d-dimensional torus of points x in $\mathbb{R}^d$ with $0 \leq x_i \leq 1$, and $T(x) = [T]x$, where all additions are modulo 1 and $[T]$ is a matrix $[T] = (t_{ij})_{i,j=1}^d$ with integer coefficients and with $\det([T]) \neq 0$ (this implies that T is surjective). The measure μ is d-dimensional Lesbesgue measure restricted to the torus. Since μ is the *unique* translation invariant probability measure on X, it follows that μ is preserved by T: define ν by

$$\nu(B) = \mu(T^{-1}(B)),$$

and for each $x \in X$ let y be such that $Ty = x$, then

$$\nu(B+x) = \nu(B+Ty) = \mu(T^{-1}(B+Ty)) = \mu(T^{-1}B+y) = \mu(T^{-1}B) = \nu(B)$$

so ν is translation invariant. See Figure 1 for the action of two examples (one invertible, one not invertible) on the two-dimensional torus.

Example 3. Rotation on the circle

Here $X = \{x \in \mathbb{C} : |x| = 1\}$, $\alpha \in [0, 2\pi]$ is fixed (the rotation angle), and T is defined by $T(x) = e^{i\alpha}x$. The measure is (normalized) arc length – measure on the circle. The behaviour of this dynamical system is rather different when α is rational or irrational, as we shall see later.

Example 4. Substitution dynamical systems

Consider an alphabet $A = \{0, 1, \ldots, r-1\}$ of r letters and the two sided shift space $A^{\mathbb{Z}}$ with the shift map T. Let u be one fixed sequence which is invariant for a substitution rule (also called inflation rule) σ, or for some power of σ.

Example. Fibonacci substitution.
Here $A = \{0, 1\}$ and σ is given by

$$0 \xrightarrow{\sigma} 01, \ 1 \xrightarrow{\sigma} 0.$$

Hence $\sigma^2(0) = 010$ and $\sigma^2(1) = 01$. There are two sequences invariant under σ^2:

$$u = \ldots 0\,1\,0\,0\,1\,\dot{0}\,1\,0\,0\,1\,0\,1\,0 \ldots$$

(the dot indicates the position of the zero'th coordinate u_0), and

$$u' = \ldots 0\,1\,0\,0\,0\,1\,0\,1\,0\,\dot{0}\,1\,0\,0\,0\,1\,0 \ldots$$

We then form the *orbit* of u consisting of all shifts of u:

$$\text{Orb}\,(u) := \{T^n(u) : n \in \mathbb{Z}\},$$

and define

$$X = \text{Closure}\,(\text{Orb}(u)).$$

The space X consists of the sequences of which any finite subsequence (also called word) will occur also as a subsequence of an element of the orbit of u.

Fact (see *e.g.* Queffélec [17]). There is a unique probability measure μ which is preserved by T on X (*i.e.*, $\mu(A^{\mathbb{Z}} \setminus X) = 0$) if we require that each letter $i \in A$ occurs in each word $\sigma^n(j), j \in A$ for some $n \geq 1$. The measure of a cylinder $[w]$ equals its limiting frequency in the σ-invariant sequence u:

$$\mu([w]) = \lim_{N \to \infty} \frac{\sharp \text{ occurrences of } w \text{ in } u_1 \ldots u_N}{N}.$$

Question: What about context-*sensitive* inflation rules?

A "simple" example: Kolakoski inflation. Here $A = \{1, 2\}$ and the rule σ is given by

$$22 \xrightarrow{\sigma} 2211, \quad 21 \xrightarrow{\sigma} 221, \quad 12 \xrightarrow{\sigma} 211, \quad 11 \xrightarrow{\sigma} 21.$$

For example if:

$$w = 2\,2\,.\,1\,1\,.\,2\,1\,.\,2\,2\,.\,1\,2$$

then $\sigma(w) = 2\,2\,.\,1\,1\,.\,2\,1\,.\,2\,2\,.\,1\,2\,.\,2\,1\,.\,1\,2\,.\,1\,1$.

There is a unique two-sided infinite sequence fixed by σ:

$$y = \ldots 1\,2\,1\,1\,2\,2\,\overset{\centerdot}{1}\,2\,2\,1\,1\,2\,1\,2\,2\,1\,2\,2 \ldots$$

Open problem: is there a probability measure μ preserved by the shift on Closure $(\mathrm{Orb}(y))$? (It is known what it should look like (see [3]).) Even the simplest recurrence questions on the sequence y are unanswered as *e.g.* "Does the beginning word 2 2 1 1 2 1 2 2 1 2 2 ever occur again?"

1.2.   Recurrence

In general there is a lot of recurrence in a dynamical system.

Poincaré Recurrence Theorem (1899)
Let $(X, T, \mathcal{B}, \mu)$ be any dynamical system. Then for each $B \in \mathcal{B}$ almost all points x of B return to B.

Here "x returns to B" means that there exists a $k \geq 1$ such that $T^k(x) \in B$, and "almost all" means that the set of points which fail to have this property has μ-measure 0.

Example. See Figure 2. As pointed out by Paul and Tatiana Ehrenfest [5] one can use Poincaré's Recurrence Theorem to justify the eventual occurrence of paradoxical events in simple models of the distribution of gas molecules. Suppose we have N molecules distributed over two interconnected volumes L and R, so that at time instants $1, 2, \ldots$ exactly one of the molecules moves from L to R with probability $\frac{i}{N}$ if there are i molecules in volume L, or from R to L with probability $1 - \frac{i}{N}$. Then almost surely there will be a time that *all* N molecules are in volume L. To see that this is implied by Poincaré's theorem, let X be the two-sided shift space over the alphabet $\{0, 1, \ldots, N\}$ (the alphabet codes the states of the system: i will mean that there are i molecules in volume R), $\mathcal{B}$ the Borel sets, T the shift (as in Ex. 1 in the previous section). We then define a measure μ on $\mathcal{B}$ *via* the transition probabilities $p_{01} = 1$, $p_{N,N-1} = 1$, and

$$p_{i,i+1} = \frac{i}{N}, \; p_{i,i-1} = 1 - \frac{i}{N},$$

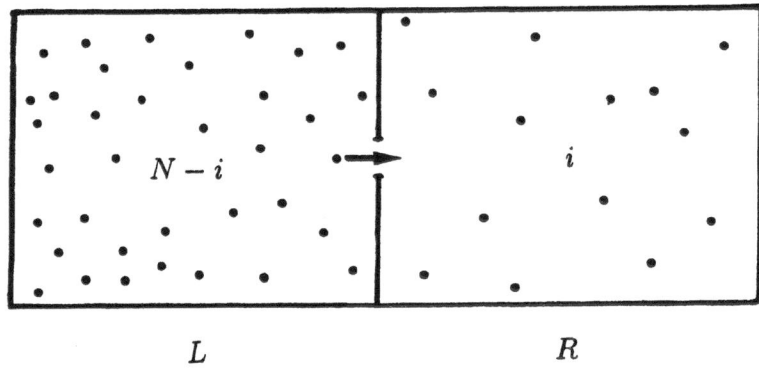

Fig. 2. — N molecules distributed over two volumes L and R.

for all $1 \leq i \leq n - 1$, and the initial distribution $\mu([i]) = 2^{-N} N!/(i!(N - i)!)$. The measure μ will be a Markov measure, which means that μ is now defined for arbitrary cylinders $[c_1, c_2, \cdots, c_m]$ by

$$\mu([c_1, c_2, \cdots, c_m]) = \mu([c_1]) p_{c_1, c_2} \cdots p_{c_{m-1}, c_m}.$$

We then apply Poincaré's Theorem to the set $B = \{x \in X : x_0 = 0\} = [0]$, which has $\mu(B) = 2^{-N} > 0$: the conclusion will be that if we start with all molecules in the left box, then with probability 1 at some moment all molecules will again be in the left box.

1.3. Ergodic theorem

It is not hard to deduce from Poincaré's theorem that almost all points of B return **infinitely often** to B. The question is: how often? *I.e.*, can we say something about the average number $\frac{1}{n}(1_B(x) + 1_B(Tx) + \cdots + 1_B(T^n x))$ of visits to B on the long term.

Birkhoff Ergodic Theorem (1931)
Let $(X, T, \mathcal{B}, \mu)$ be a dynamical system, and let f be a function integrable with respect to μ. Then there exists an integrable function $\overline{f}$ such that

$$\frac{1}{n} \sum_{k=0}^{n} f(T^k x) \to \overline{f}(x) \qquad \text{as} \quad n \to \infty$$

for almost all[4] $x \in X$. Moreover, $\overline{f}(x) = \overline{f}(Tx)$ for almost all $x \in X$, and

---

[4] *I.e.*, for all x in some set X_0 with $\mu(X_0) = 1$.

$\int_X \overline{f} d\mu = \int_X f d\mu$.

Proofs of this very important theorem have been given by Birkhoff [2], Wiener [22], Kakutani [23], Garsia [7], Kamae [10], Katznelson and Weiss [13], Kamae and Keane [11], Shields [19], ...

The most interesting case is where the limit function $\overline{f}$ is almost everywhere constant $= \int f d\mu$, $i.e.$, the case where

$$\text{``time-average''} \quad = \quad \text{``space average''}.$$

This occurs if and only if the transformation T is ergodic. A dynamical system $(X, T, \mathcal{B}, \mu)$, or simply the transformation T, is called $ergodic$ if

$$B \in \mathcal{B}, \ T^{-1}(B) = B \ \Rightarrow \ \mu(B) = 0 \text{ or } 1.$$

Reformulated in terms of integrable functions ergodicity is equivalent to

$$Tf = f \text{ almost surely} \Rightarrow f \text{ is constant a.s.}$$

Exercise. A transformation T is ergodic if and only if for all A and B with $\mu(A) > 0$ and $\mu(B) > 0$ there exists an n such that $\mu(T^{-n}A \cap B) > 0$.

We shall look at the ergodicity of our four examples.

Example 1. Coin tossing
Let $A = [v]$ and $B = [w]$ be two cylinder sets. Then, because of the product structure, $\mu(T^{-n}A \cap B) = \mu(A)\mu(B)$ as soon as n is larger than the length of v plus the length of w. Hence T is ergodic (using the exercise above).

Example 2. Linear map on the torus
Here it is known that T is ergodic if and only if the associated matrix $[T]$ has no roots of unity as eigenvalues.

Example 3. Rotation on the circle
Here it is an old result that T is ergodic if and only if the rotation angle α is irrational.

Example 4. Substitution dynamical systems
Let σ be a substitution on an alphabet $\{0, 1, \ldots, r-1\}$. Its matrix M is the $r \times r$ matrix with entries m_{ij} defined by

$$m_{ij} = \sharp \text{ occurrences of } i \text{ in } \sigma(j).$$

It is known [17] that the shift T on the closed orbit of a fixed point n of σ is ergodic if and only if there exists an n such that all entries of M^n are strictly positive – such a matrix is called $primitive$.

1.4. Unique ergodicity

Even if T is ergodic, there will be many x in X such that the time averages $\frac{1}{n}(f(x) + f(Tx) + \cdots + f(T^{n-1}x))$ do *not* converge.

Theorem. Equivalent are the following properties

(1.) $\frac{1}{n}(f(x) + f(Tx) + \cdots + f(T^n x))$ converges for *all* $x \in X$, for every continuous function $f : X \to \mathbb{R}$

(2.) There is only one μ preserved by T.

If (2.) holds, then T is automatically ergodic. To see this, let I be a measurable set such that $T^{-1}(I) = I$ almost surely, and $\mu(I) > 0$. Then we can define a measure ν by

$$\nu(B) = \frac{\mu(B \cap I)}{\mu(I)}.$$

Obviously this measure is preserved by T, so $\mu = \nu$. But then $\mu(I) = \nu(I) = 1$. Because of this property, T is called *uniquely ergodic* if property (2.) holds. The following describes the situation in the examples.

Example 1. The shift map T on $X = \{0,1\}^{\mathbb{Z}}$ is far from being uniquely ergodic: *e.g.* all the coin tossing measures with $0 \le p \le 1$ are invariant measures.

Example 2. The same holds for linear maps of the torus: a simple example of another invariant measure is the point mass in 0.

Example 3. Rotations on the circle are uniquely ergodic if and only if α is irrational [21].

Example 4. Substitution dynamical systems are uniquely ergodic as soon as the matrix M is primitive.

In Example 4 the dynamical system $X = $ closure $(\mathrm{Orb}(u))$ was obtained from a *single* sequence u (invariant for the substitution). An interesting question is whether we can "see" the unique ergodicity directly from the sequence u. The (general) answer is that this is equivalent to the existence of relative frequencies of subwords, *uniformly* in the position where one starts counting.

Theorem (Oxtoby [15])
T on Closure$(\mathrm{Orb}(u))$ is uniquely ergodic $\Longleftrightarrow$

For all $m \ge 1$, and all $c_1, \ldots, c_m \in \{0, \ldots, r-1\}$, and all $C := c_1 \ldots c_m$
$\lim_{n \to \infty} \frac{1}{n} \sum_{k=0}^{n} 1_{[C]}(T^{k+j}u)$ exists *uniformly* in $j \in \mathbb{Z}$.

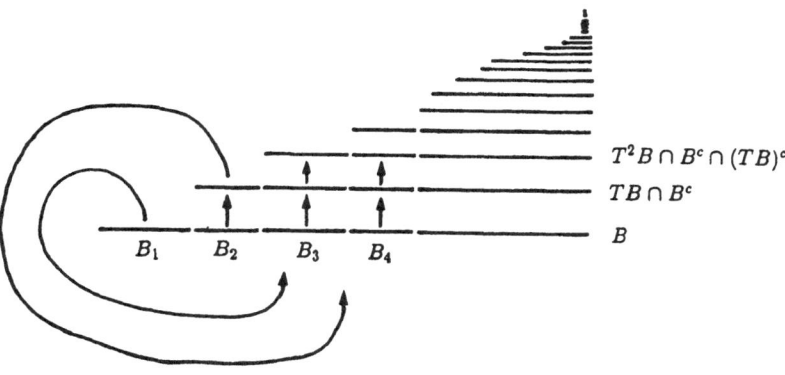

$T^2 B \cap B^c \cap (TB)^c$

$TB \cap B^c$

B

$B_1 \quad B_2 \quad B_3 \quad B_4$

Fig. 3. — The space X partioned by the B_n and their iterates.

1.5.   Expected recurrence time

We return to Poincaré's theorem, and the examples of the Ehrenfest's showing the almost sure return to a configuration where *all* molecules are in one container. One might think that the expected time till the occurrence of this strange event is infinite. There is however a result by Kac [9] that it is finite. To be specific, let us define the function n_B on X by

$$n_B(x) = n \iff x \in B, Tx \notin B, \ldots, T^{n-1}x \notin B, T^n x \in B.$$

Then it makes sense to call

$$e_B := \frac{1}{\mu(B)} \int_B n_B(x) \mathrm{d}\mu(x)$$

the expected recurrence time.

Theorem (Kac [9]). Let T be invertible and ergodic, then

$$e_B = \frac{1}{\mu(B)}$$

for all measurable B with $\mu(B) > 0$.

We give a proof by picture (Fig. 3), where intervals represent sets with a length proportional to the measure of the set.

Define

$$B_n := \{x \in B : n_B(x) = n\}.$$

We see that (because $\mu(TB_n) = \mu(T^{-1}TB_n) = \mu(B_n)$ etc.)

$$\int_B n_B(x)\mathrm{d}\mu(x) = \sum_{n=1}^{\infty} n\mu(B_n),$$

but the latter sum equals $\mu(\cup_{k=0}^{\infty} T^k B)$, which must be 1 by ergodicity of T.

2. SPECTRAL PROPERTIES OF DYNAMICAL SYSTEMS

In the sequel we consider only invertible dynamical systems (*i.e.*, there is a measurable map $T^{-1} : X \to X$ such that $T^{-1}Tx = x$ for almost all $x \in X$).

One can associate to the dynamical system $(X, T, \mathcal{B}, \mu)$ a unitary operator, which for convenience we also denote by T, on the space of square integrable functions $L^2(\mu)$ by

$$Tf(x) = f(Tx) \quad \text{for all} \quad x \in X.$$

See also the contribution of Queffélec in [1] for a more complete treatment of this subject.

2.1. The spectrum of a dynamical system

The spectrum of the operator T is called the *spectrum* of the dynamical system. In particular: a complex number λ such that

$$Tf = \lambda f$$

for some $f \not\equiv 0$ is called an *eigenvalue* of T.

Note that 1 is *always* an eigenvalue of T – its eigenspace contains the constant functions. We say that T has *continuous spectrum* if 1 is the only eigenvalue of T, with only the constant functions as eigenfunctions. On the other hand: T has *discrete* spectrum if its eigenfunctions span $L^2(\mu)$.

We discuss the nature of the spectrum in our four examples.

Example 1. Coin tossing
Coin tossing has continuous spectrum – we shall show this after we have established the connection between spectrum and mixing.

Example 2. Linear maps on the torus
These have continuous spectrum as soon as they are ergodic [21].

Example 3. Rotation on the circle
These have discrete spectrum if α is irrational. The eigenvalues are given by $\{\exp(2\pi i \alpha n) : n \in \mathbb{Z}\}$ ([21]).

Example 4. Substitution dynamical systems
Here the behaviour is quite varied. It is completely understood when the substitution σ has *constant length* l, *i.e.*, all $\sigma(i)$ have length l. Then at least the

numbers $\exp(2\pi ikl^{-n})$ with $k = 0, 1, \ldots$ and $n = 0, 1, \ldots$ will be eigenvalues. There may be (finitely many) others, and there may be a continuous part.

When σ has non-constant length (for example the Fibonacci substitution), then next to the possibilities above, there may be continuous spectrum (example: σ on two symbols given by $\sigma(0) = 001$ and $\sigma(1) = 11100$, see $e.g.$ [17]). There may also be "irrational" eigenvalues $\exp(2\pi i \tau n)$, n an integer. In the Fibonacci case there is discrete spectrum. The conjecture is that if the matrix M of the substitution has an irreducible characteristic polynomial, and its largest eigenvalue is a Pisot number then the system has discrete spectrum (see [20]).

Sequences generated by substitutions provide simple models for quasicrystals. It has been known for some time that the x-ray diffraction spectra of the quasicrystals have a close connection to the spectra of the associated dynamical systems. See Dworkin [4] and Hof [8] for an approach based on distributions. See Robinson [18] for an approach based on calculating the Fourier transform of the quasicrystal pattern in the interval $[-N, N]$ and letting $N \to \infty$. In this set up one may loose part of the spectrum, even if the eigenfunctions are continuous.

2.2. Mixing

Absence of spectrum, $i.e.$, the lack of eigenvalues indicates long range independence in the dynamical system. A weak form of long range independence already exists as soon as the transformation is ergodic. It is not hard to show (using Birkhoff's theorem) that

T is ergodic $\iff \frac{1}{n} \sum_{k=0}^{n-1} \mu(T^{-k} A \cap B) \to \mu(A)\mu(B)$ as $n \to \infty$

for all measurable A and B.

Now we define:

T is $weakly\ mixing$ if $\frac{1}{n} \sum_{k=0}^{n-1} |\mu(T^{-k} A \cap B) - \mu(A)\mu(B)| \to 0$ as $n \to \infty$

for all measurable A and B,

T is $strongly\ mixing$ if $\mu(T^{-k} A \cap B) \to \mu(A)\mu(B)$ as $n \to \infty$

for all measurable A and B.

Obviously strongly mixing T are weakly mixing, and transformations which are weakly mixing but not strongly are not easy to find. However, using the next result connecting spectrum and mixing, it follows that such an example is given by the substitution dynamical system mentioned in the previous subsection.

Theorem. Let $(X, T, \mathcal{B}, \mu)$ be an invertible dynamical system. Then T has continuous spectrum if and only if T is weakly mixing [21].

Example 1. Coin tossing

We already noted that $\mu(T^{-n} A \cap B) = \mu(A)\mu(B)$ for cylinders sets A and B as soon as n is large enough. Hence T is strongly mixing, and also weakly mixing.

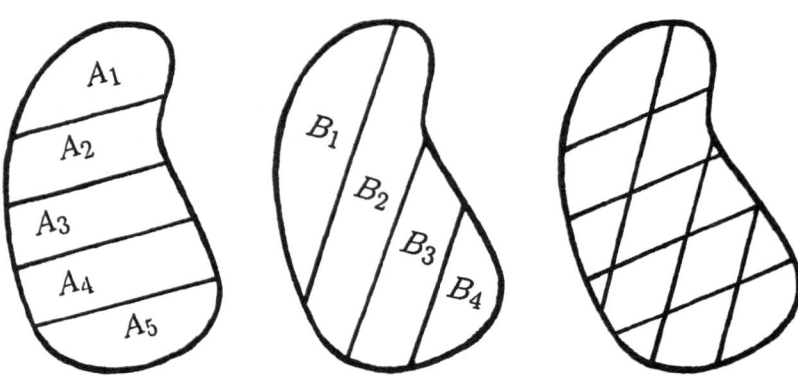

Fig. 4. — Two partitions ξ and η and their refinement $\xi \vee \eta$.

Example 2. Linear map on the torus
Here T is strongly mixing if and only if T is ergodic [21].

Example 3. Rotation on the circle
Here T has no mixing properties: it has been previously mentioned that T has discrete spectrum.

Example 4. Substitution dynamical systems
These can never be strongly mixing, but some are weakly mixing [17].

3.   ENTROPY OF DYNAMICAL SYSTEMS

Let $(X, T, \mathcal{B}, \mu)$ be a dynamical system. We let

$$\xi = \{A_1, A_r, \ldots, A_r\}$$

denote a *partition* of X with r elements, *i.e.*, the A_i are disjoint measurable sets with union X.

The *entropy of the partition* ξ is the real number   (here $0 \log 0 := 0$)

$$H(\xi) = -\sum_{i=1}^{r} \mu(A_i) \log \mu(A_i).$$

The intuitive interpretation of $H(\xi)$ is that it represents the information gained by performing the measurement which will tell us which of the A_i contains x. We have $0 \le H(\xi) \le \log r$, with $H(\xi) = \log r$ if and only if all elements A_i have the same probability $\frac{1}{r}$ (and so we obtain maximal information in this case).

Now we do another measurement on our system corresponding to a partition $\eta = \{B_1, B_2, \ldots, B_q\}$. The combined measurement corresponds to the partition

$$\xi \vee \eta = \{A_i \cap B_j : 1 \le i \le r, 1 \le j \le q\},$$

which is called the *refinement* of ξ and η.

To look where we are at the next time instant we consider the partition

$$T^{-1}\xi := \{T^{-1}A_1, T^{-1}A_2, \ldots, T^{-1}A_r\}.$$

Two successive measurements then correspond to the partition $\xi \vee T^{-1}\xi$. More generally measuring at n successive time instants is associated to

$$\xi \vee T^{-1}\xi \vee \cdots \vee T^{-n+1}\xi.$$

The *entropy of T with respect to the partition ξ* is the number

$$h(T,\xi) := \lim_{n\to\infty} \frac{1}{n} H(\xi \vee T^{-1}\xi \vee \cdots \vee T^{-n+1}\xi)$$

which corresponds to the average information gained per time unit in the long run. Here we remark that the limit always exists because the sequence (a_n) defined by $a_n = H(\xi \vee T^{-1}\xi \vee \cdots \vee T^{-n+1}\xi)$ is *subadditive*, i.e., $a_{n+m} \le a_n + a_m$ for all $n, m \ge 1$.

Now $h(T,\xi)$ is not really a property of the system, as we might do a very inefficient measurement ξ. We therefore define the *entropy* of the system, also called the entropy of T, by

$$h(T) = \sup\{h(T,\xi) : \xi \text{ a finite partition of } X\}.$$

The number $h(T)$ represents the amount of randomness in $(X, T, \mathcal{B}, \mu)$. How is $h(T)$ calculated? Here one needs Kolmogorov-Sinai's theorem which states that if we find a partition ξ which is sufficiently rich, i.e., the doubly infinite refinement $\ldots T^{-1}\xi \vee \xi \vee T\xi \vee \ldots$ equals the sigma algebra $\mathcal{B}$ modulo null sets, then $h(T) = h(T,\xi)$. Such a partition ξ is called a *generator*.

Example 1. Coin tossing
Here $\xi = \{[0], [1]\}$ is a generator. Hence

$$
\begin{aligned}
h(T) &= h(T,\xi) = \lim_{n\to\infty} \frac{1}{n} H(\xi \vee T^{-1}\xi \vee \cdots \vee T^{-n+1}\xi) = \\
&= \lim_{n\to\infty} \frac{1}{n} (H(\xi) + H(T^{-1}\xi) + \cdots + H(T^{-n+1}\xi)) = \\
&= \lim_{n\to\infty} \frac{1}{n} (nH(\xi)) = H(\xi) = -p\log p - (1-p)\log(1-p).
\end{aligned}
$$

In this computation we used the property that entropies add for independent partitions (exercise).

Example 2. Linear map of the torus
Here it is known that

$$h(T) = \sum_{|\lambda_i|>1} \log|\lambda_i|$$

where $\lambda_i, i = 1, \ldots, d$ are the eigenvalues of the matrix $[T]$ representing the linear map T. In the two examples of Figure 1 we have

$$[T_1] = \begin{pmatrix} 1 & 1 \\ 1 & 2 \end{pmatrix} \Rightarrow h(T_1) = \log(\tfrac{3+\sqrt{5}}{2}) = 0.96\ldots ,$$

$$[T_2] = \begin{pmatrix} 2 & -1 \\ 1 & 2 \end{pmatrix} \Rightarrow h(T_2) = \log(5) = 1.60\ldots$$

Example 3. Rotations of the circle
For all α one has $h(T) = 0$.

Example 4. Substitution dynamical systems
Any such system has $h(T) = 0$.

Example 5. 1-D random tilings
Suppose we have two tiles, coded 0 and 1 which can be juxtapositioned arbitrarily, except that no two type 1 tiles can occur next to each other. These tiling configurations can be represented by a subset of sequences in $\{0,1\}^{\mathbb{Z}}$, where typical sequences look like

$$\ldots 0001000101000100001001\ldots$$

It can be shown that there is a unique shift invariant probability measure μ on X which gives the shift maximal entropy $h(T)$. For the example above one has $h(T) = \log(\tau)$, where $\tau = \tfrac{1}{2}(1 + \sqrt{5})$. In ergodic theory this type of dynamical system is called a *subshift of finite type*.

3.1. Isomorphism

From the viewpoint of ergodic theory entropy $h(T)$ is mainly useful because it is an *isomorphism invariant*, i.e., if two dynamical systems $(X_1, T_1, \mathcal{B}_1, \mu_1)$ and $(X_2, T_2, \mathcal{B}_2, \mu_2)$ are isomorphic then $h(T_1) = h(T_2)$.

Here *isomorphic* means that there exists an invertible measurable map
$\varphi : X_1 \to X_2$, almost everywhere defined, such that
(i) φ preserves the measure (i.e., $\mu_2(B) = \mu_1(\varphi^{-1}B)$ for all B in $\mathcal{B}_2$),
(ii) φ preserves the dynamics (i.e., $\varphi T_1(x) = T_2\varphi(x)$ for almost all $x \in X_1$).

We conclude *e.g.* that coin tossing (for any p) can not be isomorphic to die throwing with a fair die, since $h(T_{\mathrm{die}}) = \log 6$.

Interestingly the ergodic linear maps on the torus (Ex. 2) are isomorphic to Bernoulli shifts, which are a generalisation of coin tossing (Ex. 1), i.e., product measures on a shift space $\{0, 1, \ldots, r-1\}^{\mathbb{Z}}$ [12].

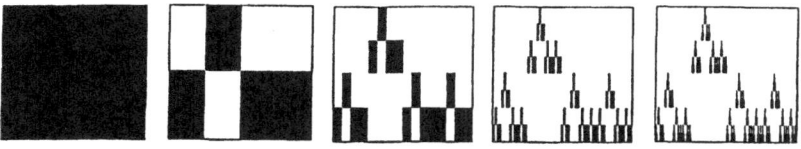

Fig. 5. — Approximants to a self affine subset of the torus.

3.2. Entropy and Hausdorff dimension

There are close relationsips between entropy and Hausdorff dimension. It is no coincidence that fair coin tossing has entropy $h(T) = \log 2$, that the scaling factor in the Cantor set C is $\frac{1}{3}$, and that the Hausdorff dimension of C equals $\log 2/\log 3$.

We do not elaborate on this (see also Furstenberg [6]), but give a more involved example. Define the *dimension* $\dim(\mu)$ *of a measure* μ as the minimal Hausdorff dimension of a Borel set B carrying μ. Let T be a linear map of the two dimensional torus (as in EXAMPLE 2) associated to the matrix

$$[T] = \begin{pmatrix} t & 0 \\ 0 & s \end{pmatrix},$$

where $t \geq s \geq 2$ are integers, and let μ be any invariant measure for T. Then the following beautiful formula relates the dimension and entropy.

Ledrappier-Young formula

$$\dim(\mu) = \frac{1}{\log t} h_\mu(T) + \left(\frac{1}{\log s} - \frac{1}{\log t} \right) h_{\pi\mu}(T_s),$$

where $h_\mu(T)$ is the entropy of the dynamical system $(\mathbb{T}^2, T, \mathcal{B}(\mathbb{T}^2), \mu)$, the map $T_s : \mathbb{T} \to \mathbb{T}$ is defined by $T_s(x) = sx \pmod 1$, and $h_{\pi\mu}(T_s)$ is the entropy of this map with respect to the measure μ projected on the second coordinate.

See the paper [14] for a short proof of this formula. The same paper extends to the d-dimensional torus results from Bedford and McMullen for the 2-torus on the Hausdorff dimension and box dimension of certain recursively defined self affine subsets. As an example, consider Figure 5 which diplays the first 5 steps in a construction where rectangles at level n are replaced by four scaled rectangles at level $n + 1$. The limit set is an self affine set, *i.e.*, a set invariant for the toral map with $[T] = \begin{pmatrix} 4 & 0 \\ 0 & 2 \end{pmatrix}$. The box dimension is $\frac{3}{2}$, but the Hausdorff-dimension is $\log(1 + \sqrt{3})/\log(2)$, which is strictly smaller.

4. EPILOGUE

These lectures are mainly based on the two excellent books by Walters [21], and Petersen [16]. Proofs, elaborations, and more examples concerning results and subjects just touched on in these notes may be found there.

REFERENCES

[1] Axel F. and Gratias D., Beyond Quasicrystals (Les Éditions de Physique and Springer, Berlin, 1995) pp. 1-619.

[2] Birkhoff G.D., Proof of the ergodic theorem, *Proc. Nat. Acad. Sci. USA* **17** (1931) 656-660.

[3] Dekking F.M., What is the long range order in the Kolakoski sequence? edited by R.V. Moody, The Mathematics of Long Range Aperiodic Order (Kluwer, Dordrecht, 1997) pp. 115-125.

[4] Dworkin S., Spectral theory and X-ray diffraction, *J. Math. Phys.* **34** (1993) 2965-67.

[5] P. and T. Ehrenfest, The conceptual foundations of the statistical approach in mechanics (Cornell University Press, Ithaca, New York, 1959) pp. 1-287.

[6] Furstenberg H., Disjointness in ergodic theory minimal sets and a problem in Diophantine approximation, *Math. Syst. Theory* **1** (1967) 1-49.

[7] Garsia A.M., A simple proof of Hopf's maximal ergodic theorem, *J. Math. Mech.* **14** (1965) 381-382.

[8] Hof A., Diffraction by Aperiodic Structures, edited by R.V. Moody, The Mathematics of Long-Range Aperiodic Order (Kluwer, Dordrecht, 1997) pp. 239-268.

[9] Kac M., On the notion of recurrence in discrete stochastic processes, *Bull. Amer. Math. Soc.* **53** (1947) 1002-1010.

[10] Kamae T., A simple proof of the ergodic theorem using nonstandard analysis, *Israel J. Math.* **42** (1982) 284-290.

[11] Kamae T. and Keane M., A simple proof of the ratio ergodic theorem, *Osaka J. Math.* **34** (1997) 653-657.

[12] Katznelson Y., Ergodic automorphisms of T^n are Bernoulli shifts, *Israel J. Math.* **10** (1971) 186-195.

[13] Katznelson Y. and Weiss B., A simple proof of some ergodic theorems, *Israel J. Math.* **42** (1982) 291-296.

[14] Kenyon R. and Peres Y., Measures of full dimension on affine-invariant sets, *Ergod. Th. & Dynam. Sys.* **16** (1996) 307-323.

[15] Oxtoby J.C., Ergodic Sets, *Bull. Amer. Math. Soc.* **58** (1952) 116-136.

[16] Petersen K., Ergodic Theory, *Cambridge Studies in Advanced Mathematics* **2** (Cambridge University Press, Cambridge, 1989) pp. 1-329.

[17] Queffélec M., Substitution dynamical systems – Spectral analysis, *Lecture Notes in Math.* **1294** (Springer Verlag, Berlin, 1987) pp. 1-240.

[18] Robinson E.A. Jr., On uniform convergence in the Wiener-Wintner Theorem, *J. London Math. Soc.* **49** (1994) 493-501.

[19] Shields P.C., The ergodic theory of discrete sample paths, *Amer. Math. Soc.* (Providence, RI, 1996) pp. 1-249.

[20] Solomyak B., Dynamics of Self-Similar Tilings, *Erg. Th. Dyn. Syst.* **17** (1997) 695-738.

[21] Walters P., An Introduction to Ergodic Theory (Springer Verlag, Berlin, 1982) pp. 1-250.

[22] Wiener N., The ergodic theorem, *Duke Math. J.* **5** (1939) 1-18.

[23] Yosida K. and Kakutani S., Birkhoff's ergodic theorem and the maximal ergodic theorem, *Proc. Imp. Acad.* **15** (1939) 165-168.

Fractality and the Kinetics of Chaos[*]

G.M. Zaslavsky

*Courant Institute of Mathematical Sciences, New York University,
251 Mercer St., New York, NY 10012, U.S.A.
and Department of Physics, New York University, 2-4 Washington Place,
New York, NY 10003, U.S.A.*

1. INTRODUCTION

A kinetic equation describes the evolution of the distribution function $F(x)$ in the space of a set of variables (x) which can be a subset of the phase space coordinates. Typically, the kinetic description of a system emerges from two basic characteristics of the dynamical process: its random features and its reduced subset $(\bar{x})$ of variables. The first characteristic, randomness, consists of

---

[*] To ease the reading of this Course, the Editors refer the non specialist readers to the following references, where they can find the basics needed together with complete definitions and examples. See also the Course by F.M. Dekking, passim, and the Glossary.
- V.I. Arnold and A. Avez, Ergodic problems of classical mechanics (W.A. Benjamin Publ., 1968).
- V.I. Arnold, Mathematical methods of classical mechanics (Springer-Verlag, 1978).
- V.I. Arnold, Ordinary differential equations (M.I.T. Press, Cambridge, MA, 1973).
- V.I. Arnold, Geometrical methods in the theory of ordinary differential equations (Springer-Verlag, 1982).
- J. Guckenheimer and P. Holmes, Non linear oscillations, dynamical systems and bifurcations of vector fields (Springer-Verlag, 1983). See *e.g.*: flows (p. 10), Ljapunov exponents (p. 283), KAM theorem and KAM tori (p. 218 sqq.), Osedelet's theorem (p. 284), Poincaré map (p. 22 sqq.) etc.
- A.J. Lichtenberg and M.A. Liberman, Regular and stochastic motion (Springer-Verlag, 1982).

a reasonable physical assumption like the Boltzmann's Stossansatz or the random phase approximation (see for example G.M. Zaslavsy, Chaos in dynamical systems, Harwood Academic publishers, New York, 1989). The second characteristic, typically, is based on a separation of slow and fast variables and averaging procedure over the latter ones.

The situation for the kinetic description of chaotic dynamics is different. Its trajectories are solutions of given dynamical equations, and there is not enough freedom to make assumptions about the random properties of the trajectories. In fact, the level of randomness strongly differs depending on the values of control parameters, type of systems, number of degrees of freedom, etc. All of this is a result of the very complicated pattern of chaotic dynamics which is a type of process whose dynamics is neither regular nor strongly random. To make this statement more "visual", consider two examples.

The first example is related to a perturbed pendulum

$$\ddot{x} + \sin x = \epsilon \sin(x - \nu t) \tag{1.1}$$

with perturbation amplitude ϵ and frequency ν. Its trajectories are solutions $x(t; x_0, \dot{x}_0)$, $\dot{x}(t; x_0, \dot{x}_0)$ which depend on the initial conditions and on the parameters (ϵ, ν) of the problem. The Poincaré map, which is defined below (see Sect. 2a.) is the most convenient way to visualize solutions in phase space $(x, \dot{x})$. Instead of plotting a whole trajectory, one can plot only a set of points $(x_n, \dot{x}_n)$ taken at the time instants

$$t_n = 2\pi n/\nu . \tag{1.2}$$

The set (1.2) represents a stroboscopic way of looking at the Hamiltonian of the system (1.1)

$$H = \frac{1}{2}\dot{x}^2 - \cos x + \epsilon \cos(x - \nu t)$$

because the perturbation has the same values for the same x if $t = t_n$. Two different cases of the Poincaré map of corresponding trajectories are shown in Figures 1a,b [1]. The difference between the values of parameters (ϵ, ν) is small but the appearance of the phase space topology is drastically different. There are many islands and island chains in Figure 1a whereas Figure 1b has no big islands. Moreoever, some parts of the phase space in Figure 1a are darker than others, which indicates nonuniformity of the distribution function. It is hard to believe that both of the two cases are described by the same kinetic equation.

The second example is more sophisticated. Consider a square billiard with absolutely ellastic collisions and a circular scatterer in the center (so-called Sinai billiard with infinite horizon). In such a billiard almost all trajectories are chaotic (Fig. 2) but they can spend a long time bouncing between two opposite sides of the square and avoiding scattering (Fig. 2a). The Poincaré map can be introduced, for example, as a set of coordinates and velocities $(x_n, \dot{x}_n)$ at the time instants t_n when trajectory hits the lower side. A Long time bounced trajectory implies scars in the phase space (Fig. 2b). There is

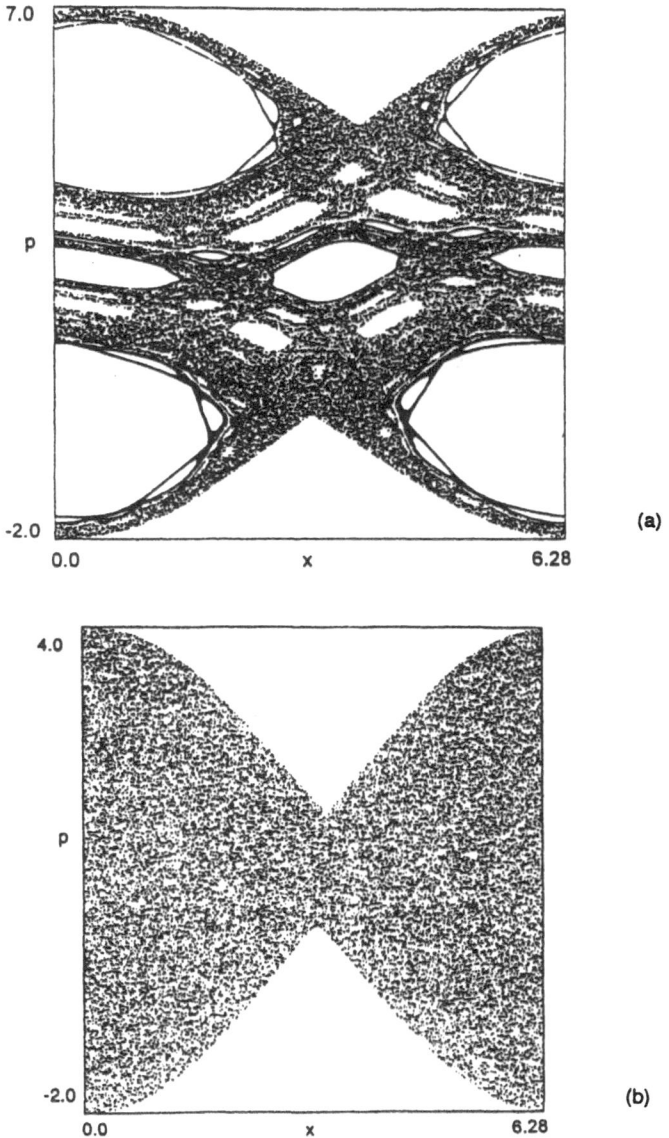

Fig. 1. — Poincaré section of an orbit for: (a) $\nu = 5.4$, $\epsilon = 0.9$; (b) $\nu = 2.05$, $\epsilon = 1$, taken with period $2\pi/\nu$, with vertical edges of the squares identified.

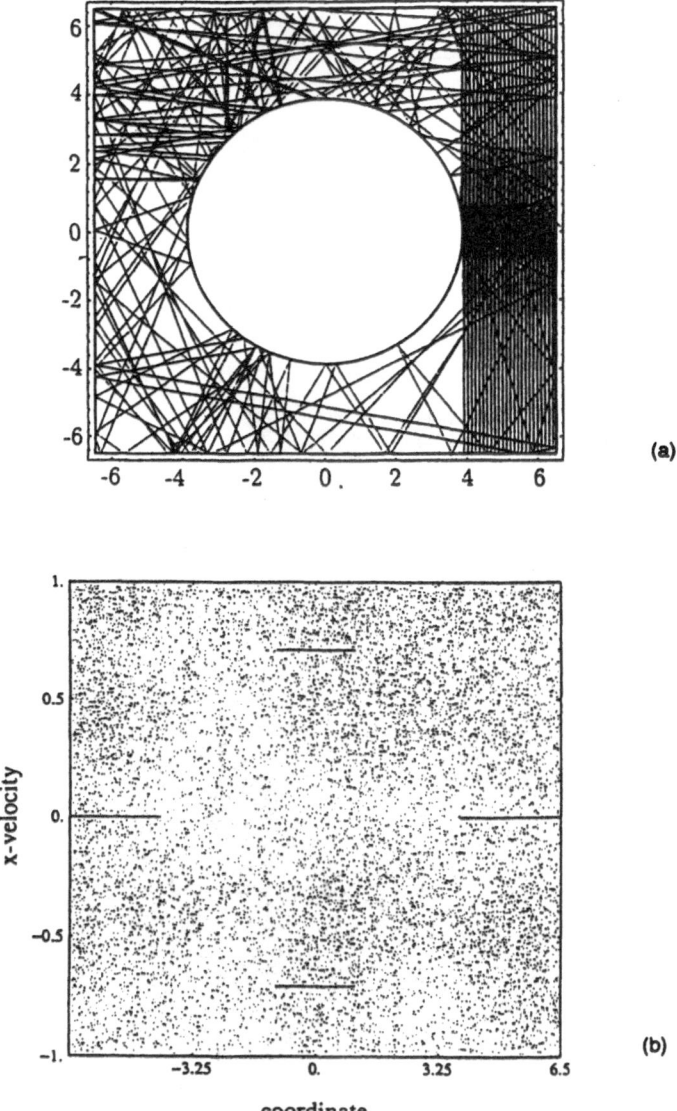

(a)

(b)

Fig. 2. — A trajectory (a) and its Poincaré section (b) for Sinai billiard.

a zero measure of them, *i.e.* a zero measure part of the phase space, which
is not included in the area of chaotic motion. The rest of the phase space of
the measure one looks uniformly filled by the trajectory points. Nevertheless,

even the existence of zero measure domains of nonergodic dynamics inside the stochastic sea influences the kinetics, making it of non-Gaussian type [2].

We can formulate a proposition: kinetic properties of chaotic dynamics strongly depend on the phase space topology, *i.e.* on the value of characteristic parameters of the system. This hypothesis was a subject of analytical and numerical investigation of recent publications [3-6]. It was concluded that a new type of kinetics can be applied to systems with chaotic dynamics, namely a fractional generalization of the Fokker-Planck-Kolmogorov equation (FFPK-equation) [7-9]. The essential point of this type of an equation

$$\frac{\partial^\beta F(x,t)}{\partial t^\beta} = \mathcal{D}\frac{\partial^\alpha F(x,t)}{\partial |x|^\alpha} \tag{1.3}$$

is that fractional exponents (α, β) depend on the phase space topology or, more precisely, on a "morphological class" of the topology for specific values of the control parameter. Actually, this loss of universality of chaotic kinetics is more complex since the form of equation (1.3) corresponds to some asymptotics for large t and $|x|$, and these asymptotics can be intermediate ones.

Even these brief introductory comments lead us to a new world of kinetics which was called "strange kinetics" [10] and which is related, to some extent, to the Lévy-type processes [11,12]. In this article we describe some reasons to introduce fractional kinetics (1.3) for the chaotic dynamics. It is conceivable that the general character of (1.3) will be revealed in its numerous applications.

2. MAPPING THE DYNAMICS

Dynamical systems can be described by the properties of their trajectories in phase space. In this article we will deal with Hamiltonian systems although some of the statements can have more general applications. Despite an arbitrary smoothness of the trajectories, they display different kinds of fractal features which are crucial for the large scale asymptotic properties of the system. In this section, we show how the fractal characteristics arise from smooth dynamics.

Typically, a trajectory in the phase space is $(\mathbf{r}(t), \mathbf{p}(t))$ with position $\mathbf{r}$ and momentum $\mathbf{p}$ at the time instant t. These functions cannot be presented in an explicit way for the generic situation because the dynamics are chaotic and the trajectories carry properties of randomness. An alternative way to describe the dynamics of a system is to map a trajectory according to certain given rules and, so doing, to reduce the desirable information about the system behavior. Here are examples of such mappings.

a. *Poincaré Map.* This is the most typical map due to its convenience and effectiveness. The Poincaré map can be defined as a set of points in phase space $\{p_j, q_j\}$ obtained from trajectory intersections with a hypersurface in the phase space at time instants $\{t_j\}$. For 1 1/2 or 2 degrees of freedom and finite motion, the closure of the set $\{p_j, q_j\}$ is an invariant curve if motion is quasi-periodic, and is a set of randomly distributed points (stochastic sea) if the motion is chaotic. For the generic situation, the Poincaré map forms a stochastic sea with implanted islands of invariant Kolmogorov-Arnold-Moser (KAM) curves (tori). Sometimes, the stochastic sea can be reduced to very thin channels of chaotic dynamics, called stochastic layers and webs.

b. *Poincaré Recurrences (or simply recurrences).* Consider a domain A and a point $(p_0, q_0) \in A$. A trajectory with initial coordinates (p_0, q_0) will definitely escape from A after a finite time and then come back again in a finite time. Such escapes and returns will repeat infinite number of times. Let $\{t_j^-\}$ be a set of the escape from A instants and $\{t_j^+\}$ be a set of the entrance to A instants. The set

$$\{\tau_j^{(\text{rec})}\}_A = \{t_j^- - t_{j-1}^-\}_A \tag{2.1}$$

is the set of recurrences. It can be fractal or multifractal [13,14]. The notion of fractal time appeared in [15] and developed in [11,16]. Numerous properties of systems can be characterized by the probability distribution function $P_{\text{rec}}(\tau; A)$.

c. *Exit Times.* Similarly to (2.1), one can introduce a set of exit (escape) times $\{\tau_j^{(\text{esc})}\}$:

$$\{\tau_j^{(\text{esc})}\}_A = \{t_j^- - t_{j-1}^+\}_A \tag{2.2}$$

and the corresponding probability distribution function $P_{\text{esc}}(\tau; A)$. The properties of the distributions of recurrences and escapes are strongly correlated [17,18].

d. *Zeno Map.* The Zeno map is a set of positions $\{x_j\}$ and the corresponding instants t_j attributed to the Zeno paradox about Achilles who could never overtake a turtle. Let $\{x_j\}$ be the turtle coordinate at time instant t_n. The trick of the Zeno paradox is that the set of time instants is specially taken as

$$t_{n+1} = t_n + (x_n - x_{n-1})/v_a,$$

while the turtle's coordinate is

$$x_{n+1} = x_n + v_t\, t_{n+1}, \quad (n \geq 0),$$

where v_a and v_t are velocities of Achilles and the turtle. Then, for initial conditions, let us put $t_0 = 0$, $x_0 = 0$, $x_1 = \ell$, *i.e.* $t_1 = \ell/v_a$, $x_1 = \ell v_t/v_a$, ... There is always some distance between the turtle and Achilles, *i.e.* $t_{n+1} > t_n$ although the term $x_{n+1} - x_n \to 0$. The example demonstrates that not any map generates a sufficient information to solve a problem.

e. *Roulette Map.* The roulette is an example of random number generator. A set of time instants $\{t_j\}$ can be taken in the same way as numbers from the roulette when it stops. Sometimes, physical events (for example, collisions) follow the same order as the roulette numbers order.

One should consider different ways of mapping dynamical systems while keeping in mind the final goal, *i.e.* obtaining the necessary probabilistic properties about the system's evolution. In this article we focus on kinetics and transport, *i.e.* time evolution of moments. We can expect that, after choosing a specified mapping method, a fractal or multifractal support arises for the system's trajectories in the phase space. The important questions are: how intrinsic are fractal properties of the selected way of mapping? Why do we need this complication and is it unavoidable? In the next section we demonstrate that natural chaotic trajectories of real dynamical systems possess space-time (multi)fractal structure for a typical situation. Here "real" means typical physical systems rather than specially prepared models; "space" means the phase space; and "fractal" situation means an actually multifractal one with very narrow spectral function of dimensions.

3. TOPOLOGICAL ZOO (SINGULAR ZONES)

Here we want to provide a motivation for introducing a fractal in the space-time of the support for chaotic trajectories. As it was mentioned above, a typical structure of the phase space of a Hamiltonian system with finite chaotic motion consists of the so-called stochastic sea and islands imbedded into the sea. In fact, there are also other structures, such as cantori or stable/unstable manifolds. As a result, we see what is presented in Figures 3 and 4 [5] where we plot the mapping of trajectories for the web-map

$$u_{n+1} = v_n, \quad v_{n+1} = -u_n - K \sin v_n, \quad (u_n, v_n = \mathrm{mod}\ 2\pi) \qquad (3.1)$$

for two different values of the parameter K. Figure 3 simply demonstrates how complicated the phase portrait of the system can be. In contrast to Figure 3, Figure 4 presents an elusive simplification. The dark area near the islands in Figure 4 corresponds to the stickiness, the theory of which does not yet exist. This zone can be considered a singular one, in which a particle is trapped for a long time. The "trap" has a fine structure which displays a new set of islands with again sticky boundaries. This hierarchy of islands continues to infinity, creating a fractal or multifractal structure of the trap. The remaining part of the stochastic sea outside the islands appears uniform, although it is only a problem of resolution to see more structures.

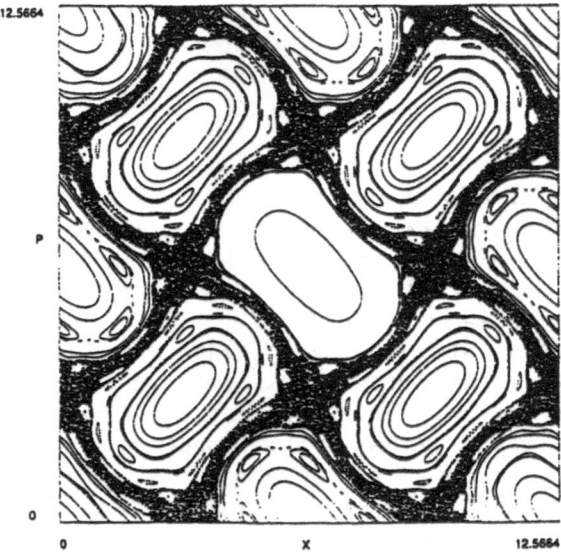

Fig. 3. — Phase portrait for the web-map with $K = 1.5$.

The second example for the perturbed pendulum model (1.1) was shown in Figure 1 where one can see similar trapping zones around certain islands. There is a fast propagation of particles along the dark strips, which we call a flight.

Despite the given two examples corresponding to two different models, all of them can occur in any model depending on parameter values. Dark domains in Figures 1 and 4 correspond to a nonuniformity of the phase space and, consequently, to a nonuniformity of the level of randomness of different parts of trajectories. This situation leads to the absence of uniformity of the kinetics for different time scales and to different strategies in the statistical description of chaotic dynamics. We expect a loss of universality of kinetics for chaotic dynamical systems, in general, and a "restricted universality" for systems within the same topological class of dynamics. A topological class is defined for systems with topologically equivalent phase spaces.

Let us discuss, in a qualitative way, how one can distinguish different topological classes of dynamics. Depending on the duration of observation of the system (say 10^{11}), we can neglect certain singular zones, *i.e.* the zones of stickiness that have too small probability to capture a trajectory. Only a few singular zones are responsible for the observable asymptotics of the trajectories and their characteristic time windows can overlap. Assuming, for example, that the dynamics in each singular zone can be characterized by a kind of self-similarity, one can consider a general situation of motion as multi-fractal in space and time. To avoid so high a level of complexity of the motion, let us

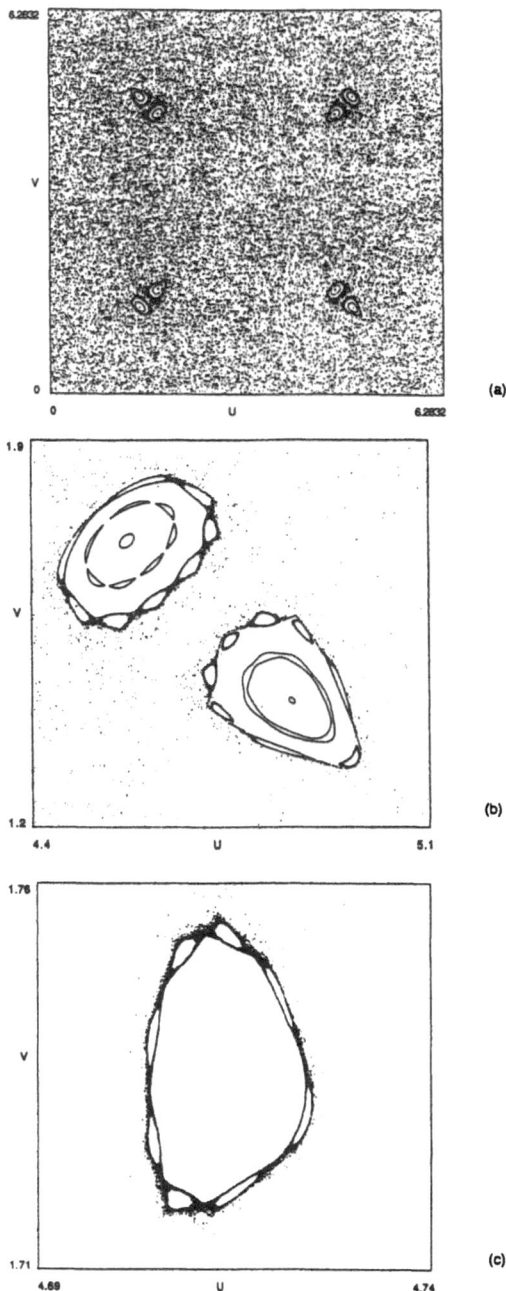

Fig. 4. — Phase portrait for the web-map with $K = 6.349972$ (a) and two consequent magnifications of the islands (b,c).

introduce parameter windows in such a way that for a given parameter window there are well separated time scales of the particle residence in singular zones, *i.e.* characteristic times of trapping into different domains of the phase space strongly differ. This gives an opportunity to select a parameter window and a corresponding time window for which intermediate asymptotics can be considered, and the fractional kinetic equation can be introduced. Such a case is similar to fractal rather than multi-fractal dynamics. Some of the windows are described in [5,6,19].

4. SELF-SIMILAR HIERARCHY OF ISLANDS

In Figure 4 of Section 3 we show an example of the hierarchical structure of islands for the web-map (3.1). A corresponding value of K, found in [20], is

$$K^* = 6.349972\ldots \tag{4.1}$$

the value (4.1) provides an infinite set of islands-around-islands with the proliferation number $q = 8$ (number of daughter-islands around the mother-island). Let the sequence of the islands be written in the form of a "word"

$$\{m_j\} = \{m_1, m_2, \ldots\}$$

where m_j is the number of islands of the j-th generation around an island of the $(j-1)$-th generation. Typically, the sequence $\{m_j\}$ depends on the value of K and all possibilities can occur in (4.2) up to some minor restrictions. For the case (4.1) all m_j are equal $m^* = 8$.

Let us introduce an area ΔS_j of an j-th generation island and a period T_j of the last invariant curve for the same island. It was shown in [20] that there exists a self-similarity

$$T_{k+1} = \lambda_T T_k , \qquad \lambda_T > 1$$
$$\delta S_{k+1} = \lambda_S \delta S_k , \qquad \lambda_S < 1 \tag{4.2}$$

where

$$\delta S_k = q \Delta S_k \tag{4.3}$$

and λ_T, λ_S are some scaling parameters. The relationship between λ_T, λ_S can be established for a given dynamical system.

The situation of the existence of equations (4.2), which appear for a special choice of the parameter $K = K^*$, is nontrivial. Nevertheless, one can consider different regular sequences $\{m_j\}$ for different values of K and determine a spectral function of scalings $F(\lambda)$ in analogy to the multifractal spectral function [21]. Then, the sequence of periods $\{T_j\}$ will have a much more complicated form than (4.2). At the same time, sequence $\{T_j\}$ defines how long a trajectory spends in the vicinity of the boundary of an island when it is being trapped in the singular zone. This brief comment explains the origin of (multi)fractal dynamics in the set of boundary layers of the sequence of islands $\{m_j\}$. To

Fig. 5. — Scheme of the phase space for islands surrounded by subislands.

simplify the situation, it is convenient to select a "window" of K, which corresponds to the case of constant values of λ_T, λ_S as in (4.1, 4.2). Then we have to introduce a kinetic equation for specific dynamics when trajectories randomly switch from islands of one hierarchy to islands of another. This indicates the support of dynamics is fractal in space and time simultaneously. In Figure 5, we show a structure of the islands hierarchy described above. It somewhat resembles the Sierpinsky carpet. The crucial difference between this structure and the Sierpinsky carpet is that each circle (island) in Figure 5 carries an additional parameter which is the time span that trajectories spend inside the circle.

5. QUASI-TRAPS

In this section we formalize some statements of the previous one using the notions of fractal time and quasi-traps (since traps do not exist in Hamiltonian dynamics).

The concept of fractal time is based on the fractal properties of a time set $\{t_j\}$ where the instants t_j are related to a characteristic set of events, such as the length of steps in a random walk [11,16]. The application of the fractal time concept to Hamiltonian chaotic dynanics needs to be modified [2,5,13,14].

Let the finite phase space Γ_0 (square in Fig. 6) be filled by a chaotic trajectory almost uniformly, except for a relatively small domain $\Gamma_1 \subset \Gamma_0$ (a smaller square

Fig. 6. — A sketch of the quasi-trap.

in Fig. 6) where the trajectory is almost uniform except on a smaller domain $\Gamma_2 \subset \Gamma_1$ and so on. We assume the existence of the sequence of sets

$$\Gamma_0 \supset \Gamma_1 \supset \dots \tag{5.1}$$

with a space-scaling constant λ_Γ:

$$\lambda_\Gamma = \Gamma_{n+1}/\Gamma_n < 1, \qquad (\forall n). \tag{5.2}$$

It follows from (5.2) that

$$\Gamma_n = \lambda_\Gamma^n \Gamma_0. \tag{5.3}$$

Now introduce a time T_0 that the trajectory spends in $\Gamma_0 - \Gamma_1$, a time T_1 that the trajectory spends in $\Gamma_1 - \Gamma_2$, and so on. Assume the existence of a time-scaling constant λ_T:

$$\lambda_T = T_{n+1}/T_n > 1, \qquad (\forall n) \tag{5.4}$$

which means that the smaller the domain Γ_j, the larger the residence time of a particle in the domain. This is a way to construct a quasi-trap, which corresponds to the possibility to approach a singular zone (or a set in the phase space) where the Lyapunov exponent approaches zero but is never zero. Particularly, the sequence for the islands hierarchy of the web-map with parameter (4.1) can be considered as an example to apply the described construction.

Let us make some simple transformations. It follows from (5.4) that

$$T_n = \lambda_T^n T_0 \tag{5.5}$$

or

$$n = \ln(T_n/T_0)/\ln\lambda_T. \tag{5.6}$$

The substitution of (5.6) into (5.3) gives

$$\Gamma_n/\Gamma_0 = (T_0/T_n)^{\mu_\Gamma} \tag{5.7}$$

with

$$\mu_\Gamma = |\ln\lambda_\Gamma|/\ln\lambda_T. \tag{5.8}$$

Consider an asymptotic formula for the escape probability density

$$\psi(t;\Delta\Gamma) =\sim 1/t^\gamma \tag{5.9}$$

where we introduce a probability density $\psi(t;\Delta\Gamma)$ for the particle to stay time t in a volume element $\Delta\Gamma$ with an exponent γ. The location in phase space of the domain $\Delta\Gamma$ will be specified below. The phase volume G_n of particles that pass the domain Γ_n during the escape time $t > T_n$ is of the order [18]:

$$G_n \sim \int_{T_n}^\infty t\psi(t;\Gamma_n)\mathrm{d}t \sim 1/T_n^{\gamma-2} \tag{5.10}$$

in accordance with (5.9). These particles guarantee that recurrence occurs after T_n. The expression (5.10) should be of the same order Γ_n up to a small correction of order $\Gamma_{n+1} \ll \Gamma_n$. Comparing (5.10) and (5.7) with $G_n \sim \Gamma_n$ we get in the limit $n \to \infty$

$$\gamma = 2 + \mu_\Gamma = 2 + |\ln\lambda_\Gamma|/\ln\lambda_T. \tag{5.11}$$

This result coincides with the result in [5] obtained by the Renormalization Group approach. A qualitative way of obtaining the expression (5.11) shows how the quasi-trap can work in Hamiltonian systems. The described construction of the quasi-trap corresponds to the situation considered in [5] when chain of islands make a fractal type hierarchy.

6. BOUNDARY LAYER AS A QUASI-TRAP

Consider now the situation when the quasi-trap is of fairly general type which we call "boundary layer". The dynamics in the quasi-trap can be considered in detail if one can specify its internal structure.

Our first step is to define a phase volume

$$\Delta\Gamma_{\max} = \Delta x \cdot \Delta p \tag{6.1}$$

of particles that enter the $\Delta\Gamma_{\max}$ region where they spend a fairly long time due to the intermittent and sticky properties of the domain. Let us assume that the stickiness appears because of the vicinity of the island boundary to the destroyed separatrix. Then slow dynamics in the boundary layer can be described by a typical Hamiltonian

$$H_{\text{ef}} = \frac{1}{2}\{(\Delta p)^2 - (\Delta x)^2 + (\Delta x)^3\} \tag{6.2}$$

where we renormalize all constants. The first two terms describe the motion near a saddle point and the third one is the first nonlinear term which defines a potential well for the trapping. We also assume that Δx_{max} is the length of a step. The corresponding value Δp_{max} can be obtained from (6.2) in a straightforward manner:

$$\Delta p_{\text{max}} \sim (\Delta x_{\text{max}})^{3/2} \tag{6.3}$$

and from (6.1) and (6.3), we have the phase volume of escaping particles

$$\Delta \Gamma_{\text{max}} = (\Delta x_{\text{max}})^{5/2}. \tag{6.4}$$

The necessary time to escape is

$$t \geq t_0 = N_0 \tau_0 \tag{6.5}$$

where τ_0 is the time to perform a step and N_0 is number of steps. Let a be the characteristic length of the trap, $i.e.$ the length of the boundary layer. Then

$$N_0 = a/\Delta x_{\text{max}} \tag{6.6}$$

and it follows from (6.4–6.6) that

$$\Delta \Gamma_{\text{max}} = \frac{(a\tau_0)^{5/2}}{t^{5/2}}. \tag{6.7}$$

From a definition similar to (5.9) we have

$$\psi(t) \sim P_{\text{esc}}(t) \sim \Delta \Gamma_{\text{max}} \sim (a\tau_0)^{5/2}/t^{5/2} \tag{6.8}$$

$i.e.$

$$\gamma = 5/2. \tag{6.9}$$

A similar value for γ was obtained in [22] for a different model of the boundary layer. Actually, all estimations near the saddle point can be up to a logarithmic factor.

7.   FRACTAL AND MULTIFRACTAL SPACE-TIME OF KINETICS

In this section we look at the time sets $\{\tau^{(\text{rec})}\}$, $\{\tau^{(\text{esc})}\}$ of (2.1, 2.2) in a more detailed way. Each interval $\{\tau_j^{(\text{rec})}\}$ is called a Poincaré cycle. The distribution $P(\tau; \Delta \Gamma)$ of cycles can be normalized as

$$\frac{1}{\Delta \Gamma} \int_0^\infty P(\tau; \Delta \Gamma) \, d\tau = 1 \tag{7.1}$$

and the limit

$$P(\tau) = \lim_{\Delta\Gamma\to 0} P(\tau;\Delta\Gamma)/\Delta\Gamma \qquad (7.2)$$

exists if there exists ergodicity and phase space compactness [17]. Moreover, if the measure of the chaotic orbits is nonzero, then

$$\langle\tau\rangle \equiv \tau_0 < \infty \qquad (7.3)$$

i.e. the mean recurrence time is finite [17] (see also [18]). Using these results, it follows from the asymptotic formula

$$P(\tau) \sim \tau^{-\gamma}, \qquad \tau \to \infty \qquad (7.4)$$

that there exists a restriction

$$\gamma > 2 . \qquad (7.5)$$

We can obtain additional information considering distributions of the escape times from different boundary layer zones [5,20] by introducing a time delay distribution. Let $\psi(t;\Delta\Gamma)$ (compare to (5.9)) be the probability density to escape from $\Delta\Gamma$ during the time interval $(t, t + dt)$. Then the probability to escape from $\Delta\Gamma$ in a time not less than t is

$$P_{\text{esc}}(t;\Delta\Gamma) = \int_0^t d\tau\ \psi(\tau;\Delta\Gamma). \qquad (7.6)$$

The corresponding probability of survival until time t is

$$\Psi(t;\Delta\Gamma) = 1 - P_{\text{esc}}(t;\Delta\Gamma) = 1 - \int_0^t d\tau\ \psi(\tau;\Delta\Gamma) \qquad (7.7)$$

with the normalization condition

$$P_{\text{esc}}(t \to \infty;\Delta\Gamma) = 1. \qquad (7.8)$$

The mean time of trapping to the domain $\Delta\Gamma$ is

$$t_{\text{s}}(\Delta\Gamma) = \int_0^\infty d\tau \cdot \tau\psi(\tau;\Delta\Gamma). \qquad (7.9)$$

In the case where there is only one "leading" exponent in (7.4) we can assume that the same has happened with the escape distribution function $\psi(\tau;\Delta\Gamma)$, *i.e.*

$$\psi(\tau;\Delta\Gamma) \sim \tau^{-\gamma} \qquad (7.10)$$

as in (5.9).

The chaotic dynamics can be considered as normal if they have a Poissonian distribution

$$P(\tau) = \frac{1}{\langle\tau\rangle} \exp(-\tau/\langle\tau\rangle) \qquad (7.11)$$

or any similar distribution with all finite moments. Such a situation was described in [23]. In fact, simulations show that the distribution (7.6) is not typical for generic Hamiltonian chaotic dynamics, although it can be considered as a good approximation at least for a finite time interval. Situation for (7.4) looks more acceptable from the simulation, but it should also be improved by introducing different exponents γ for different time and parameter windows. That is the way how we arrive to the necessity to introduce multifractal distributions for Poincaré recurrences and exit times [13,14].

Consider the space-time partitioning that was introduced in [9] and resembles the Sierpinsky carpet (Fig. 5). Let the central square be an island of zero-generation. Surround the island by an annulus which represents the boundary island layer. It consists of g_1 ($g_1 = 8$ in Fig. 5) sub-islands of the first generation (dashed small islands in Fig. 5). We can partition the annulus into g_1 domains so that each of them includes exactly one island of the first generation, then we surround each of the first generation island by an annulus of the second generation and repeat the process. At the n-th step the structure can be described by a "word"

$$w_n = w(g_1, g_2, \ldots, g_n). \tag{7.12}$$

The full number of islands at the n-th step is

$$N_n = g_1 \ldots g_n \tag{7.13}$$

and any island belonging to the n-th generation can be labeled by

$$u_i^{(n)} = u(i_1, i_2, \ldots, i_n), \qquad 1 \leq i_j \leq g_j, \ \forall\, j. \tag{7.14}$$

Let us now the time that a particle spends in the boundary layer of an island. This time

$$T_i^{(n)} = T\left(u_i^{(n)}\right) \tag{7.15}$$

carries all the information about the n-th generation islands (7.12–7.14). By introducing a residence time for each island boundary layer, we have a new situation comparing to the plain Sierpinsky carpet or plain fractal situation because of the nontriviality of the space-time coupling. In fact, we are attaching an additional parameter, responsible for the temporal behavior, to a geometric construction which is similar to the Cantor-set.

A simplified situation corresponds to the exact self-similarity of the construction described above, i.e.

$$S_i^{(n)} = S^{(n)} = \lambda_S^n \cdot S^{(0)}, \qquad \forall\, i$$

$$T_i^{(n)} = T^{(n)} = \lambda_T^n \cdot T^{(0)}, \qquad \forall\, i \tag{7.16}$$

where $S_i^{(n)}$ is area of an island $u_i^{(n)}$ and $T_i^{(n)}$ is introduced in (7.15). The expressions in (7.16) correspond to equal areas and residence times for all islands of the same generation. The two scaling parameters, λ_S and λ_T, represent the existence of exact self-similarity in space and in time correspondingly. Precisely such a situation was described in [5,20] for the standard map with

$$\lambda_S < 1 , \qquad \lambda_T > 1. \tag{7.17}$$

In addition to (7.16) there is a self-similarity in the islands' proliferation, $i.e.$

$$g_n = \lambda_g^n g_0 , \qquad \lambda_g \geq 3. \tag{7.18}$$

It follows from (7.13) and (7.14) that

$$N_n = \lambda_g^n g_0 = \lambda_g^n \tag{7.19}$$

if we start from the only island ($g_0 = 1$). It is useful to introduce the "residence frequency"

$$\omega_i^{(n)} = 1/T_i^{(n)} \tag{7.20}$$

with the self-similarity property

$$\omega_i^{(n)} = \omega^{(n)} = \lambda_T^{-n} \omega^{(0)} . \tag{7.21}$$

This expression completes a simplified version of the space-time partition given by (7.16, 7.18, 7.21). For this particular partition it was found in [5] that

$$\gamma = 2 + \mu = 2 + |\ln \lambda_S| / \ln \lambda_T \tag{7.22}$$

where μ is the transport exponent. For the web map (3.1) μ is defined by

$$\langle u^2 + v^2 \rangle \sim t^\mu . \tag{7.23}$$

Thus, the expression (7.22) connects the recurrence distribution (7.4) to the self-similar structure of the space-time.

8. DIMENSION SPECTRUM OF THE MULTIFRACTAL SPACE-TIME

In analogy with [21,24-26], we can introduce a dimension spectral function to describe the self-similar properties of a system. In our case we are dealing with a new type of object that reveals its multifractal properties in space and time simultaneously.

Following a way usual in statistical mechanics, let us introduce a partition function of the form [14]

$$Z_{\text{discr}}^{(n)}\{\lambda_T, \lambda_S; q\} = \sum_{i_1, i_2, \ldots, i_n} \left(\omega_i^{(n)} S_i^{(n)} \right)^{\gamma q} . \tag{8.1}$$

Here we use the space-time bin probability

$$P_i^{(n)} = \omega_i^{(n)} S_i^{(n)} \tag{8.2}$$

i.e. the probability for a particle to stay in the domain $u_i^{(n)}$ of area $S_i^{(n)}$ during the time interval $T_i^{(n)}$. Considering a multi-scaling situation, we assume that a real elementary probability to occupy a bin should have the same scaling dependence as in (8.2) up to a power γ, and there are different values of γ in the sum. With exponent q we can consider different moments of the elementary bin probability. In particular, for $q = 0$ we obtain simply the number of bins. Let us replace the summation by an integration and write

$$Z^{(n)}\{\lambda_T, \lambda_S; q\} = \int d\gamma' \rho(\gamma') [\omega^{(n)} S^{(n)}]^{-f(\gamma') + \gamma' q} \qquad (8.3)$$

where the density of the space-time bins is introduced as

$$dN^{(n)}(\gamma') = d\gamma' \, \rho(\gamma') [\omega^{(n)} S^{(n)}]^{-f(\gamma')}. \qquad (8.4)$$

The function $f(\gamma')$ is a spectral function of the space-time dimension-like characteristics, or simply, dimensions. The distribution density $\rho(\gamma')$ is a slow function of γ'. To be more accurate, we should assume that the bin-probability $\omega^{(n)} \cdot S^{(n)}$ also depends on γ' because for different island-sets the bins have different structures. Nevertheless, the dependence of $\omega^{(n)} \cdot S^{(n)}$ on γ' is slow in comparison to the exponential law in (8.4).

Using (7.16, 7.21) and (7.22, 8.3) transforms into

$$Z^{(n)}\{\lambda_T, \lambda_S; q\} = \int d\gamma' \rho(\gamma') \exp\{-n[\gamma' q - f(\gamma')](|\ln \lambda_S| + \ln \lambda_T)\}, \quad (8.5)$$

where λ_S and λ_T are slow functions of γ'. For $n \to \infty$ the standard steepest descent procedure gives

$$Z^{(n)}\{\lambda_T, \lambda_S; q\} \sim \exp\{-n[\gamma_0 q - f(\gamma_0)] \cdot (|\ln \lambda_S| + \ln \lambda_T)\} \qquad (8.6)$$

with the equation to determine $\gamma_0 = \gamma_0(q, \lambda_S, \lambda_T)$

$$q = f'(\gamma_0) . \qquad (8.7)$$

In (8.6) λ_S and λ_T should be taken at the saddle point γ_0. From another side recall that for $q = 0$ expression (8.1) defines $Z_{\text{discr}}^{(n)}\{\lambda_T, \lambda_S; 0\}$ as a number of bins. For the one-scale situation we have this number from (7.19) as λ_g^n. For the multi-fractal case we can write a power of λ_g^n by introducing a generalized dimension D_q in the analogy to [27].

$$Z^{(n)}\{\lambda_S, \lambda_T; q\} \sim \lambda_g^{-n(q-1)D_q} \sim \exp\{-n \ln \lambda_g \cdot (q-1)D_q\} \qquad (8.8)$$

and by considering λ_g as some characteristic constant.

The comparison of (8.8) and (8.6) gives

$$(q-1)D_q \cdot \ln \lambda_g = [\gamma_0 q - f(\gamma_0)](\ln \lambda_T + |\ln \lambda_S|). \qquad (8.9)$$

For some extreme cases λ_g should satisfy the conditions

$$\lambda_g = \begin{cases} 1/\lambda_S, & \text{if } \lambda_T = 1 \\ \lambda_T, & \text{if } \lambda_S = 1. \end{cases} \tag{8.10}$$

For the general situation $\lambda_g, \lambda_T, \lambda_S \neq 1$ and we can rewrite (8.9) in the final form

$$(q-1)D_q = \frac{\ln \lambda_T}{\ln \lambda_g}(1+\mu)[\gamma_0 q - f(\gamma_0)] \tag{8.11}$$

where the parameter μ is the same as in (7.22).

It follows from (8.11) for $q = 0$

$$D_0 = \frac{\ln \lambda_T}{\ln \lambda_g}(1+\mu) \cdot f(\gamma_0) , \tag{8.12}$$

i.e. there is no simple connection between the dimension D_0 and the spectral function. The regular formula

$$D_0 = f(\gamma_0) \tag{8.13}$$

appears only in the cases (8.10) when one has a multifractal structure only in space or in time. For $q = 1$ using (8.7) and (8.11) we obtain

$$D_1 = \gamma_0(1) \cdot \frac{\ln \lambda_T}{\ln \lambda_g}(1+\mu) \tag{8.14}$$

where the value $\gamma_0(1) = \gamma_0(q = 1)$ can be obtained from the equation (8.7): $f'(\gamma_0(q = 1)) = 1$.

In (8.9) we expressed the generalized dimension D_q through the spectral function $f(\gamma')$ as in [21,24-26]. Nevertheless formulas (8.9, 8.11, 8.12), and (8.14) show that the knowledge of the spectral function is not sufficient for a typical dynamical system, and some additional information is necessary about the system's structure in space and time.

9. FRACTIONAL KINETICS

Using the kinetic equation with fractional derivatives is a convenient way to take into account such properties as power tails of the exit and recurrence time distributions. There are fairly detailed articles about the derivation of fractional kinetics [5,7-9,28] and we will refer the reader to them for more details. Here is a brief description of the way to get equation (1.3). Let $P(x,t)$ be the p.d.f. to find a particle at time t with a variable value x which can be the coordinate or the momentum. The presence of singular zones can be used to split the original $P(x,t)$ into two parts, one of which describes the singularity:

$$P(x,t) = P_n(x,t) + P_s(x,t) \tag{9.1}$$

where n corresponds to the normal part and s to the singular one. The same splitting can be performed in q-space

$$P(q,t) = \int e^{iqx} P(x,t) \mathrm{d}x = P_n(q,t) + P_s(q,t). \tag{9.2}$$

Considering the asymptotics

$$|x| \to \infty, \qquad t \to \infty, \tag{9.3}$$

i.e. $q \to 0$ in (9.2), we can assume that there exists an expansion

$$P(q,t) = \sum_{m=0}^{n} \frac{1}{m!} q^m \left. \frac{\partial^m P(q,t)}{\partial q^m} \right|_{q=0} + A_m |q|^\alpha + \text{remainder} \tag{9.4}$$

where A_m is a constant and

$$n < \alpha < n+1 \tag{9.5}$$

with the definition

$$P_s(q,t) = A_m |q|^\alpha. \tag{9.6}$$

The normal part of (9.4) can be rewritten using definition (9.2):

$$P_n(q,t) = \sum_{m=0}^{n} \frac{(iq)^m}{m!} \langle x^m \rangle. \tag{9.7}$$

This expression, together with (9.6) and (9.5) simply states that $P(x,t)$ possesses finite first n moments and infinite $(n+1)$-th moment because of

$$\left. \frac{\partial^{n+1} P(q,t)}{\partial q^{n+1}} \right|_{q=0} = i^{n+1} \langle x^{n+1} \rangle \tag{9.8}$$

$$= A_m \alpha(\alpha-1)\dots(\alpha-n)|q|^{\alpha-1-n} = \infty, \qquad q = 0.$$

As a result of this simple introduction we can consider the distribution function with an arbitrary power-wise tail as a function with corresponding infinite moments, and use for such a function expansion of the (9.4)-type.

Denote $\ell = |x|$ and assume that

$$P_s(x,t) = P(\ell,t) \tag{9.9}$$

where the subscript s is omitted for simplicity. Consider an infinitesimal change in time of the distribution function

$$\delta_t P(\ell,t) \equiv P(\ell, t+\Delta t) - P(\ell;t) \tag{9.10}$$

where the notation (9.9) has been used. This deviation can be expressed through the all the available paths with their probabilities. Thus

$$\delta_t P(\ell,t) = \sum_{\Delta \ell} \{ \overline{P(\ell+\Delta\ell, t) - P(\ell,t)} \} \equiv \delta_\ell P(\ell,t) \tag{9.11}$$

where the sum over $\Delta\ell$ means a summation over all such paths within an infinitesimal interval $\Delta\ell$ that gives the infinitesimal evolutional change $\delta_t P$, and the bar means averaging over the paths. As an example one can remind the conventional averaging over phases. Assuming that there exists a balance equation for the particles, *i.e.* that (9.11) and (9.12) are equal, we can write

$$(\Delta t)^\beta \frac{\partial^\beta P}{\partial t^\beta} = \sum_{\Delta\ell} \frac{\partial^\alpha}{\partial \ell^\alpha} (\overline{(\Delta\ell)^\alpha \mathcal{A}'} \, P) \qquad (9.12)$$

where $\mathcal{A}'$ is some expansion constant and choice of a term with power α depends on the first nonvanishing value after averaging over phases in the singular part of the expansion (9.4).

Here the fractional derivative with respect to ℓ should be considered on the basis of Fourier transform (9.2). Namely, let $g(x)$ be an arbitrary function for which the derivative exists, and $g(q)$ is its Fourier transform

$$g(x) \xrightarrow{F} g(q).$$

The following formulas are valid

$$\frac{d^\alpha}{dx^\alpha} g(x) \xrightarrow{F} (-iq)^\alpha g(q),$$

$$\frac{d^\alpha}{d(-x)^\alpha} g(x) \xrightarrow{F} (iq)^\alpha g(q). \qquad (9.13)$$

Now introduce the operator of the symmetrized fractional derivative using the Fourier transforms (9.13)

$$\frac{d^\alpha}{d|x|^\alpha} g(x) \xrightarrow{F} -|q|^\alpha g(q). \qquad (9.14)$$

Then, combining (9.13) and (9.14), one obtains

$$\frac{d^\alpha}{d|x|^\alpha} = -\frac{1}{2\cos(\pi\alpha/2)} \left[\frac{d^\alpha}{dx^\alpha} + \frac{d^\alpha}{d(-x)^\alpha} \right], \qquad (\alpha \neq 1). \qquad (9.15)$$

The last expression (9.15) defines the fractional derivative with respect to ℓ. The derivative with respect to t is slightly different since $t > 0$. For example, for $0 < \beta < 1$ and an arbitrary function $f(t)$ for which the derivative exists, we have

$$\frac{d^\beta}{dt^\beta} f(t) = \frac{1}{\Gamma(1-\beta)} \int_{-\infty}^{t} \frac{df(\tau)}{d\tau} \frac{d\tau}{(t-\tau)^\beta} \qquad (9.16)$$

(see more in [28]).

The important part of the derivation of a fractional kinetic equation is the assumption that for infinitesimal Δt (comparing to the observed time of evolution of $P_s(\ell, t)$) the expression

$$\mathcal{A} = \sum_{\Delta\ell} \frac{\overline{(\Delta\ell)^\alpha}}{(\Delta t)^\beta} \mathcal{A}' \qquad (9.17)$$

is finite. Then, assuming that $\mathcal{A}$ is a constant, the kinetic equation (9.11) takes the form

$$\frac{\partial^\beta P}{\partial t^\beta} = \mathcal{A}\frac{\partial^\alpha P}{\partial \ell^\alpha} \qquad (9.18)$$

that coincides with (1.3).

Equation (9.18) corresponds to a specific window of parameters and time, where corresponding values α, β are applicable. For different windows, one can expect different pairs (α, β). Each of α and β is proportional to the generalized fractal dimension of the corresponding support. This information is sufficient to get a transport exponent. Multiplying (9.18) by ℓ^α and integrating it, we obtain the moment equation

$$\langle \ell^\alpha \rangle \sim \mathcal{A}t^\beta \qquad (9.19)$$

or, approximately,

$$\langle \ell^2 \rangle \sim \langle |x|^2 \rangle \sim t^{2\beta/\alpha} = t^\mu. \qquad (9.20)$$

This is an asymptotic result that is valid for

$$t^\beta/\ell^\alpha \ll 1 \qquad (9.21)$$

(see [28]).

Parameters (α, β) should be taken from the dynamical properties of the trajectories. The approach, described above, implies that there exists a self-similarity of the dynamics in space and time, and this property indicates the way to find (α, β) from the invariants of the scaling transformations [9]. Particularly, for a Gaussian process $\alpha = 2$, $\beta = 1$, while for the case of hierarchical island chains

$$2\beta/\alpha = \mu = |\ln \lambda_S|/\ln \lambda_T. \qquad (9.22)$$

For the case of the stickiness in a boundary layer (Sect. 6), we have to put

$$\beta = \gamma - 1 = 3/2\,, \qquad \alpha \sim 2 \qquad (9.23)$$

which gives

$$\mu \sim 3/2. \qquad (9.24)$$

Simulation shows [29] that the value of α is not exactly 2 and that the p.d.f. of displacements has a power-like tail that permits to evaluate all exponents.

10.   CONCLUSIONS

Chaotic trajectories of generic Hamiltonian systems with 1 1/2 or 2 degrees of freedom are not filling phase space uniformly. In their avoiding of islands on their way, the trajectories can be trapped for a while into specific zones near the island boundaries. Such dynamics lead to the (multi)fractal structure of space-time partitioning of the distribution function along a trajectory. A few examples developed here and their simulations have shed a new light on the fine features of these chaotic dynamics. At the same time, these new properties of chaos reveal such important applications as anomalous transport.

Acknowledgements

This work was supported by the U.S. Department of Navy, Grants N00014-93-1-0218, N00014-97-1-0426, and by the U.S. Department of Energy, Grant DE-FG02-92ER54184. Computations were performed at National Energy Research Scientific Computing Center on CRAY J90.

REFERENCES

[1] S.S. Abdullaev and G.M. Zaslavsky, *Phys. Rev. E* **51** (1995) 3901.
[2] G.M. Zaslavsky and M. Edelman, *Phys. Rev. E* **56** (1997) 5210.
[3] A.A. Chernikov, R.Z. Sagdeev and G.M. Zaslavsky, in *Topological Fluid Dynamics*, edited by H.K. Moffatt and A. Tsinober (Cambridge Univ. Press, 1990) p. 45.
[4] G.M. Zaslavsky, D. Stevens and H. Weitzner, *Phys. Rev. E* **48** (1993) 1683.
[5] G.M. Zaslavsky, M. Edelman and B.A. Niyazov, *Chaos* **7** (1997) 159.
[6] S. Benkadda, S. Kassibrakis, R.B. White and G.M. Zaslavsky, *Phys. Rev. E* **55** (1997) 4909.
[7] G.M. Zaslavsky, in *Topological Aspects of The Dynamics of Fluids and Plasmas*, edited by H.K. Moffatt, G.M. Zaslavsky, P. Compte and M. Tabor (Kluwer, Dordrecht, 1992) p. 481.
[8] G.M. Zaslavsky, *Physica D* **76** (1994) 110.
[9] G.M. Zaslavsky, *Chaos* **4** (1994) 25.
[10] M.F. Shlesinger, G.M. Zaslavsky and J. Klafter, *Nature* **363** (1993) 31.
[11] E.W. Montroll and M.F. Shlesinger, in *Studies in Statistical Mechanics*, edited by J. Lebowitz and E.W. Montroll (North-Holland, Amsterdam, 1984) p. 1.
[12] M. Shlesinger, G.M. Zaslavsky and U. Frish, in *Lévy Flights and Related Topics in Physics* (Springer, 1995).
[13] V. Afraimovich, *Chaos* **7** (1997) 12.
[14] V. Afraimovich and G.M. Zaslavsky, *Phys. Rev. E* **55** (1997) 5418.
[15] J.M. Berger and M. Mandelbrot, *IBM Journal* **7** (1963) 224.
[16] M. Shlesinger, *Ann. Rev. Phys. Chem.* **39** (1988) 269.
[17] M. Kac, in *Probability and Related Topics in Physical Sciences, Proc. of Summer Seminar, Boulder, Colorado, 1957* (Interscience, NY, 1958).
[18] J. Meiss, *Chaos* **7** (1997) 39.
[19] R.B. White, S. Benkadda, S. Kassibrakis and G.M. Zaslavsky, to be published.
[20] G.M. Zaslavsky and B.A. Niyazov, *Physics Reports* **283** (1997) 73.
[21] H.G.E. Hentschel and I. Procaccia, *Physica D* **8** (1983) 435; P. Grassberger and I. Procaccia, *Physica D* **13** (1984) 34.
[22] S.C. Venkataramani, T.M. Aantonsen Jr. and E. Ott, *Phys. Rev. Lett.* **78** (1997) 3864.

[23] G.A. Margulis, *Funkshional Anal. i Prilozh.* **3** (1969) 80; *ibid.* **4** (1970) 62.

[24] U. Frisch and G. Parisi, in *Turbulence and Predictability of Geophysical Flows and Climate Dynamics*, edited by M. Ghill, R. Benzi and G. Parisi (North Holland, Amsterdam, 1985).

[25] M.H. Jensen, L.P. Kadanoff, A. Libshaber, I. Procaccia and J. Stavans, *Phys. Rev. Lett.* **55** (1985) 439; T.C. Halsey, M.H. Jensen, L.P. Kadanoff, I. Procaccia and B.I. Schraiman, *Phys. Rev. A* **33** (1986) 1141.

[26] G. Paladin and A. Vulpiani, *Phys. Rep.* **156** (1987)147.

[27] A. Rényi, in *Probability Theory* (North-Holland, Amsterdam, 1970).

[28] A.I. Saichev and G.M. Zaslavsky, *Chaos* **7** (1997) 753.

[29] G.M. Zaslavsky and M. Edelman, unpublished.

Long-Tailed Distributions in Physics

Given by M.F. Shlesinger

Written by M.F. Shlesinger[1], J. Klafter[2] and G. Zumofen[3]

[1]*Office of Naval Research, Physical Sciences Division,*
Arlington, VA 22217-5660, U.S.A.
[2]*School of Chemistry, Tel-Aviv University, Ramat-Aviv 69978, Israel*
[3]*Laboratorium fur Physikalische Chemie, ETH-Zentrum,*
8092 Zurich, Switzerland

1. INTRODUCTION

It is not difficult to produce random variables whose probability densities have long tails with infinite moments. We give a few example below.

If the distribution of a random variable peaks at the origin, then perhaps the distribution for the inverse of this variable will have a slow decay at large values of the variable. For example, let the random variable x have a Gaussian probability density $g(x)$, with zero mean and unit variance.

$$g(x) = (2\pi)^{1/2} \exp(-x^2/2). \tag{1}$$

Then $y = 1/x$ will have the probability density $f(y)$ with

$$f(y)\mathrm{d}y = g(x)\mathrm{d}x = (2\pi)^{-1/2} \exp(-1/2y^2)(\mathrm{d}y/y^2). \tag{2}$$

This is similar to the Cauchy probability, in that $f(y)$ goes to zero at large y as $1/y^2$. This long tail insures that $f(y)$ will have an infinite variance.

The Gaussian distibution arises as the probability limit distribution for a sum of identically distributed random variables. The lognormal distribution arises as a probability distribution for a product of identically distributed random variables. That is the lognormal is a Gaussian distribution for the variable

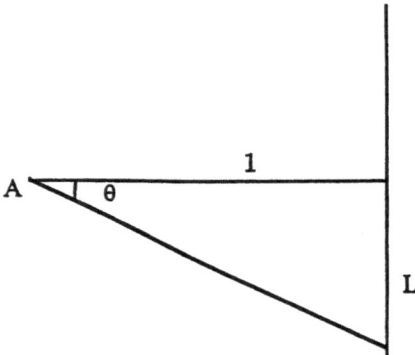

Fig. 1. — For a ray shot out at a random angle from point A, the probability that it hits the wall a distance L away in the vertical direction will be the Cauchy distribution.

$\log(x)$. If $m(x)$ is the lognormal distribution (with x measured in terms of $\langle x \rangle$ and with unit variance) and $g(x)$ is the Gaussian then

$$m(x)\mathrm{d}x = g(\log x)\mathrm{d}(\log x) = g(\log x)\frac{\mathrm{d}x}{x}$$

so

$$(3)$$

$$m(x) = \frac{1}{x}\frac{1}{\sqrt{2\pi}}\exp\left(-\frac{(\log x)^2}{2}\right).$$

There will be some intermediate range of x values where $m(x)$ looks like it it is slowly decaying as $1/x$. The lognormal is one of those distributions all of whose moments are finite, but the n-th moment divided by n, diverges as n goes to infinity. This implies that the lognormal function is not uniquely expressed by its moments, *i.e.*, other distributions can have all of the same moments.

Next consider a light source at point A which releases photons which travel in a straight lines through angles θ and strike an infinite wall one unit away in the x-direction, at distances L units away in the y-direction. from the point where the perpendicular to the wall passing through the point A crosses the wall (see Fig. 1).

If the angle θ has a distribution $f(\theta)$ uniformly randomly distributed between $-\pi/2$ and $\pi/2$ so $f(\theta) = 1/\pi$, the distribution of angles θ, will induce a distribution of hitting points $g(L)$ along the wall. Since

$$f(\theta)\mathrm{d}\theta = \frac{1}{\pi}\mathrm{d}\theta = g(L)\mathrm{d}L$$

and

$$\begin{aligned} \theta &= \tan^{-1}(L) \\ \mathrm{d}\theta/\mathrm{d}L &= 1/(1+L^2) \end{aligned} \qquad (4)$$

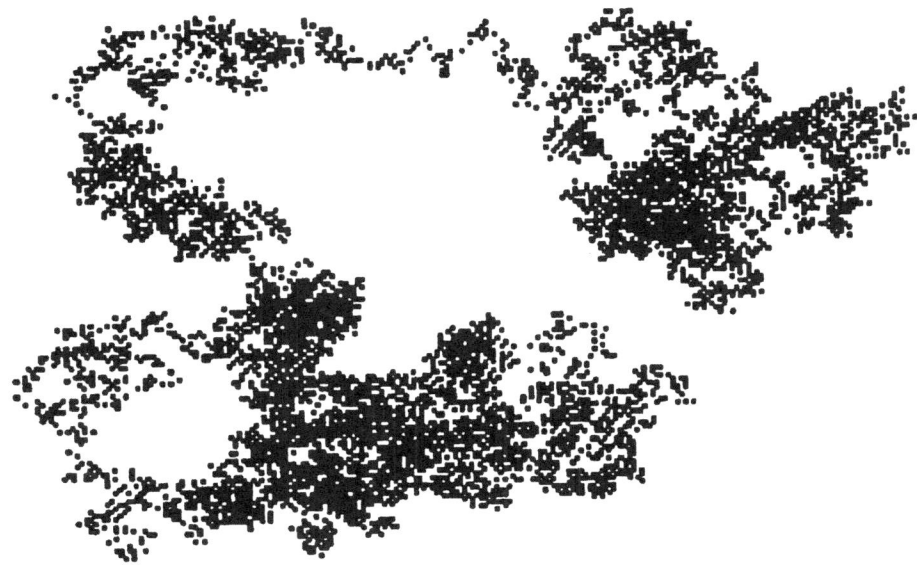

Fig. 2. — A typical Browian motion random walk with jumps on a square lattice to nearest neighbor sites. This should be contrasted with the Lévy flight path in Figure 2.

which allows one to solve for $g(L)$ as

$$g(L) = \frac{1}{\pi} \frac{1}{1 + L^2}$$

which is the Cauchy distribution which has an infintie second moment.

As a first example of a long tailed distribution in time, consider a particle in a potential well of height E. Suppose the time it spends in the well is governed by the Arrhenius law

$$\tau = \nu \exp(E/kT) \tag{5}$$

where ν is a frequency, k is Boltzmann's constant and T is the temperature. Let the barrier height, E, be a random variable governed by the probability distribution $f(E)$, where

$$f(E) = \exp(-E/E_0) \tag{6}$$

i.e., E_0 is the width of the distribution.

This distribution of barrier heights induces a distribution of trapping times,

$$\psi(\tau) = f(E) \frac{\mathrm{d}E}{\mathrm{d}\tau}. \tag{7}$$

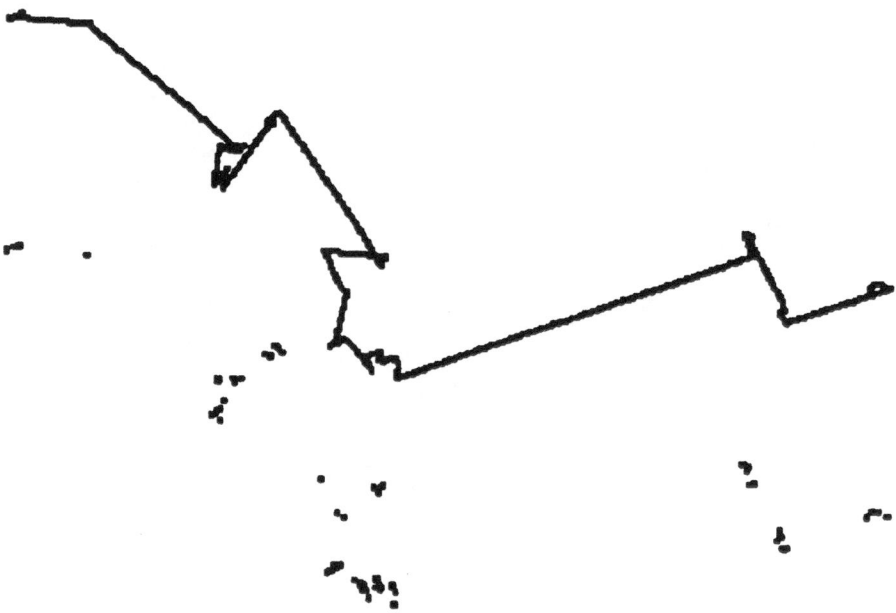

Fig. 3. — A Lévy flight is shown above, and the fractal set of turning points in the trajectory is shown below. If a velocity is ascribed to each segment of the flight then the mean square displacement of the flight becomes a function of time, even when the mean square flight segment size is infinite.

Using equations (5) and (6) one finds the distribution of trapping times to be [1]

$$\psi(\tau) = (kT\nu^{\beta})/\tau^{1+\beta}, \quad \text{where} \quad \beta = kT/E_0. \tag{8}$$

We are interested in the long time behavior of $\psi(\tau)$. If there is a minimum barrier height greater than zero, then τ will always be greater than zero, and we can ignore the short time singularity in equation (5). The value of β is greater than zero which insures that the probability is normalized. When β is less than one, the mean time for hopping over a barrier is infinite. This does not mean that every hop has an infinite waiting time. It just means that waiting times of all scales arise, such that, none of them dominates, so one cannot calculate a characteristic time. The times of jumps marked on a time axis will be a fractal point set of fractal dimension β. When β is greater than or equal to one the jump events occur with a finite average frequency. The dimension of time than stick to the value of one as an upper critical dimension.

Historically, in the early 1700's the St. Petersburg Paradox was designed to yield a probability distribution with an infinite first moment. The paradox involved calculating the average winnings in a game of chance. The game was to flip a coin until a head was achieved for the first time. One could get N tails prior to the first head, and this case would be rewarded with a win of 2^N coins. The probability for this event is $(1/2^{N+1})$. So one can win one coin with probabiltiy 1/2, 2 coins with probabiltiy 1/4, etc. One can see a scaling law develop where winning an order of magnitude more occurs with an order of magnitude less probability. The mean winning is $1 \times 1/2 + 2 \times 1/4 + 4 \times 1/8 + ...$ which is just adddding up 1/2 an infinite number of times. This was treated as a paradox because the notion of probability distribution with an infintie first moment seemed ill posed. But this is precisely the situation we illustrated above for thermal jumps over random barriers.

2. FRACTAL TIME [2]

Let us look at the case of $\langle \tau \rangle$ being infinite in a more general context. Consider the Laplace transform of the waiting-time density,

$$\psi^*(s) \equiv \int_0^\infty \psi(t) \exp(-st) \mathrm{d}t = \sum_{n=0}^\infty (-1)^n s^n \langle t^n \rangle \tag{9}$$

where $\langle t^n \rangle$ is the n-th moment of $\psi(t)$. This expansion only holds if all the moments of $\psi(t)$ are finite. Let us consider a case where this does not occur. Choose $\psi(t)$ as,

$$\psi(t) = \frac{1-q}{q} \sum_{j=1}^\infty q^j \lambda^j \exp(-\lambda^j t) \qquad (\lambda < q < 1). \tag{10}$$

This has a scaling form, with an order of magnitude longer waiting time (in base $1/\lambda$) occuring an order of magnitude less often (in base $1/q$). Our waiting-time distribution is easily checked to be normalized to one, but the condition that $\lambda < q < 1$ leads to $\langle t \rangle$ being infinite, i.e.,

$$\langle t \rangle = \frac{1-q}{q} \sum_{j=1}^\infty \left(\frac{q}{\lambda}\right)^j = \infty. \tag{11}$$

How then can one proceed with a power series expansion of $\psi^*(s)$? We write $\psi^*(s)$ as,

$$\psi^*(s) = \frac{1-q}{q} \sum_{j=1}^\infty \frac{(q\lambda)^j}{[s + \lambda^j]}. \tag{12}$$

The scaling relation,

$$\psi^*(s) = q\psi^* \left(\frac{s}{\lambda}\right) + \frac{\lambda(1-q)}{(s + \lambda)} \tag{13}$$

allows one to determine the non-integer exponent in the expansion of $\psi^*(s)$ in terms of powers of s. Any non-integer exponent must come from the homogeneous part of the equation, $\psi^*(s) = q\psi^*\left(\frac{s}{\lambda}\right)$, which has a solution of the form,

$$\psi^*(s) \approx s^\beta, \qquad \text{with} \quad \beta = \frac{\ln(q)}{\ln(\lambda)}. \tag{14}$$

The asymptotic behavior for small s and large t are,

$$\psi^*(s) \approx 1 - s^\beta + O(s), \text{ as } s \to 0$$

and, as in equation (4)   (15)

$$\psi(t) \approx t^{-1-\beta}, \text{ as } t \to \infty.$$

If β is greater than unity, then $\langle t \rangle$ is finite, and the s term will dominate the s^β term, for small s, so the standard Poisson behavior returns when $\langle t \rangle$ is finite. This is similar to the Central Limit Theorem where the important spatial quantity was $\langle r^2 \rangle$. If this was finite, Gaussian behavior ensued. The condition , $\lambda < q < 1$, which makes $\langle t \rangle$ infinite, also keeps $\beta < 1$. When β equals one, jumps occur at a well-defined constant rate and time flows forward (as measured by the number of jumps) as a one dimensional process. The above $\psi(t)$, with β less than one, is called a fractal time random process because if one made points on a time axis when jumps occur, the set of points would look like a random Cantor set with fractal dimension β. The jumps do not occur at a well defined rate, but in a hierarchical patchy manner.

This mechanism of fractal time was used to describe charges moving in glassy materials, specifically in thin films used in xerography. In an electric field the charges tend to move in one direction. Each jump has a mean distance $\langle d(E) \rangle$ which we assume is proportioanl to the electric field E. The distance covered after a time t can be calculate for fractal time governed by an exponent α to be

$$\langle l(t) \rangle \approx \langle d(E) \rangle t^\alpha \approx E t^\alpha. \tag{16}$$

If the thin film has a length L, then at some time T, the mean position of a packet of charges will be equal to L. At time T,

$$L \approx E T^\alpha$$

or   (17)

$$\frac{1}{T} \approx \left(\frac{E}{L}\right)^{\frac{1}{\alpha}}$$

for normal transport the velocity V is usually written as $V = \mu E$ where μ is the charge mobility, and V is a constant equal to L/T. This gives $(1/T)$ proportional (E/L). The exponent α in the fractal time case seems to imply that the mobility μ should scale as $\mu \propto \left(\frac{E}{L}\right)^{1-\frac{1}{\alpha}}$.

3. SLOW RELAXATIONS [3]

Consider the problem of dielectric relaxation of glasses involving a frozen-in dipole which can be relaxed only when it is hit by a mobile defect. This problem involves the first passage time of random walkers (defects) to reach the origin (frozen-in dipole). Slower than exponential decay processes have been called slow relaxations. The mathematical analysis of the problem is in the form of letting there be V lattice sites and letting N walkers be initially randomly distributed among these sites, not including the origin. The probability $\Phi(t)$ that none of the walkers has reached the origin by time t is given by,

$$\Phi(t) = \left[1 - \frac{1}{V}\sum_r \int_0^t F(\mathbf{r},\tau)\mathrm{d}\tau\right]^N \tag{18}$$

where $F(\mathbf{r},\tau)$ is the probability density that a walker starting at site $\mathbf{r}$ will reach the origin for the first time at time τ. The integral allows for a first passage of a walker to the origin in the interval $(0,t)$. The $1/V$ enters as the probability of a walker starting at a site, and a sum over all possible starting points for a walker is performed. The bracket calculates the probability that a particular walker has not reached the origin and it is raised to the N-th power for the probability that none of the walkers has yet reached the origin. The problem is easier in the limit $N \to \infty$, $V \to \infty$ but with the ratio remaining constant $N/V = c$. In this limit, with $S(t)$ being the number of distinct sites a walker visits in time t,

$$\Phi(t) = \exp\left\{-\left(c\sum_\ell \int_0^t F(\ell,\tau)\mathrm{d}\tau\right)\right\} = \exp(-cS(t)). \tag{19}$$

The simplifiication involving $S(t)$ was accomplished by noting that any of the sites from which a walk can reach the origin in a time t are exactly the same sites a walker starting at the origin can reach in a time t. $S(t)$ has the following form for the random walk jumps governed by $\langle t \rangle$ finite ($\beta = 1$) and infinite ($\beta < 1$),

$$S(t) \approx \exp(-\text{const.}t^\beta). \tag{20}$$

We have found that the stretched exponential (the $\beta < 1$ case) is a probability limit distribution.

4. FRACTAL SPACE PROCESSES

Consider random walks with infinite mean jump lengths. Let S_N be the sum of N identically distributed random variables with zero mean and variance σ^2, i.e., $S_N = X_1 + ... + X_N$. For walks with finite jump variances,

$$p(x) = \lim_{N\to\infty} \text{Prob}\left[x < S_N/\sqrt{N} < x + \mathrm{d}x\right] \tag{21}$$

$$= (2\pi N\sigma^2)^{-1/2} \exp(-x^2/2N\sigma^2).$$

Lévy [4] asked the question of how to proceed when σ^2 is infinite. If the variance of each jump is infinite, then the variance of N jumps is also infinite. This in turn implies that the distribution of the sum of N steps should have similar properties to a single step. This is basically the question of fractals of when does the whole (the sum) look like its parts. Lévy found the answer that the Fourier transform of $p(x)$ was of the form $\tilde{p}(k) = \exp(-|k|^\beta)$ with $\beta \leq 2$. The Gaussian is the case $\beta = 2$. The Cauchy is the case $\beta = 1$. This is reminiscent of our analysis of fractal time where non-integer exponents entered and signaled self-similar behavior and infinite moments of a probability distribution. This time the moments are spatial, instead of temporal.

Let us look at a particular pedagogical case, called the Weierstrass random walk [5], which illuminates the above discussion. Begin with a random walk on a lattice with the following probabilities for jumps of length r,

$$p(r) = \frac{q-1}{2q} \sum_{j=0}^{\infty} q^{-j} (\delta_{r,b}^j + \delta_{r,-b}^j). \tag{22}$$

In this scheme jumps an order of magnitude farther (in base q) occur an order of magnitude less often (in base b). About q jumps of size unity are made, and form a cluster of q sites, before a jump of length b occurs, and about q such clusters of visited sites are formed. Before a jump of length b^2 occurs, and so on until an infinite fractal hierarchy of clusters is formed. See Figure 1, for a 2D Weierstrass random walk. To see the mathematical condition necessary for this occur, examine the Fourier transform of $p(r)$ to obtain,

$$\tilde{p}(k) = \frac{q-1}{q} \sum_{j=0}^{\infty} q^{-j} \cos(kb^j). \tag{23}$$

This is the famous Weierstrass function which is everywhere continuous, but nowhere differentiable when $b > q$. Note that,

$$\langle r^2 \rangle = \sum_{r=-\infty}^{\infty} r^2 p(r) = \frac{q-1}{q} \sum_{j=0}^{\infty} \left(\frac{b^2}{q} \right)^j = -\frac{\partial^2}{\partial k^2} \tilde{p}(k)_{|k=0} = \infty$$

when $b^2 > q$. This divergence of the second moment (absence of a scale) is a sign of fractal self-similar properties. We can write a power series expansion of $\tilde{p}(k)$, but the coefficient of the $\langle r^2 \rangle$ term is infinite. As in the case of a fractal time distribution, a scaling equation is employed,

$$\tilde{p}(k) = \frac{1}{q}\tilde{p}(bk) + \frac{q-1}{q} \cos(k) \tag{24}$$

whose homogeneous part has a solution of the form k^β with $\beta = \frac{\ln(q)}{\ln(b)}$. The full solution behaves for small k (large distances r) as,

$$\tilde{p}(k) \approx \begin{cases} 1 - \text{const.}|k|^\beta + O(k^2) & \approx \exp(-|k|^\beta) \text{ for } \beta < 2 \\ 1 - \frac{1}{2}\langle r^2 \rangle k^2 & \approx \exp(-|k|^2) \text{ for } \beta > 2. \end{cases} \tag{25}$$

Lévy's solution treats a more general case allowing for a bias and additional parameters, but the above discussion captures the general spirit. Although $\tilde{p}(k)$ has a simple form, explicit forms for $p(x)$ directly in terms of analytic functions are not derivable, in general.

Despite the beauty of Lévy flights they have been largely ignored in the physics literature because of their infinite moments. This infinity can be tamed by associating a velocity with each flight trajectory segment [6,7]. Still in the continuous-time random walk framework we consider $\Psi(r,t)$ to be the probability density to make a jump of displacement r, in a time t. We write [8],

$$\Psi(r,t) = \psi(t|r)p(r)$$

or   (26)

$$\Psi(r,t) = p(r|t)\psi(t)$$

where $p(r)$ and $\psi(t)$ have the same meaning as before and $\psi(t|r)$ and $p(r|t)$ are conditional probabilities for a jump taking a time t given it is of distance r, and for a jump being of distance r, given it took a time t. A simple choice is

$$\psi(t|r) = \delta\left(t - \frac{r}{V(r)}\right).$$ (27)

In the study of turbulent diffusion Kolmogorov assumed a scaling law implying $V(r) \sim r^{1/3}$. For Kolmogorov's $V(r)$ employed together with

$$p(r) \approx |r|^{1+\beta}$$ (28)

with small enough values of β so $\langle r^2 \rangle = \infty$ one finds for the mean square displacement [7],

$$\langle r(t) \rangle^2 = \begin{cases} t^3 & \text{for} & \beta \leq 1/3 \\ t^{2+\frac{3}{2}(1-\beta)} & \text{for} & 1/3 \leq \beta \leq 5/3 \\ t & \text{for} & \beta \geq 1/3. \end{cases}$$ (29)

The t^3 case is called Richardson's law of turbulent diffusion. It corresponds to a Lévy walk with Kolmogorov scaling for $V(r)$ combined with a β such that?the mean time spent in a segment of the trajectory is infinite. The reason we did not obtain the usual Lévy flight result of $\langle r^2(t) \rangle = \infty$ is that with the coupled space-time memory $\Psi(r,t)$ one does not calculate $\langle r^2 \rangle = -\frac{\partial^2}{\partial k^2}\tilde{p}(k=0)$ which would be infinite, but one uses $\int \exp(ikr)\psi(s|r)p(r)dr$ instead of $\tilde{p}(k)$. In effect, one is asking how far a walker has gone in a time t, instead of how long is the flight segment. This type of random walk process has been used to model turbulent pipe flow [9].

5. NONLINEAR DYNAMICS

The same type of fractal space and time statistics we have discussed above in random systems, also appears in dynamical systems, both dissipative and Hamiltionian [10-15].

The Lévy flight case of a long-tailed distribution for dynamical phase rotation, with a constant rotation rate was first presented by Geisel *et al.* [4]. Their investigations involved the study of chaotic phase diffusion, in a Josephson junction, *via* a dynamical map. With a constant voltage across the junction the phase rotates at a fixed rate, but can change direction intermittently. N complete clockwise rotations correspond to a random walk with a jump of N units (a laminar phase or flight) to the right. Analysis of data led to a consideration of a map of the form $x_{t+1} = (1 + \varepsilon)x_t + ax_t^z - 1$, with ε small, The iterations of the map tend to approach and cluster near $x = 0$. This can be interpreted as the system remaining in the same state, *i.e.*, CW or CCW rotations, for long times. The number of iterations clustered in a series near zero depends on the initial condition. Nearby initial conditions can lead to vastly differnet cluster sizes, such that a long-tailed distribution arises. The statistical analysis, averaging over initial conditions, finds that [4] $\langle r^2(t) \rangle \approx t^2$, if $z = 2$; t^γ with $\gamma = 3 - 1/(z - 1)$ if $3/2 < z < 2$; $t \ln t$ if $z = 3/2$; and t if $1 < z < 3/2$. This corresponds to a constant velocity Lévy flight with a long-tailed flight time distribution between reversals of motion of $\psi(t) \approx t^{-1-\beta}$ with $\beta = 1/(z - 1)$.

In the above example, one can go directly from the exponent of the map, "z" to the temporal exponent "$\beta = 1/(1 - z)$" in the mean square displacement. In most nonlinear dynamical systems the connection between the equations and the kinetics is not obvious or simple. In fact, say for the standard map,

$$
\begin{aligned}
x_{n+1} &= x_n + K \sin(2\pi\vartheta_n) \\
\vartheta_{n+1} &= \vartheta_n + x_{n+1}
\end{aligned}
\tag{30}
$$

there is no apparent exponent in sight, but non-integer exponents abound in the description of the orbit kinetics, hinting at a complex dynamics [16].

In a plot of x *vs.* ϑ, say, with $K = 1.1$ some orbits are of the simple closed circular type, while others wander chaotically. Still other orbits exhibit periodicity, *i.e.*, $|x_p - x_0| = \ell$ where a particle moves along the x axis ℓ units in p iterations, with the velocity ℓ/p. Sticking close to these orbits provides for the laminar segments of Lévy flights. The fractal nature of these island hierarchies (called Cantori as they form a Cantor-like set of tori; see Fig. 4) creates complex kinetics with Lévy statistics which have been dubbed "strange kinetics" [10]. The probability densities for spending time in laminar states near period 3 and period 5 orbits are long-tailed and have the Lévy distribution algebraic asymptotic form, $\psi_{\text{period }3}(t) \approx t^{-2.2}$ and $\psi_{\text{period }5}(t) \approx t^{-2.8}$. A finer resolution of the period 3 orbit uncovers a period 7 orbit, and a finer resolution of one of these would find ten sub-islands for each period 7 orbit. One can keep

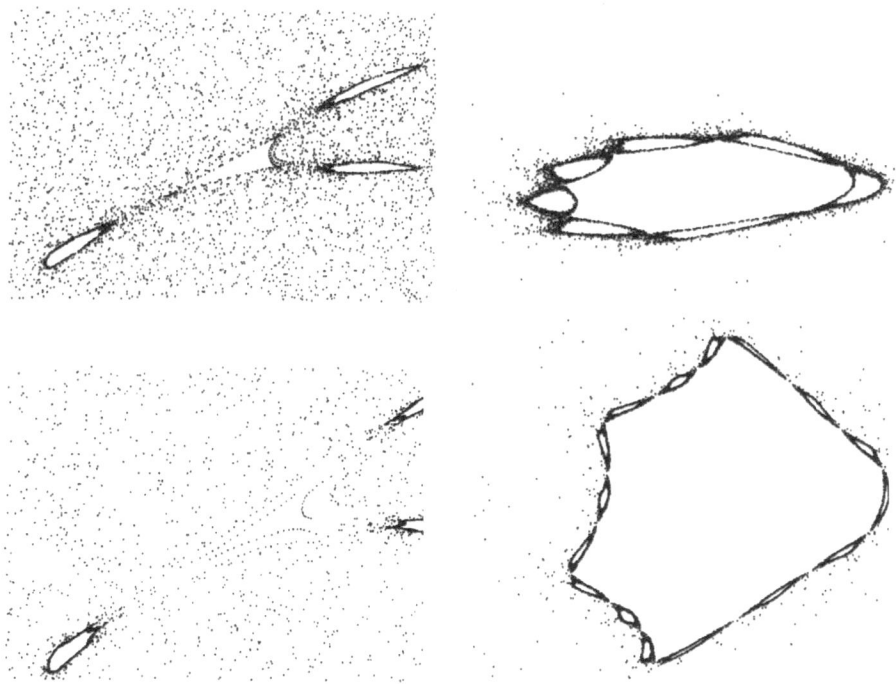

Fig. 4. — a) In the standard map for $K = 1.1$ a period 3 orbit is shown surrounded by a chaotic sea. b) under higher resolution each of these period 3 orbits is found to consist of 7 sub-islands. c) for the slightly different value of $K = 1.1015$ the period 3 orbit is shown again. d) under higher resolution a different hierarchy of islands is seen. A period 16 orbit is found instead of the period 7 orbit for the $K = 1.1$ case.

finding that islands are composed of sub-island down to the resolution of the computation.

To predict these exponents, for these long tailed distributions, is difficult. For example if we choose a K value of 1.1015 rather than 1.1 the period 3 orbit upon further resolution goes into a period 16 orbit, rather than the period 7 orbit (Figs. 4a-d). Slightly different K values can give rise to different exponents because the precise island structure changes in a sensitive manner to the energy "K" of the system. Also at a different energy the new orbits do not necessarily traverse the hierarchy in the same order. Escape times from a level of the hierarchy is not a simple function of distance from the island [16].

Another example of Lévy flights in dynamical systems are the orbits in the Zaslavsky map [10,14]

$$u_{n+1} = (u_n + K\sin(\nu_n))\cos(\alpha) + \nu_n\sin(\alpha) \qquad (31)$$
$$\nu_{n+1} = -(u_n + K\sin(\nu_n))\sin(\alpha) + \nu_n\cos(\alpha)$$

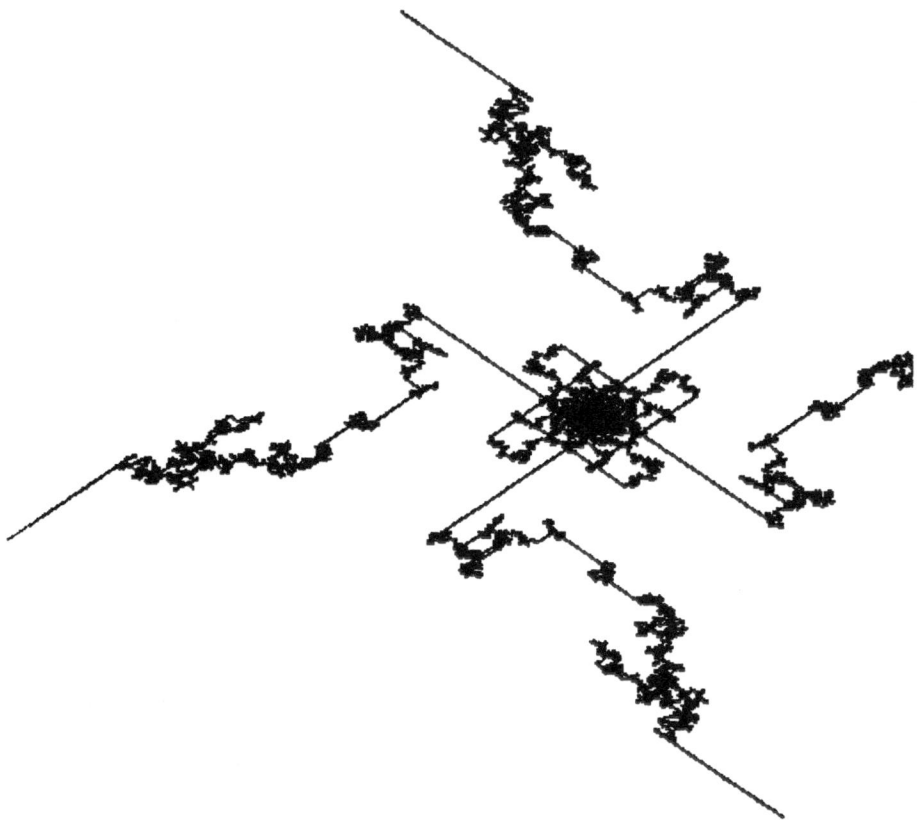

Fig. 5. — A Lévy-like trajectory with a 4-fold symmetric ($q = 4$) is found for the Zaslavsky map when the value $K = 6.349972$ is chosen.

where $\alpha = 2\pi/q$, which form an intricate fractal web, with q-fold symmetry, throughout a 2D (u, ν) phase space. This map only has simple trigonometric nonlinearities, but like the standard map, it requires non-integer exponents for its characterization of trajectories.

As a last example, a quasi two-dimensional rotating laboratory flow was experimentally investigated by Solomon *et al.* [12]. In their experiment, a fluid filled annulus rotates as a rigid body. The flow is established by pumping fluid through holes in the bottom of the annulus. Since their flow was nearly two-dimensional the stream function, given by a solution of the viscous Navier-Stokes equations, is a Hamiltonian for neutrally buoyant tracer particles in the flow. Even though the velocity field is laminar, Lévy flight advection of neutrally buoyant tracer particles is not ruled out. In fact, direct measurements were made of Lévy flights and enhanced diffusion in this system. An instability of the axisymmetrically pumped fluid led to a stable chain of six vortices. Tracer particles were followed and statistics were collected on the probability distribution for the time for a tracer particle to be to confined to a vortex, and the "distance" (number of vortices passed) by a tracer particle "in flight" before it is trapped again in a vortex.

When a tracer particle leaves a vortex and then passes r vortices in a counter-clockwise (clockwise) fashion before it is trapped again in a vortex, this counts as a random walker jumping r units to the right (left) at a constant velocity. In Lévy walk notation, the following distributions were measured

$$\psi_{\text{flight}}(t|r) = \delta(t - |r|/v) \tag{32}$$
$$p_{\text{flight}}(r) \approx |r|^{-2.3} = |r|^{-1-\beta}.$$

For this constant magnitude of velocity, v, flight motion, one has

$$\langle r^2(t) \rangle \approx \begin{cases} t^2 & \text{for} \quad 0 < \beta < 1 \\ t^2/\ln(t) & \text{for} \quad \beta = 1 \\ t^{3-\beta} & \text{for} \quad 1 < \beta < 2 \\ t\ln(t) & \text{for} \quad \beta = 2 \\ t & \text{for} \quad \beta > 2. \end{cases} \tag{33}$$

A long time tail was found in the distribution for the times a particle is trapped in a vortex $\psi_{\text{trapped}}(t) \approx t^{-2.6}$. The mean trapped time is finite and so the trapping exponent will not effect the exponent for the mean squared displacement. Under these conditions Lévy flight theory implies $\langle (r(t) - \langle r(t) \rangle)^2 \rangle \approx t^\gamma$, with $\gamma = 3 - \beta = 1.7$ which is consistent with the experimental data and also presents a beautiful example of enhanced diffusion.

REFERENCES

[1] G. Pfister and H. Scher, *Adv. Phys.* **27** (1974) 1474.
[2] M.F. Shlesinger, *Ann. Rev. Phys. Chem.* **39** (1988) 269.
[3] H. Scher, M.F. Shlesinger and J.T. Bendler, *Physics Today* (1991) 26-34.
[4] P. Lévy, Théorie de l'addition des variables aléatoires (Gauthier-Villars, Paris, 1937).
[5] B.D. Hughes, M.F. Shlesinger and E.W. Montroll, *Proc. Nat. Acad. Sci. (USA)* **78** (1981) 3287-3291.
[6] T. Geisel, J. Nierwetberg and A. Zacherl, *Phys. Rev. Lett.* **54** (1985) 616.
[7] M.F. Shlesinger, B.J. West and J. Klafter, *Phys. Rev. Lett.* **58** (1987) 1100-1103.
[8] M.F. Shlesinger, J. Klafter and Y.M. Wong, *J. Stat. Phys.* **27** (1982) 499.
[9] F. Hayot, *Phys. Rev. A* **43** (1991) 806.
[10] M. Shlesinger, G. Zaslavsky and J. Klafter, *Nature* **263** (1993) 31.
[11] J. Klafter, G. Zumofen and M. Shlesinger, *Fractals* **1** (1994) 389.
[12] T.H. Solomon, E.R. Weeks and H.L. Swinney, *Phys. Rev. Lett.* **71** (1993) 3575.

[13] T. Geisel, J. Nierwetberg and A. Zacherl, *Z. Phys. B* **71** (1988) 117.
[14] D.K. Chaikovsky and G.M. Zaslavsky, *Chaos* **1** (1991) 463.
[15] J. Klafter and G. Zumofen, *Phys. Rev. E* **49** (1994) 4873.
[16] J. Klafter, M.F. Shlesinger and G. Zumofen, *Physics Today* (1996) 33-39.

Distribution of Galaxies: Scaling *vs.* Fractality

R. Balian

Service de Physique Théorique, Centre de Saclay,
91191 Gif-sur-Yvette, France

---

1. INTRODUCTION

The distribution of matter in the Universe is probably the oldest paradigm of fractality. In the major part of the literature devoted to this theme, which is already ancient, models of fractal structures are first constructed, then checked against observations. This short course will present a *converse approach*. The data about the actual distribution of galaxies in the 10 to 100 Mpc range are now rather precise and reliable, and we can thus start from them so as to *derive* the fractal properties of the set. We shall give below in a self-contained way the main ideas of this construction; the figures, equations, detailed derivations, discussions and references can be found in two articles [1,2] denoted here as I and II. We shall focus less on astrophysical implications than on general methods which may be of interest for other problems.

We have to reconcile two evidences (II, Sect. 2). On the one hand, the Universe is *homogeneous at very large scales*, with a finite density; this agrees with the cosmological theory of the expansion of the Universe, as well as with the count of galaxies according to their apparent magnitude (II, Fig. 2). On the other hand, the galaxies are clustered at the 10 Mpc scale in (possibly branched) filaments or walls, separated by extremely large voids (II, Fig. 16); this suggests a *fractal structure* on this scale, with dimension between 1 and 2.

We shall use three complementary tools to describe the distribution of galaxies: (i) two-point, three-point, ... *correlation functions*. (ii) *Counts in cells*, which provide the probabilities $P_N(v)$ that N galaxies (with $N = 0, 1, 2, ...$) lie within a region of space having given volume v. (iii) Fractal or *multifractal exponents*, depending on the scale.

Observations yield statistical data about *correlation functions* and about *counts in cells*. On the other hand *theoretical studies* of the dynamics of galaxies may provide information on the correlation functions. The data about correlations and counts in cells are complementary, and we give in Section 2 a formalism which encompasses both sets of quantities. We then show (Sect. 3) that a scaling hypothesis for correlation functions, supported by observational data, gives rise to a rich structure for $P_N(v)$. Finally (Sect. 4) we derive therefrom the rather exotic fractal properties of the set. We conclude (Sect. 5) on the relationship between scaling and fractality.

2. ALGEBRA OF POINT DISTRIBUTIONS

We regard the galaxies as a set of indistinguishable points in a Euclidean 3-dimensional space. Their statistical distribution can be characterized by various functions that we shall relate to one another.

2.1. Densities and correlations

The analysis of the relative positions of galaxies, taken two by two, three by three, ..., characterizes their statistical distribution by means of correlation functions. The formalism, recalled below, closely follows that of classical statistical mechanics [3], except for the replacement of phase space by ordinary space.

The most direct description relies on the *global probability densities* $D_M(\mathbf{r}_1...\mathbf{r}_M)$. Given a large volume V, the largest one about which we may get data, we denote (as in statistical mechanics) by $D_M(\mathbf{r}_1...\mathbf{r}_M)d\tau_M$ the probability that the number of points lying in V is M, that the first of these M points lies within the volume element $d^3\mathbf{r}_1$ around $\mathbf{r}_1$, the second point within $d^3\mathbf{r}_2$ around $\mathbf{r}_2$,..., and the last point within $d^3\mathbf{r}_M$ around $\mathbf{r}_M$. The points being indistinguishable, it is convenient to normalize the measure $d\tau_M$ as $(M!)^{-1}d^3\mathbf{r}_1...d^3\mathbf{r}_M$. If the positions of all points are known, M takes a single value and D_M is a product of M δ-functions, symmetrized by permuting the labels of the points. Otherwise the overall density is $n = V^{-1}\sum_M \int d\tau_M D_M M$. The set D_M can be characterized by its generating functional

$$\mathcal{D}\{\mu(\mathbf{r})\} \equiv \sum_M \int d\tau_M D_M \mu(\mathbf{r}_1)...\mu(\mathbf{r}_M), \tag{1}$$

depending on an arbitrary test function $\mu(\mathbf{r})$.

The *reduced densities* $F_N(\mathbf{r}_1...\mathbf{r}_N)$ refer to a subset of N points, N being small compared to nV. The quantity $F_N(\mathbf{r}_1...\mathbf{r}_N)d^3\mathbf{r}_1...d^3\mathbf{r}_N$ is defined as the joint probability that some point lies within $d^3\mathbf{r}_1$ around $\mathbf{r}_1$, some other point within $d^3\mathbf{r}_2$ around $\mathbf{r}_2$, etc. In particular, F_0 equals 1 in agreement with the normalization of D, and $F_1(\mathbf{r})$ is the local density at $\mathbf{r}$. The reduced densities are expressed in terms of the probabilities D_M by

$$F_N(\mathbf{r}_1...\mathbf{r}_N) = \sum_M \int d\tau'_M D_{N+M}(\mathbf{r}_1...\mathbf{r}_N\mathbf{r}'_1...\mathbf{r}'_M). \tag{2}$$

Hence, their generating functional, defined as in (1), is found to be

$$\mathcal{F}\{\nu(\mathbf{r})\} \equiv \sum_N \int d\tau_N F_N \nu(\mathbf{r}_1)...\nu(\mathbf{r}_N) = \mathcal{D}\{\mu(\mathbf{r}) = \nu(\mathbf{r}) + 1\}. \tag{3}$$

Indeed, the replacement of μ by $\nu + 1$ in (1) produces the coefficients (2); the normalization chosen for $d\tau_N$ takes care of the combinatory factors. Conversely (3) implies

$$D_M(\mathbf{r}_1...\mathbf{r}_M) = \sum_N (-)^N \int d\tau'_N F_{M+N}(\mathbf{r}_1...\mathbf{r}_M\mathbf{r}'_1...\mathbf{r}'_N). \tag{4}$$

The *connected N-point functions* G_N are recursively defined as the cumulants of the set F_N, namely,

$$
\begin{aligned}
F_1(\mathbf{r}_1) &= G_1(\mathbf{r}_1), \\
F_2(\mathbf{r}_1\mathbf{r}_2) &= G_1(\mathbf{r}_1)G_1(\mathbf{r}_2) + G_2(\mathbf{r}_1\mathbf{r}_2), \\
F_3(\mathbf{r}_1\mathbf{r}_2\mathbf{r}_3) &= G_1(\mathbf{r}_1)G_1(\mathbf{r}_2)G_1(\mathbf{r}_3) + G_1(\mathbf{r}_1)G_2(\mathbf{r}_2\mathbf{r}_3) \\
&\quad + G_1(\mathbf{r}_2)G_2(\mathbf{r}_1\mathbf{r}_3) + G_1(\mathbf{r}_3)G_2(\mathbf{r}_1\mathbf{r}_2) + G_3(\mathbf{r}_1\mathbf{r}_2\mathbf{r}_3), \\
F_4 &= G_1G_1G_1G_1 + \sum G_1G_1G_2 + \sum G_2G_2 \\
&\quad + \sum G_3G_1 + G_4, \quad \text{etc.,}
\end{aligned}
\tag{5}
$$

with the convention $G_0 = 0$. The set of equations (5) is equivalent to the relation

$$\mathcal{F}\{\nu(\mathbf{r})\} = \exp \mathcal{G}\{\nu(\mathbf{r})\} \tag{6}$$

between the generating functionals of the two sets F_N and G_N.

For a statistically homogeneous distribution, the density $n = F_1 = G_1$ is uniform, and it is convenient to introduce the dimensionless *correlation functions*

$$\xi_N(\mathbf{r}_1...\mathbf{r}_N) \equiv n^{-N} G_N(\mathbf{r}_1...\mathbf{r}_N), \tag{7}$$

which are proportional to the connected cumulants, with $\xi_0 = 0, \xi_1 = 1$. For $N \geq 2$, these correlation functions depend only on the differences of coordinates; in particular the two-point function ξ_2 depends only on $|\mathbf{r}_1 - \mathbf{r}_2|$ if the distribution is statistically isotropic. The generating functional for the set ξ_N is

$$\sum_N \int d\tau_N \xi_N \nu(\mathbf{r}_1)...\nu(\mathbf{r}_N) = \mathcal{G}\{n^{-1}\nu(\mathbf{r}\}. \tag{8}$$

The distribution of points can thus equivalently be characterized by the set D_M, the set F_N, the set $G_N(N \geq 1)$ and the set n, ξ_N $(N \geq 2)$, which are related to one another through (2, 6) and (8).

2.2. Counts in cells

An alternative means for characterizing the distribution of points is provided by *counts in cells*. Consider a test cell of volume $v = l^3$, with either *cubic* or *spherical* shape. At the scale l, the statistical distribution may conveniently be described by the probability $P_N(l)$ that a number N of points lie within the volume $v = l^3$. In practice, $P_N(l)$ is found by dividing the Universe into identical cells and by counting the number of galaxies within each cell. An important special case is $P_0(l)$, the *void probability*, which is the probability for a volume $v = l^3$ to be empty.

Let us express the set P_N in terms of the correlation functions. The probability P_N is first related to the global probability density by

$$P_N = \sum_M \int_v d\tau_N \int_{Cv} d\tau'_M D_{N+M}(\mathbf{r}_1...\mathbf{r}_N\mathbf{r}'_1...\mathbf{r}'_M), \qquad (9)$$

where the first N coordinates are integrated inside v, the last M ones outside v. The generating function

$$\mathcal{P}(\lambda) \equiv \sum_N P_N \lambda^N \qquad (10)$$

is then obtained by integration of (9) over all points, with a weight $\mu(\mathbf{r})$ equal to λ for each $\mathbf{r} \in v$, to 1 for each $\mathbf{r} \notin v$. Hence, using the definition (1) and the relation (3), we find

$$\mathcal{P}(\lambda) = \mathcal{D}\{\mu(\mathbf{r})\} = \mathcal{F}\{\nu(\mathbf{r})\}, \qquad (11)$$

where we should set $\nu(\mathbf{r}) = \lambda - 1$ inside v, $\nu(\mathbf{r}) = 0$ outside. The relations (6) and (8) then allow us to express the void probability P_0 in terms of the density n and the correlation functions ξ_N as

$$P_0 = \mathcal{P}(0) = \exp \sum_{N=1}^{\infty} (-n)^N \int_v d\tau_N \xi_N(\mathbf{r}_1...\mathbf{r}_N). \qquad (12)$$

More generally the generating function $\mathcal{P}(\lambda)$ is

$$\mathcal{P}(\lambda) = \exp \sum_N (\lambda n - n)^N \int_v d\tau_N \xi_N(\mathbf{r}_1...\mathbf{r}_N), \qquad (13)$$

which implies

$$P_N = \frac{(-n)^N}{N!} \left(\frac{d}{dn}\right)^N \mathcal{P}(0). \qquad (14)$$

The counts in cells of size l are thus directly related to integrals of the correlation functions over distances shorter than l. However, their expression involves correlation functions of all orders.

Observations yield reliable information, on the one hand about n, ξ_2, and with less accurency ξ_3 and ξ_4, on the other hand about $P_N(v)$ for some ranges of N and v, in particular the void probability as a function of l. Such pieces of information are *complementary*, and equations (12–14) are useful to merge them so as to account simultaneously for them as exemplified in the next section.

3. SCALE INVARIANCE

3.1. Scaling of correlations

The first four correlation functions of galaxies exhibit scaling properties for distances in the range of 0.1 to 10 or possibly 100 h^{-1} Mpc. The two-point function ξ_2, which depends only on the distance $r = | \mathbf{r}_1 - \mathbf{r}_2 |$, behaves there as

$$\xi_2(\mathbf{r}_1\mathbf{r}_2) \sim \left(\frac{r_0}{r}\right)^\gamma, \tag{15}$$

with an exponent $\gamma \simeq 1.8$ and a coefficient $r_0 \simeq 5$ h^{-1} Mpc. The data on the three-point function agree in the same range with a hierarchical model $\xi_3(\mathbf{r}_1\mathbf{r}_2\mathbf{r}_3) \simeq \xi_2(\mathbf{r}_1\mathbf{r}_2)\xi_2(\mathbf{r}_1\mathbf{r}_3)+$ permutations, or at least with a scale invariance involving the exponent -2γ.

More generally, the scaling assumption

$$\xi_N(k\mathbf{r}_1, k\mathbf{r}_2...k\mathbf{r}_N) = k^{-(N-1)\gamma}\xi_N(\mathbf{r}_1\mathbf{r}_2...\mathbf{r}_N) \tag{16}$$

for any $N \geq 2$ is compatible with observational data for $2 \leq N \leq 4$, and with numerical simulations of galaxy dynamics; it is also compatible with the BBGKY dynamics for $\Omega = 1$, as well as for $\Omega < 1$ in the limit of large times.

3.2. The void probability

From (12) and (16) we find that the void probability has the form

$$P_0 = \exp\left[-nv + \frac{1}{2}n^2v^2\bar{\xi} + \sum_{N\geq 3} \frac{(-nv)^N}{N!}S_N\bar{\xi}^{N-1}\right], \tag{17}$$

where $\bar{\xi}$ is the average of ξ_2 over the cell v, namely from (15),

$$\bar{\xi} = 4.4\left(\frac{r_0}{l}\right)^\gamma \equiv \left(\frac{l_0}{l}\right)^\gamma \tag{18}$$

for a sphere of volume $v = l^3$, and where the S_N are c-numbers depending only on the form of the functions ξ_N and on the shape of the cell. For an uncorrelated set of points, we would get only the first term, $P_0 = e^{-nv}$. The second term describes Gaussian statistics, and the 3-point, 4-point, ... correlation functions provide deviations from Gaussian statistics.

By defining $S_1 \equiv S_2 \equiv 1$ and by introducing the function

$$\varphi(y) \equiv \sum_{N=1}^{\infty} \frac{(-)^{N-1}}{N!} S_N y^N, \tag{19}$$

we can rewrite (17) under the scaling form

$$P_0 = \exp\left[-\varphi(N_c)/\bar{\xi}\right], \tag{20}$$

where the cell size l appears through $\bar{\xi}$ (with the exponent $-\gamma$) and through

$$N_c(l) \equiv n v \bar{\xi} = n l_0^3 (l/l_0)^{3-\gamma} \tag{21}$$

(with the exponent $3-\gamma$). The density n appears through N_c, and the constant l_0 is about 10 h^{-1} Mpc.

Accordingly we get from (13) the scaling behaviour of the generating function $\mathcal{P}(\lambda)$ by replacement of n by $n(1-\lambda)$, which yields

$$\mathcal{P}(\lambda) = \exp\left\{-\varphi[N_c(1-\lambda)]/\bar{\xi}\right\}. \tag{22}$$

3.3.   Scaling of counts in cells

In spite of the simple form of (22), the probabilities P_N obtained by expanding it in powers of λ according to (10) have a surprisingly rich scaling structure. We first note that $n l_0^3$ is of the order of 10^3, much larger than 1. Hence $\mathcal{P}(\lambda)$ will be governed by the few following properties of the function $\varphi(y)$:

(i) Its power expansion near $y = 0$ reads

$$\varphi(y) \underset{y \to 0}{\approx} y - \frac{y^2}{2} + \frac{S_3 y^3}{3!} - \frac{S_4 y^4}{4!}, \tag{23}$$

where S_3 can be estimated as $S_3 \simeq 3$ and S_4 as $S_4 \simeq 12$ from the available data on ξ_3 and ξ_4.

(ii) It behaves for large y as

$$\varphi(y) \underset{y \to \infty}{\sim} a y^{1-\omega}, \tag{24}$$

with $\omega \simeq 0.7$ and $a \simeq 1.5$, as can be shown by adjusting the observed probability P_0 of large voids ($N_c(l) > 1$) upon the scaling form (20) (see II, Fig. 4). The information (23) given by correlation functions and (24) given by the void probability are clearly complementary, as expected.

(iii) Adjustement of the curve $P_0(l)$ for smaller values of l, such that $N_c(l) < 1$, shows the existence of a singularity near $y \simeq -0.1$ in the function $\varphi(y)$ extrapolated towards $y < 0$. The coefficients S_N in (19) thus roughly increase as $10^N N!$ for large N, indicating that the N-point correlations are large for increasing N. The statistics of galaxies is far from being Gaussian, in which case we would have $\varphi(y) = y - \frac{1}{2}y^2$.

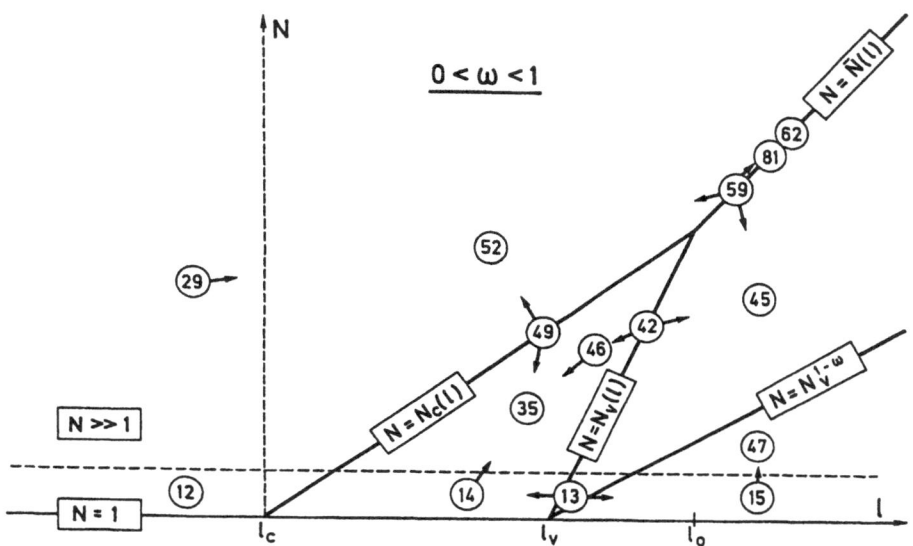

Fig. 1. — The various forms of $P_N(l)$ in the (l, N) plane, in logarithmic scales. The circled numbers refer to equations in I, Appendix B, and arrows indicate simplifications of analytic forms.

From the scaling hypothesis (16) and from the above empirical properties of $\varphi(y)$, we can derive, using (22), the behaviour of $P_N(l)$ as a function of N and of the cell size $v = l^3$, and then check these results against actual counts in cells. We refer to I for this derivation and for a detailed discussion, and give here only a few qualitative indications. The property $nl_0^3 \gg 1$ entails the existence of numerous, rather well separated regimes for $P_N(l)$, depending on regions in the (l, N) plane (Fig. 1). These regions are separated by several lines, namely:

(i) $N = \bar{N}(l)$, where

$$\bar{N}(l) \equiv nl^3 \tag{25}$$

is the average number of galaxies in the cell of volume $v = l^3$;

(ii) $N = N_c(l)$, where $N_c(l)$ is defined by (21);

(iii) $N = N_v(l)$, where we introduce

$$N_v(l) \equiv a^{1/(1-\omega)} nl^3 (l/l_0)^{\gamma\omega/(1-\omega)}; \tag{26}$$

and

(iv) $N = N_v(l)^{1-\omega}$. Three different length scales also occur:

   (i) the largest one is l_0;

   (ii) the smallest one, l_c, defined by the condition $N_c(l_c) = 1$, describes a *clustering length*, *i.e.*, the size of a cell centered on a galaxy and such that the closest galaxy lies on its boundary;

   (iii) the intermediate length, l_v, defined by the condition $N_v(l_v) = 1$, describes a typical *void size*.

For the small scales $l < l_c$, $\ln P_N$ decreases linearly with N, as for Poisson statistics but with a cross-over between two different slopes (I, Fig. 6).

For larger scales, $l_c < l < l_v$, P_N behaves as a power law $P_N \propto N^{-(2-\omega)}$ up to $N \simeq N_c(l)$, except for the first few values of N where discreteness matters; it decreases exponentially beyond the cut-off $N \simeq N_c$ (I, Fig. 7). Moreover, when l varies in this range, the non-trivial dependence of $P_N(l)$ on l and N comes out through the combination $N/l^{3-\gamma} \propto N/N_c(l)$, in the form

$$P_N(l) = \frac{\bar{N}(l)}{N^2} h\left(\frac{N}{N_c(l)}\right), \tag{27}$$

where the function $h(x)$ is related to $\varphi(y)$ through the Laplace transform

$$\varphi(y) = \int_0^\infty dx (1 - e^{-xy}) h(x). \tag{28}$$

The universal scaling form (27) for $P_N(l)$ in the region $l_c < l < l_v, N \gg 1$ agrees with observations on the statistics of objects of various sizes, galaxies, groups of galaxies, rich clusters (I, Fig. 10).

Finally, for the very large scales $l > l_v$, P_N again satisfies the power law $P_N \propto N^{-(2-\omega)}$, but now only for N between $N_v(l)$ and $N_c(l)$; it is much smaller, both for $N < N_v(l)$ and for $N > N_c(l)$ (I, Fig. 8). Here again, we find scaling forms involving either $N/N_c(l)$ as in (27) or $N/N_v(l)$.

These scaling properties of $P_N(l)$, more complicated than the underlying scaling property (16) of the correlation functions from which they emerge, suggest the existence of a scale-dependent fractal structure, that we now discuss.

4.   FRACTALITY

4.1.   Correlation dimension

Consider a galaxy lying at the center of a spherical cell, of volume $v = l^3$. The average number of its neighbours, those which lie within the sphere, is obtained from the two-point correlation function as

$$\bar{N}_{\text{nei}}(l) = n \int_v d^3\mathbf{r} [1 + \xi_2(0\ \mathbf{r})]. \tag{29}$$

The scaling (15) of ξ_2 yields

$$\bar{N}_{\rm nei}(l) \sim nl^3 + \left(\frac{4\pi}{3}\right)^{\gamma/3} \frac{3nr_0^\gamma}{3-\gamma} l^{3-\gamma}. \tag{30}$$

This expression agrees with the countings of galaxies as a function of their apparent luminosity, which is related to their distance from our own galaxy, taken to lie at the center of the cell. It displays a *cross-over* between two regimes (II, Fig. 1). For $l < l_0 \simeq 10 \ h^{-1}$ Mpc, the correlation dimension $D_{\rm corr}$, defined as the exponent of l in (30), equals

$$D_{\rm corr} = 3 - \gamma. \tag{31}$$

It reflects the scaling property (15).

For $l > 5 \ l_0$, the expression (30) is dominated by its first term. The correlation dimension $D_{\rm corr}$ is then the trivial dimension 3. This reflects the *statistical homogeneity* of the Universe at very large scales.

In a purely fractal set, we would have $\bar{N}_{\rm nei}(l) \propto l^{3-\gamma}$ at all scales, which implies (for $l \to \infty$) that the overall density of the Universe would vanish. The cosmological evidence for a finite density n and for a translational invariance of the statistical distribution is reflected in the cross-over of the fractal correlation dimension from $3 - \gamma$ to 3. Even if the *scaling property* (15) *of ξ_2 holds at all scales*, in which case (30) is exact, the *fractality* (31) *of the set holds only up to a maximum range l_0*. The scale invariance of correlations may generate fractality, but it also sets up limits on the validity domain of fractality. Scaling thus appears as a more general property than fractality. We shall moreover see below that a single type of scaling may lead to different, complex types of fractality.

4.2. Hausdorff dimension for occupied cells

Consider a large fixed volume V, divided into cells of variable size l, with volume l^3. The total number of boxes is Vl^{-3}. Hence the number of empty boxes is $Vl^{-3}P_0(l)$, where P_0 is the void probability, and the number of occupied boxes reads

$$C_{\rm occ}(l) = Vl^{-3}[1 - P_0(l)]. \tag{32}$$

We can define in some range of l a dimension D_0 associated with the occupied cells if in this range the number $C_{\rm occ}(l)$ behaves as a power law l^{-D_0}. In such a case the scaling form (20) of P_0 is reflected by a fractal property.

For $l \ll l_c$, we thus find a trivial dimension $D_0 = 0$, associated with the *discreteness* of the galaxy distribution. Indeed, when the cells are small, the occupied ones nearly always contain a single galaxy, so that $C_{\rm occ} = nV$. Another trivial dimension, $D_0 = 3$, occurs for $l \ll l_v$, in which case P_0 nearly vanishes. Being much larger than the void size, the cells are then *all occupied*.

However, for $l \ll l_v$, *most cells are empty*, so that $C_{\rm occ}(l) \sim Vl^{-3}\varphi(N_c)\bar{\xi}$. We recover $C_{\rm occ} = nV$ for $N_c \ll 1$, but find a non-trivial power-law when

$l_c \ll l \ll l_v$, in which case $\varphi(N_c)$ takes its asymptotic form corresponding to $N_c \gg 1$. We then obtain a Hausdorff dimension

$$D_0 = (3 - \gamma)\omega \tag{33}$$

that differs from D_{corr} since $\omega \simeq 0.7$. Actually this dimension (33) holds in a range $l_c \ll l \ll l_v$ much more narrow than the range $l < l_0$ for (31).

The difference between the two non-trivial dimensions (31) and (33) arises from the fact that D_{corr} tests only the *two-point correlation* ξ_2 while D_0 tests only the *void probability* P_0. This suggests that by testing the whole set P_N we should find multifractality.

4.3. Renyi index

Models of multifractality for the distribution of galaxies have been proposed long ago by Benoit Mandelbrot, as he recalled in his course at this school. Our purpose here is to find which type of multifractality emerges from the scaling form (14, 22) of the counts in cells, a form itself supported by observational data.

As in the Section 4.2 above, we consider the partition of the volume V into a number Vl^{-3} of cells with size l, but we now *weight these cells* differently, depending on the number of galaxies that they contain (instead of distinguishing only the occupied cells from the empty ones). Denoting by N_i the number of galaxies which lie in each occupied cell i, we define the weighted number of occupied cells as

$$C_q(l) \equiv \sum_i \left(\frac{N_i}{N_{\mathrm{tot}}}\right)^q, \tag{34}$$

when $N_{\mathrm{tot}} = nV$ is the total number of galaxies in V and q is an arbitrary *testing exponent*. For $q = 0$, $C_0(l)$ reduces to the number (32) of occupied cells. As q increases we focus more and more on the dense regions where N_i is large. We can express (34) in terms of the probabilities $P_N(l)$ as

$$C_q(l) = Vl^{-3} \sum_{N=1}^{\infty} \left(\frac{N}{N_{\mathrm{tot}}}\right)^q P_N(l) = \frac{Vl^{-3}}{(nV)^q} \langle N^q \rangle. \tag{35}$$

For given q, if $C_q(l)$ behaves in some range of l as a power law

$$C_q(l) \propto \left(\frac{l}{\Lambda}\right)^{\tau(q)}, \tag{36}$$

the *Renyi index* D_q is defined in terms of the exponent $\tau(q)$ of (36) as

$$D_q = \frac{\tau(q)}{q - 1} = \frac{1}{q - 1} \frac{d(\ln C_q)}{d(\ln l)}. \tag{37}$$

The validity of the scaling (36) in some range corresponds to the constancy in this range of the logarithmic derivative in (37).

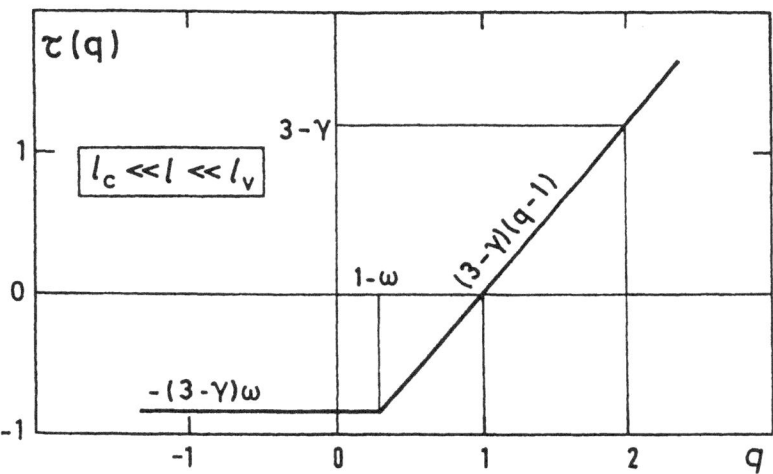

Fig. 2. — The exponent $\tau(q)$ for scales $l_c \ll l \ll l_v$.

For $q = 0$, we recover $C_0 = C_{\text{occ}}$, so that D_0 is the *Hausdorff dimension* of Section 4.2. For $q = 2$, we note that

$$\langle N^2 \rangle = nv(nv + 1) + n^2 \int_v 2d\tau_2 \xi_2, \tag{38}$$

which yields, for volumes such that $nv \gg 1$,

$$C_2(l) = \frac{l^3}{V}\left(1 + \frac{1}{nl^3} + \bar{\xi}\right) \simeq \frac{l^3}{V}\left[1 + \left(\frac{l_0}{l}\right)^\gamma\right]. \tag{39}$$

This expression behaves as (30), so that $\tau(2) = D_2$ is identical with the *correlation dimension* of Section 4.1. For $q = 1$, the quantity

$$D_1 = l\frac{d}{dl}\frac{1}{N_{\text{tot}}}\sum_i N_i \ln N_i = l\frac{d}{dl}\frac{V}{l^3 N_{\text{tot}}}\langle N \ln N \rangle \tag{40}$$

defines an *entropy dimension* in ranges where it does not vary with l.

More generally, we find from the scaling of $P_N(l)$ *two ranges* of l in which (35) behaves as a non-trivial power law (36).

(i) For *large scales $l_c \ll l \ll l_v$*, the exponent $\tau(q)$ and hence D_q take different values (Fig. 2) depending on q, the signature of multifractality. However, $\tau(q)$ is made of two straight lines, which exhibits the *bifractal* nature of the galaxy distribution at such scales.

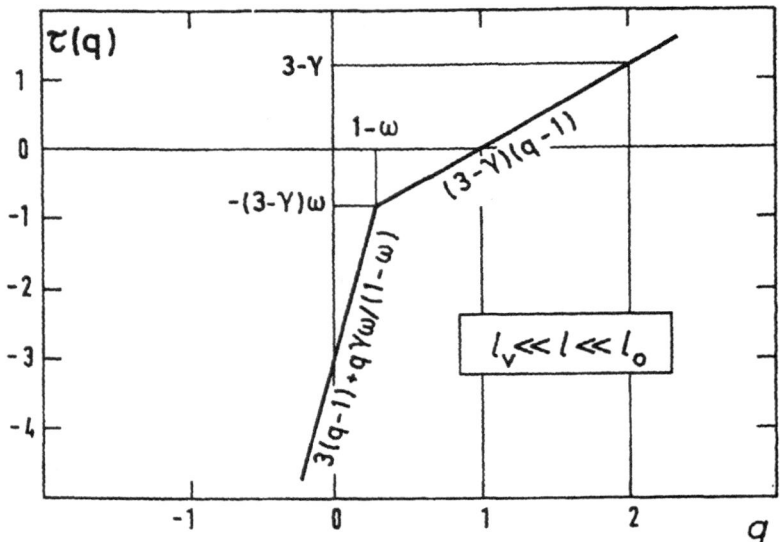

Fig. 3. — The exponent $\tau(q)$ for scales $l_v \ll l \ll l_0$.

(ii) For the *very large scales* $l_v \ll l \ll l_0$, we find a similar behaviour, again exhibiting bifractality (Fig. 3), but with different slopes for $\tau(q)$. The Renyi index is simply given by $D_q = 3 - \gamma$ for $q \geq 1 - \omega$ in both ranges, but D_q equals $(3 - \gamma)\omega/(1 - q)$ for $q \leq 1 - \omega$ and $l_c \ll l \ll l_v$, while D_q equals $3 - q\gamma\omega/(1 - \omega)(1 - q)$ for $q \leq 1 - \omega$ and $l_v \ll l \ll l_0$.

Note that D_q is an *increasing* function of q for the *large scales* $l_c \ll l \ll l_v$, a *decreasing* function of q for the *very large scales* $l_v \ll l \ll l_0$. This difference is associated with the fact that the *reference length* Λ occuring in (36) is equal to l_0 in the first case, to l_c in the second case. The dimensionless *scaling variable* l/Λ of (36) is therefore much *smaller than* 1 for $l_c \ll l \ll l_v$, much *larger than* 1 for $l_v \ll l \ll l_0$, so that we encounter two opposite types of fractality (with $l \to 0$ and $l \to \infty$) in these two ranges.

4.4. Multifractal dimension

Depending on the local density of galaxies, the number N_i of galaxies contained in a cell i, which is a function of the size of this cell, may scale differently with l. We thus introduce a *local scaling exponent*

$$\alpha_i \equiv \frac{\mathrm{d}(\ln N_i)}{\mathrm{d}l}. \tag{41}$$

For a simple monofractal set of points with dimension D, the exponent α_i would equal D with little statistical fluctuation. Here, $\alpha \equiv \alpha_i$ is a *random variable*, which describes how the underlying random variable N_i increases with the size of the boxes, and which may display large fluctuations. Its statistical distribution is characterized by the *number $C(\alpha, l)$ of cells i* such that N_i *scales as l^α* for small changes of l.

The multifractal dimension $f(\alpha)$ associated with the exponent α is then defined as the Hausdorff *dimension of the set $C(\alpha, l)$*, provided that $C(\alpha, l)$ scales as a function of l as

$$C(\alpha, l) = B(\alpha) \left(\frac{l}{\Lambda} \right)^{-f(\alpha)} \tag{42}$$

in the considered range of l. In (42), Λ is the same upper or lower characteristic length as in Section 4.3 while $B(\alpha)$ is some slowly varying coefficient. Note that $\sum_\alpha C(\alpha)$ is the number C_{occ} of occupied cells. For a monofractal set with dimension D, the quantity $f(\alpha)$ is defined only for $\alpha = D$ since $N_i \propto l^D$, and its value at this point is also $f(D) = D$ since $C_{\text{occ}} \propto l^{-D}$.

The multifractal dimension $f(\alpha)$ can be related to the Renyi index of Section 4.3 by noting that, if N_i scales locally as

$$N_i \sim A_i \left(\frac{l}{\Lambda} \right)^{\alpha_i}, \tag{43}$$

where the coefficient A_i depends on α_i less than exponentially, the expression (34) reads

$$C_q(l) = \sum_i \left[\frac{A_i}{N_{\text{tot}}} \left(\frac{l}{\Lambda} \right)^{\alpha_i} \right]^q = \frac{1}{N_{\text{tot}}^q} \sum_\alpha C(\alpha) \langle A^q(\alpha) \rangle \left(\frac{l}{\Lambda} \right)^{\alpha q}. \tag{44}$$

By using (42) we find

$$C_q(l) = \frac{1}{N_{\text{tot}}^q} \sum_\alpha B(\alpha) \langle A^q(\alpha) \rangle \left(\frac{l}{\Lambda} \right)^{\alpha q - f(\alpha)}. \tag{45}$$

Since $B(\alpha)$ and $\langle A^q(\alpha) \rangle$ are assumed to vary slowly, this sum is dominated for $l \ll \Lambda$ by the *smallest exponent $\alpha q - f(\alpha)$*, for $l \ll \Lambda$ by the *largest*. The results described by the end of Section 4.3 thus imply, in the two non-trivial scaling ranges, that

$$\tau(q) = \text{Min}_\alpha[\alpha q - f(\alpha)] \text{ for } l_c \ll l \ll l_v, \tag{46}$$

$$\tau(q) = \text{Max}_\alpha[\alpha q - f(\alpha)] \text{ for } l_v \ll l \ll l_0. \tag{47}$$

Thus the Renyi index $\tau(q)$ is the Legendre transform of the multifractal dimension $f(\alpha)$. The two different forms (46) and (47) are consistent with the *different convexities* of the curves of Figures 2 and 3. However $f(\alpha)$ cannot be completely deduced from $\tau(q)$: it is a more detailed property.

A complete study of the multifractal index $f(\alpha)$, as resulting from the above formalism, can be found in II. Bifractality is again exhibited in each of the

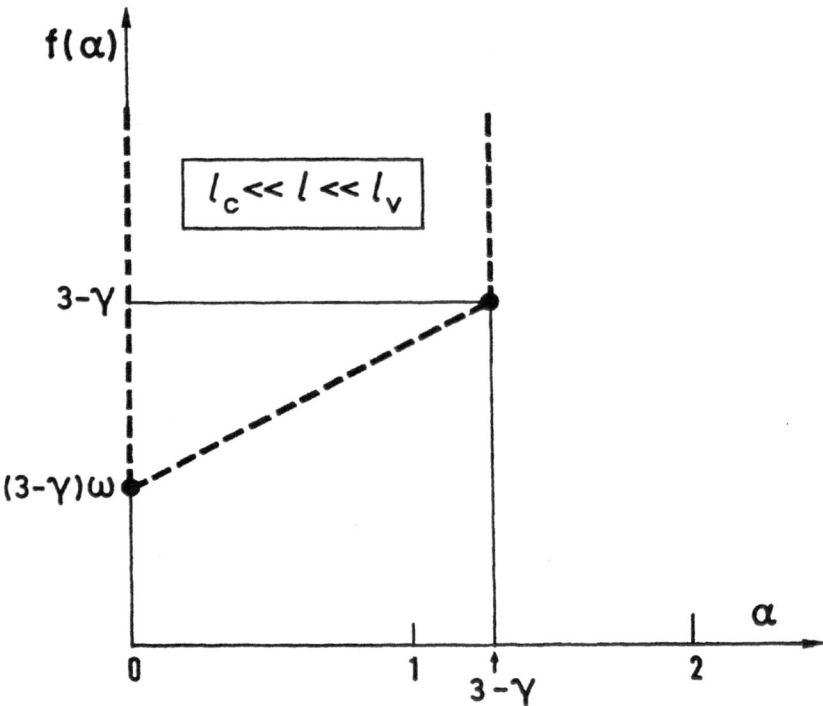

Fig. 4. — The multifractal dimensions for scales $l_c \ll l \ll l_v$.

two scaling regions. For $l_c \ll l \ll l_v$, the *two values for α and for $f(\alpha)$* that come out are $\alpha = 0$ for which $f(0) = (3 - \gamma)\omega$, and $\alpha = 3 - \gamma$ for which $f(3 - \gamma) = 3 - \gamma$. For $l_v \ll l \ll l_0$, the two values are again $\alpha = 3 - \gamma$ for which $f(3 - \gamma) = 3 - \gamma$, and $\alpha = 3 + \gamma\omega/(1 - \omega)$ for which $f(\alpha) = 3$. The shapes of the two curves (Figs. 4 and 5) are inverted, in agreement with the fact that the scaling variables are $l/l_0 \ll 1$ and $l/l_c \gg 1$, respectively.

A qualitative understanding for this behaviour can be found by noting that the actual distribution of galaxies (II, Fig. 16) exhibits *dense clusters* as well as *large voids*. Most galaxies lie in clustered regions, where their relative distances are of order l_c but they occupy a small fraction of space. The *dimension $3 - \gamma$*, which arises throughout the range $l_c \ll l \ll l_0$, is *associated with these clusters*, where the distribution behaves as an ordinary fractal set with dimension $\alpha = f(\alpha) = 3 - \gamma$. However, most cells are much less occupied, and they behave differently. For $l_c \ll l \ll l_v$, the number of galaxies in most occupied cells is rather small, and it thus remains *nearly constant* as l increases; this corresponds to $\alpha = 0$. The number $C(\alpha = 0) \simeq C_{occ}$ of such cells scales with the exponent (33), so that the dimension $f(\alpha = 0)$ of these underdense regions

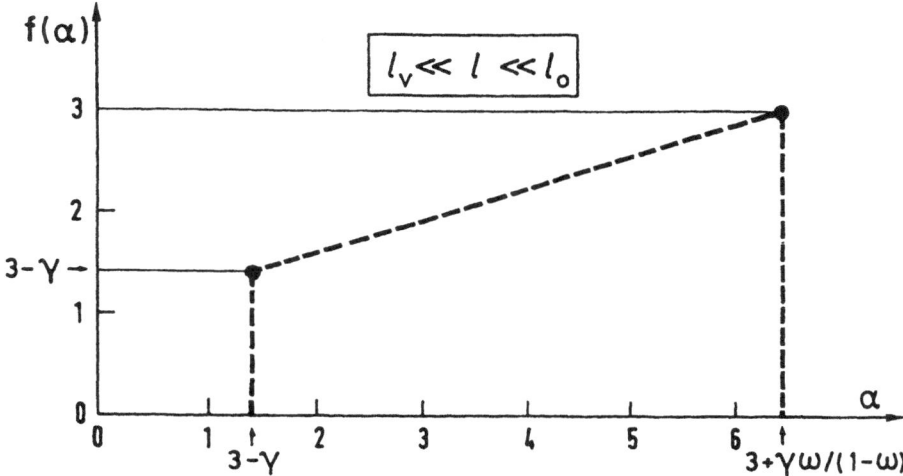

Fig. 5. — The multifractal dimensions for scales $l_v \ll l \ll l_0$.

is the *Hausdorff dimension* $(3 - \gamma)\omega$. For $l_v \ll l \ll l_0$, the cells are larger than the void size, and they are practically never empty, yielding $f(\alpha) = 3$. Apart from the small proportion of cells which lie in overdense regions, most of them contain a *number of galaxies of order* $N_v(l)$. This number, given by (26), rises *faster than the cell volume* ($\alpha = 3 + \gamma\omega/(1 - \omega) > 3$) because a typical cell, lying in underdense regions, may be surrounded by clusters. The corresponding value $f(\alpha) = 3$ is associated with the *restoration of homogeneity* at scales $l \gg l_v$.

5. CONCLUSION

Apart from their relevance to astrophysics, the above results may provide insight on the relationship existing between a *scale invariance* of a set of points and its *fractality*. Our *only hypothesis*, the scale invariance (16) of the set of correlation functions ξ_N, together with the finiteness of the overall density n, has led to a *rather complex structure* for the counts in cells $P_N(l)$, which are generated by the expressions (13, 22). The fractal properties of the set, which directly result from the probabilities $P_N(l)$, were in turn derived and are also complicated, existing only for some scales. Scale invariance thus appears as a *more general concept* than fractality.

Although the original scaling property of the set ξ_N involves directly only a single length l_0, several other *characteristic lengths*, namely the clustering length l_c, the average distance $\bar{l}$ and the void size l_v, come out in the derivation. They separate different scales, in which the probabilities $P_N(l)$ behave differently (Fig. 1), and in most of which some or other type of fractality occurs. The cross-over between them is rather narrow because the parameter nl_0^3 is large. It is remarkable that *two different*, well separated non-trivial *fractal structures* emerge at *two different scales* (Figs. 2 to 5). Indeed, a distinctive feature of the present approach is the contrast between the simplicity of the underlying scaling property (16) and the resulting occurence of many different regimes for the counts in cells and for the fractality.

The Figures 2 to 5 exhibit the *bifractality* of the set, in *both* non-trivial fractal regimes. Indeed the interpretation of bifractality given at the end of Section 4 shows that our single scaling hypothesis entails a rather sharp distinction between two regions of space, in which the galaxies are either underdense (voids) or overdense (clusters).

The large y behaviour of the function $\varphi(y)$, which was defined by (19) in terms of the integrals S_N of the N-point correlation functions, depends on the way these correlations decrease with N. We have used astrophysical information to determine this large y behaviour (24). Indeed, the information given by ξ_N for small N and by P_N for small N are *complementary*, and (24) directly results from the probability $P_0(l)$ for large voids, while the small y behaviour of φ results from the first few N-point correlations. However, the above formalism based on *generating fuctionals* is general, and it can readily be adapted, beyond astrophysics, to other behaviours of $\varphi(y)$, and even to other types of scaling. It thus provides a convenient mathematical framework to *build new types of fractal sets* of points, by starting from various hypotheses on the correlation functions of these points.

REFERENCES

[1] R. Balian and R. Schaeffer, *A&A* **220** (1989) 1-29, here referred to as I.

[2] R. Balian and R. Schaeffer, *A&A* **226** (1989) 373-414, here referred to as II.

[3] see, for instance, R. Balian, From Microphysics to Macrophysics (Springer-Verlag, Berlin 1991) Section 2.3.

An extensive bibliography can be found in I and II.

Glossary

Abelian group: A group with the property that its elements commute. Example $\mathbf{Z}$ (the integers under addition).

Absolutely continuous: See measure, see also Ergodic theory – Introductory lectures, Springer. Lecture Notes, Math. 458 (1975).

Acceptance Domain: Cross-section of the window that yields a quasicrystal in the projection method. See also model set.

Almost periodic (in the sense of Bohr): A bounded, continuous function $f(x)$ on $\mathbf{R}$ is called (uniformly) almost periodic if any subset of the set of its possible shifts $f_\xi(x) = f(x + \xi)$ contains a sequence that converges uniformly on the entire axis. The set $\tau_\varepsilon = \left\{ \xi \mid \sup_{x \in \mathbf{R}} |f_\xi(x) - f(x)| < \varepsilon \right\}$ is said set of ε-periods of f and is relatively dense if $f(x)$ is uniformly almost periodic. The closure of the set of shifts $\{f_\xi(x) : \xi \in \mathbf{R}\}$ in the uniform convergence topology is called the hull Ω_f of the function f. The hull Ω_f has a natural structure of compact topological group.

An almost periodic function is quasiperiodic if its hull Ω_f is a n-dimensional torus T^n. Then it can be written as $f_\xi(x) = F(\alpha_1 x, ..., \alpha_n x)$, where $F(x_1, ..., x_n)$ is periodic of period 1 on every variable, and the numbers $\alpha_1, ..., \alpha_n$ are rationally independent.

An almost periodic function $f(x)$ is determined by its Fourier-Bohr coefficients c_λ, calculated as the limits

$$
\begin{aligned}
c_\lambda &= \lim_{T \to +\infty} \frac{1}{T} \int_{a-T}^{a} f(x) e^{ix\lambda} dx \\
&= \lim_{T \to +\infty} \frac{1}{T} \int_{a}^{a+T} f(x) e^{ix\lambda} dx,
\end{aligned}
$$

for any $a \in \mathbf{R}$.

Almost-Mathieu equation: The equation $c_{n+1} + c_{n-1} + 2\lambda \cos(2\pi\alpha n + \theta)c_n = Ec_n$ which is a discrete Schrödinger equation.

Alphabet: A finite set $\mathcal{A}$, the elements of which are called "letters". From the letters are formed words (or blocks) by juxtaposition just as in the current language.

Anderson localisation: Mechanism of electronic localisation due to multiple interference effects induced by scattering or disorder. This results in a shrinkage of the electronic wave function which becomes exponentially localized in space. A finite temperature conduction may occur through variable range hopping (see variable range hopping).

Anomalous transport: Transport which cannot be described by the diffusional equation.

Aperiodic crystal: A solid with Bragg peaks in its diffraction pattern, but without lattice periodicity.

Approximants: Structure obtained from a quasiperiodic one by a (small) phason strain. See also periodic approximant.

Atomic cluster: See local environments.

Atomic measure: See measure.

Atomic surface: The closure of the set of points in the unit cell in superspace which are equivalent by lattice translations with atomic positions in physical space.

Automorph (of an integral quadratic form): It is defined *e.g.* in Dickson, Introduction to the Theory of Numbers, Dover, 1957 p. 72 and in Buell, Binary Quadratic Forms, Springer, 1989 p. 31. It is obtained as an integral solution of the Pell (plus) equation and corresponds to a positive unit of the associated quadratic field.

Band: Continuous electronic energy range which allows for Bloch type boundary conditions over a (unit) cell.

Band germ: Energy region where band conditions are fulfilled on a finite (sub-) string.

Band superband: Bands in periodic structures which are glued without gaps and overlaps.

Bernoulli measure: The simplest example of a Bernoulli measure is the measure associated to the infinitely lasting game of tossing a unbiased coin, *i.e.*, $\mu = *_1^\infty \left(\frac{\delta_0}{2} + \frac{\delta_1}{2} \right)$.
More generally, one calls Bernoulli measure any convergent infinite product of discrete convolution measures having finite support.

Bernouilli shift: A shift space equipped with a Bernouilli measure (*i.e.* product measure of a measure on the alphabet).

Bifractality: A special type of multifractality in which the multifractal dimension $f(\alpha)$ associated with the exponent α takes only two values, corresponding to two values of α.

Billiard: It is a dynamical system consisting of a particle that moves freely in space until encountering with a billiard's border. Typically, collisions of the particle with borders are elastic.

Binary integral quadratic forms: Binary quadratic forms; Quadratic fields; Pell equation(s): these concepts are discussed in practically every textbook on number theory. See in particular:

1. D.A. Buell, Binary Quadratic Forms, Springer, 1989.

2. L.E. Dickson, Introduction to the Theory of Numbers, Dover, 1957.

3. H. Cohn, A Second Course in Number Theory, Wiley, 1963.

4. H. Hasse, Vorlesungen uber Zahlentheorie, Springer 1964.

Block frequency: Consider an infinite sequence u taking its values in a finite alphabet $\mathcal{A}$. Let w be a word appearing in u. For all N, define N_w the number of appearances of w in the first N terms of the sequence u. If $\lim\limits_{N \to \infty} \frac{N_w}{N}$ exists, it represents the frequency of appearance of w in u.

Caution: The limit may not exist. For instance, if $\mathcal{A} = \{0, 1\}$ and $u = 0\ 11\ 0000...\ 1^{2^n}\ 0^{2^{n+1}}\ .....$ the frequency of 0 or 1 does not exist.

Bloch germ: Set of Bloch states on a finite patch.

Bloch label: Irreducible representation label for a discrete translation group.

Bloch state: From a group representation point of view, state transforming irreducibly under a discrete translation group.

Borel function: A function f is called a Borel function if and only if $f^{-1}[(a, b)]$ is a Borel set for all a, b.

Borel set: The class of "Borel sets" on $\mathbf{R}$ is somehow the class of sets that can be obtained from repeated denumerable unions or denumerable intersections of intervals. There is no description of Borel sets. A "length" can be defined for these sets. See measure.

Borel support: A Borelian measure μ on $\mathbf{R}$ admits B for its Borel support if $\mu(B^c) = 0$ where $B^c = \mathbf{R} \backslash B$. It is clear that it is not defined in a unique way since, if N is negligible for μ, $B \cup N$ and $B \cap N^c$ are still Borel supports of μ.

Boundary layer: A narrow strip in the stochastic sea near an island boundary.

Bounded measure: See measure.

Box dimension or Bouligand-Minkowski dimension: Let E be a bounded subset of $\mathbf{R}^n$. Consider $N_\varepsilon(E)$ the minimum number of elements of coverings of E by balls of diameter ε. The Bouligand-Minkowski dimension of E is the number $\Delta(E) = \limsup\limits_{\varepsilon \to 0} \frac{\log N_\varepsilon(E)}{-\log \varepsilon}$.

Brillouin zone: (First Brillouin zone) Wigner-Seitz primitive cell of the reciprocal lattice. The Wigner-Seitz cell about one point is the region of space that is closest to that point than to any other lattice point. This definition can be extended to quasiperiodic lattices, with the restriction than the so called "pseudo-Brillouin zone" is only defined at the origin of the reciprocal space. The points in reciprocal space are then defined by the diffusion vectors of the most intense diffraction peaks.

Cantor function (see Cantor set): The Cantor function (or "devil staircase") $\alpha(x)$ is constructed in the following way:

$$\alpha(x) \;=\; \frac{1}{2} \;\text{on}\; \left(\frac{1}{3}, \frac{2}{3}\right),$$

$$\alpha(x) \;=\; \frac{1}{2^2} \;\text{on}\; \left(\frac{1}{3^2}, \frac{2}{3^2}\right),$$

$$\alpha(x) \;=\; \frac{3}{2^2} \;\text{on}\; \left(\frac{2}{3} + \frac{1}{3^2}, \frac{2}{3} + \frac{2}{3^2}\right)$$

and so on.

Extend α to $[0,1]$ by making it continuous. Then $\alpha(x)$ is a nonconstant continuous function such that $\alpha'(x)$ exists almost everywhere (with respect to the Lebesgue measure) and is zero almost everywhere. The Stieltjes measure, defined by $\mu_\alpha((a,b)) = \alpha(b-0) - \alpha(a+0)$, is continuous and concentrated on the Cantor set C, since $\mu_\alpha(S) = 0$. Therefore μ_α and the Lebesgue measure live on completely different sets: μ_α is singular continuous with respect to the Lebesgue measure (See also Cantor measure).

Cantor measure: It is the Bernoulli measure $*_1^\infty \frac{1}{2}(\delta_0 + \delta_{2.3^{-n}})$ or the measure defined by the following self similarity property:

$$\int_{[0,1]} h(x)\mathrm{d}\mu(x) = \frac{1}{2}\int_{[0,1]} h\left(\frac{x}{3}\right)\mathrm{d}\mu(x) + \frac{1}{2}\int_{[0,1]} h\left(\frac{x+2}{3}\right)\mathrm{d}\mu(x)$$

for any positive measurable h on $[0,1]$.

Cantor set: Let S be the subset of $[0,1]$

$$S = \left(\frac{1}{3}, \frac{2}{3}\right) \cup \left(\frac{1}{3^2}, \frac{2}{3^2}\right) \cup \left(\frac{2}{3} + \frac{1}{3^2}, \frac{2}{3} + \frac{2}{3^2}\right)$$

$$\cup \left(\frac{1}{3^3}, \frac{2}{3^3}\right) \cup \left(\frac{2}{3^2} + \frac{1}{3^3}, \frac{2}{3^2} + \frac{2}{3^3}\right) \cup ...,$$

that is the set obtained by removing the middle third of what is not in S at each stage and add it to S. The Lebesgue measure of S is $S = \frac{1}{3} + \frac{2^2}{3^3} + ... = 1$. The (ternary) Cantor set C is the complement of S in $[0,1]$: $C = [0,1]\backslash S$. It has the Lebesgue measure 0. $x \in C$ if and only if the base 3 expansion of x has no 1's. C is an uncountable set of measure 0. More generally, 2^{n+1} closed intervals $I_{n+1,i}$, $i = 1, ..., 2^{n+1}$, are obtained from 2^n closed intervals

I_{ni}, $i = 1, ..., 2^n$, by removing open intervals of length ε_{n+1}, lying in their middles. The general Cantor set is $C = \bigcap_{n=1}^{\infty} I^{(n)}$ where $I^{(n)} = \bigcup_{i=1}^{2^n} I_{ni}$.

The general Cantor set is homeomorphic to the ternary set.

Cartan matrix: See root matrix.

Cesarò mean: A sequence (u_n) $n \in \mathbf{N}$ is said to converge by Cesarò mean if $\bar{u}_N = \frac{1}{N} \sum_1^N u_n$ converges. If a sequence converges then it converges by Cesarò mean. The converse is not true.

Closure: The closure $\bar{Y}$ of a subset Y of a topological space X is the smallest closed set containing it. If this is X itself then Y is dense in X. Well known examples are the rational numbers in $\mathbf{R}$ and the ring $\mathbf{Z}[\tau] := \mathbf{Z} + \tau\mathbf{Z}$ in $\mathbf{R}$ where $\tau = (1 + \sqrt{5})/2$. The interior $\text{int}(Z)$ of a subset Z of X is the set of those points x for which an open set U exists with $x \in U \subset Z$. Thus Z is an open set if and only if it is its own interior. The boundary of Z is the complement of its interior inside its closure: $\bar{Z}\backslash\text{int}(Z)$. A neighbourhood of a point x in a topological space is any subset containing x in its interior.

Coboundary: See cohomology.

Cocycle: See cohomology.

Cohomology: Let X be a topological space and $\mathcal{U} = \{U_\alpha\}_{\alpha \in A}$ a covering of X with open sets U_α. A cochain of dimension n with coefficients in an

abelian group Π is an alternating function C: $\underbrace{A \times A \times ... \times A}_{(n+1)} \rightarrow \Pi$

with definition set

$$\{(\alpha_0, \alpha_1, ..., \alpha_n) | U_{\alpha_0} \cap U_{\alpha_1} \cap U_{\alpha_2} ... \cap U_{\alpha_n} \neq \emptyset\}.$$

The set of n-dimensional cochains is a group denoted by $C^n(\mathcal{U}, \Pi)$. The coboundary operator (or derivation) d sends C^n into C^{n+1} in the following way

$$\text{d}c(\alpha_0, ..., \alpha_{n+1}) = \sum_{i=0}^{n+1} (-1)^i c(\alpha_0, ..., \hat{\alpha}_i, ..., \alpha_{n+1})$$

(where $\hat{\alpha}_i$ means that α_i is omitted). The cochain dc is called coboundary of c. If d$c = 0$, then c is named cocycle. Let $Z^n(\mathcal{U}, \Pi)$ (resp. $B^n(\mathcal{U}, \Pi)$) denote the set of all n-dimensional cocycles (resp. coboundaries). Since $d^2 = 0$, we have the inclusion $B^n \subset Z^n$.

The coset $H^n(\mathcal{U}, \Pi) = Z^n(\mathcal{U}, \Pi)/B^n(\mathcal{U}, \Pi)$ is a group called cohomology group of dimension n (according to Čech) for the covering $\mathcal{U}$.

Compact: A topological space X is compact if whenever a family $\mathcal{F}$ of open subsets covers it (*i.e.* the union of the sets of $\mathcal{F}$ is all of X) then there is a finite number of elements of $\mathcal{F}$ that still covers X. A subset of X is compact if it is compact in the induced topology. Some authors (*e.g.* Bourbaki) also require compact spaces to be Hausdorff. In the case of Euclidean spaces, compactness is equivalent to closed and bounded. A subset of X is compact if it is compact in the induced topology. A topological space is locally compact if every point has a compact neighbourhood. Typical examples are real and complex spaces.

Complete: If (X, d) is a metric space then X is complete if every Cauchy sequence in it converges. Each metric space can be embedded in a so-called completion which is complete and in which X is dense. The reals are the completion of the rationals in this way.

Composite structure: A crystal phase consisting of two or more subsystems which are displacively modulated incommensurate crystal phases, whereas the basic structures are mutually incommensurate. See T. Janssen, A. Janner, Adv. Phys. 36, 519 (1987).

Conductivity: Response function to a thermal or electrical current flux. In particular κ and σ are the linear response coefficient, respectively $\kappa \nabla T = \theta$ and $\sigma \nabla U = j$, with T the temperature, U the voltage, j the electrical and θ the thermal current flux.

Connected: A subset of a topological space is connected if it is not the union of two disjoint open sets. Given a point x of a topological space, the union of all the connected subsets containing it is also connected. This unique maximum connected subset containing x is the connected component of x. A topological space X is totally disconnected if each point of X is its own connected component. Examples of these are discrete sets, the rational numbers under the topology induced from the reals, and the p-adic topologies.

Connected densities: The connected N-point densities are the cumulants of the reduced densities F_N. See cumulants reduced densities.

Continuous mapping: A mapping $f: X \to W$ of one topological space into another is continuous if the inverse image of any open set in W is open in X. If it is also invertible with continuous inverse, then f is a homeomorphism.

Continuous measure: See measure.

Continuous singular: See measure.

Continuous spectrum: See energy spectrum or spectrum.

Convex hull: Convex hull of vectors: $x_1, ..., x_k \in R^n$ is a set of all their convex combinations, *i.e.* elements of the form $x = \sum_{i=1}^{k} \alpha_1 x_1$, such that $\sum_{i=1}^{k} \alpha_1 = 1$ and $\alpha_1 \geq 0$ for all $i = 1, ..., k$.

Correlation dimension: If the average number of points of a random set which lie at a distance smaller than ℓ from a given point scales as ℓ^{D_c} in some range of ℓ, D_c is the correlation dimension of the set in this range.

Correlation functions: For a translationally invariant random set of points with overall average density n, the N-point correlation function is the connected N-point density divided by n^N.

Correlation measures: See measure.

Counts in cells: They are characterized by the probability $P_N(\ell)$ that N particles lie in a cell of volume ℓ^3.

Coxeter group: A group generated by reflections. See also root systems.

Critical states: Electronic eigenstates supposed found in quasiperiodic structures. These are neither extended nor localized, they present a complicated structure roughly described by the fact that they decay like a power law with distance from a given site.

Cross-over: The transition between two different scaling regimes.

Cumulants (for random variables): Let X be a discrete random variable with generating function $g(z) = \sum_{k \geq 0} P(X = k) z^k$.
The coefficients K_n in the following expansion $\log g(e^t) = \sum_{n \geq 0} \frac{K_n}{n!} t^n$ are called the cumulants of X.

Cumulant (for reduced densities): Let $\{F_n(\mathbf{r}_1 \mathbf{r}_2 ... \mathbf{r}_n)\}$ the set of reduced densities (see reduced densities). The cumulants $G_n(\mathbf{r}_1 \mathbf{r}_2 ... \mathbf{r}_n)$ of the $\{F_n\}$ are recursively defined as follows:

$$F_1 = G_1,$$

$$F_2 = G_1 G_1 + G_2, ... \; F_N = \underbrace{G_1 G_1 + G_1}_{N} + \sum \underbrace{G_1 G_1 ... G_1 G_2}_{N-1} ...$$

Cut and projection (also called strip/projection method): See model set.

Cylinder set: A subset $[C]$ of shift space X defined by fixing a finite number of coordinates: If $C = C_1 ... C_m$, then $[C] = \{x \in X; x_1 = c_1, ..., x_m = c_m\}$.

Decagonal phase: Phase which exhibits a tenfold symmetry axis, incompatible with periodic translation. The known decagonal phases are quasicrystalline in planes and periodic along the third direction (10 fold axis direction) perpendicular to them.

Delaunay property, Delaunay set: See Delone property.

Delone property: A set $\Lambda \subset \mathbf{R}^n$ is called Delone if it satisfies two conditions: (i) Λ is uniformly discrete, *i.e.* the distances between any pair of points in Λ are greater than a fixed $\varepsilon > 0$, (ii) Λ is relatively dense, *i.e.* there exists $r > 0$, such that $\mathbf{R}^n$ is covered by balls of radius r centered at points of Λ.

Dense point spectrum: See spectrum.

Density of states: $N(E)$: Number of electronic orbitals or states per energy unit, as a function of energy. For instance in a free electron model (*i.e.* no interaction with atomic potential), $N(E)$ varies as the square root of the energy. $N(E)$ gives the number of accessible states for an electron excited at energy E. At zero temperature (no excited states), the electrons of the system (finite number) fill the states from the lowest energy up to a maximum energy called the Fermi level E_F. The Fermi surface is the iso-energetic surface at $E = E_F$ in the electron wave vector k-space. The specific heat can give the value of $N(E_F)$, which is proportional to the linear temperature term. The Einstein relation relates the electrical conductivity σ to the number of available states and the mobility of these states through the electronic diffusion constant D: $\sigma = e^2 D N(E)$, with e the electronic charge.

Dia-(para) magnetism: For an introduction see N.W. Ashcroft & N. Mermin, Solid State Physics, HRW International Edition (1976), chapter 31.

Diffraction pattern: A positive measure based on the Fourier transform of the autocorrelation of a point set. The pure point part of this measure is the Bragg spectrum.

Dihedral group: the group of symmetries of a n-gon.

Diffusion constant: For a random walk motion, the mean travelling distance L from the origin after a time t is $L = \sqrt{Dt}$, with D the diffusion constant, which is related to the mobility of the particle. The electrical conductivity σ (see conductivity) is proportional to D: $\sigma = e^2 D N(E_F)$, with $N(E_F)$ the electronic density of states (see density of states).

Dimension of a curve: See M. Mendes-France in Beyond Quasicrystals, F. Axel and D. Gratias eds. Les Éditions de Physique/Springer Verlag, p. 318.

Dimension of measure: The minimal Hausdorff dimension of sets having the measure as support.

Dirac mass (or Bragg peak): See measure, diffraction.

Discrete measure: See measure.

Discrete Perp. Space: A subspace of perp. space in which the projected basis vectors of the hyperlattice take on discrete values.

Discrete point spectrum: See measure.

Discrete space: The space X is discrete if every point is an open set (in which case every subset is open). A subset Z of X is discrete if it is discrete in the induced topology, which amounts to saying that it is possible to put an open set of X around each point of X which contains no other point of Z. This is to be distinguished from the stronger property: A subset Z of a metric space (X, d) is uniformly discrete if there is a positive r with $d(x, y) > r$ for all distinct x, y in Z.

Discrete spectrum: See spectrum.

Dynamic structure factor: Differential cross section for inelastic scattering as a function of energy and momentum transfer. (See: S.W. Lovesey: Neutron Scattering, Oxford University Press, Oxford).

Dynamical system: See page 19 of P. Walters, An introduction to ergodic theory, Grad. Texts in Math. 79, Springer, New York, 1981. See also J. Guckenheimer, P. Holmes: Non linear oscillations, dynamical systems and bifurcations of vector fields, Springer-Verlag 1983 and A.J. Lichtenberg, M.A. Liberman, Regular and stochastic motion, Springer Verlag 1982.

Electron electron interaction: See quantum interference effects.

Electronic mean free path: The average travelling distance for an electron before being scattered, either elastically (no loss of energy in the scattering) or inelastically. Usual values for the elastic mean free path are interatomic distances in amorphous phases, and up to tens of micrometers in high purity metal elements.

Elliptic matrix: See linear map.

Energy gauge: Function on an energy range for a patch, with value 0 in band germs and value 1 otherwise.

Energy spectrum: The spectrum of a Hamiltonian operator H, *i.e.* the values λ for which the operator $H - \lambda E$ is singular. It defines a spectral measure that can be continuous or discrete in the measure theoretical sense (see measure). These are called the continuous and discrete spectrum, respectively.

Entropy of a dynamical system: See P. Walters, An introduction to ergodic theory, Grad. Texts in Math. 79, Springer, New York, 1981.

Ergodic, Ergodicity: See page 27 of P. Walters, An introduction to ergodic theory, Grad. Texts in Math. 79, Springer, New York, 1981. A measure μ is called $\underline{T\text{-ergodic}}$ or $\underline{\text{ergodic}}$ if the T-invariant sets in A have either μ measure $\overline{0 \text{ or } 1}$. See measure.

Faber-Ziman picture: Mechanism proposed to account for the negative temperature dependence of the electronic conductivity in simple liquid metals and amorphous phases. It is based on diffraction effects.

Fermi surface, Fermi level: See density of states.

Fibonacci sequence: The sequence LSLLS... generated by starting with the seed S and then repeatedly applying the substitution rule $L \rightarrow LS$; $S \rightarrow L$.

Finite local complexity: A countable set of points Λ in $\mathbf{R}^d$ such that for each compact set C there are, up to translation, only finitely many C-patches.

Flip: A minimal rearrangement of tiles in a tiling.

Fourier module: An infinite set of vectors in reciprocal space that are linear combinations with integer coefficients of a discrete set of basis vectors.

Fourier transform:

- The Fourier transform of a function $u(x)$, $x \in \mathbf{R}^n$, is defined by

$$\mathcal{F}u(y) = \hat{u}(y) = \int_{\mathbf{R}^n} u(x)e^{-2i\pi x \cdot y}\mathrm{d}x.$$

The linear map $\mathcal{F}$ is well-defined on the Lebesgue space $L^1(\mathbf{R}^n)$ of integrable function $u(x)$, $\int_{\mathbf{R}^n} |u(x)|\mathrm{d}x < \infty$.

- The Fourier transform $\mathcal{F}$ is a linear bijection (automorphism) of the Schwartz space $S(\mathbf{R}^n)$ (see functional spaces) such that $\hat{\hat{u}}(x) = u(-x)$, i.e. $u(x) = \int_{\mathbf{R}^n} \hat{u}(y)e^{2i\pi x \cdot y}\mathrm{d}y$.
  The Gaussian $e^{-\pi|x|^2}$ is a fixed point of this automorphism.

- The Fourier transform $\mathcal{F}$ is a unitary operator on the Hilbert space $L^2(\mathbf{R}^n)$:

$$\int_{\mathbf{R}^n} |u(x)|^2\mathrm{d}x = \int_{\mathbf{R}^n} |\hat{u}(y)|^2\mathrm{d}y \quad \text{(Plancherel formula)}.$$

Fourier-Bohr coefficient: See almost periodic.

Fractal atomic surfaces: Atomic surface with a Hausdorff dimension that is not an integer, and smaller than the dimension of the internal space.

Fractal time: A time series with fractal properties.

Fractional kinetics: Kinetics that describes trajectories with fractal space and (or) time properties.

Free group: Given a set of symbols $s_1, ..., s_n$, to each symbol s_i we assign another symbol s_i^{-1}. A word is a sequence of symbols s_i and s_j^{-1} in any order, for example $s_4 s_2 s_4^{-1} s_1 s_1$.
The product of two words V and W is the word obtained by writing W after V, and then cancelling out all adjacent pairs of s_i and s_j^{-1} until we get a reduced word. The set of reduced words with the

operation forms a group, called free group with generators $s_1, ..., s_n$, and denoted $\mathcal{S}_n$. The identity element e is the empty word and the inverse of a word is the word in which the symbols are written out in the opposite order, with s_i replaced by s_i^{-1} and s_i^{-1} by s_i. Every group G is a homomorphic image of a free group $G = \mathcal{S}_n/N$, where N is a normal subgroup generated by the so-called defining relations.

Free group automorphism: A bijective free group endomorphism.

Free group endomorphism: A map from a free group to itself which preserves the group operation.

Functional spaces (usual): They are spaces of functions $u(x)$, $x \in \mathbf{R}^n$, with values in $\mathbf{C}$.

- $C(\mathbf{R}^n)$: Space of continuous functions.

- $C^r(\mathbf{R}^n)$: Space of functions that are r-times continuously differentiable.

- $C^\infty(\mathbf{R}^n)$: Space of indefinitely differentiable functions.

- $S(\mathbf{R}^n)$: Schwartz space of rapidly decreasing indefinitely differentiable functions:

$$S(\mathbf{R}^n) = \left\{ u \in C^\infty(\mathbf{R}^n) : \forall \alpha, \beta \in \mathbf{N}^n, \ \sup_{x \in \mathbf{R}^n} \ \left| x^\alpha \partial^\beta u(x) \right| < \infty \right\},$$

where

$$\begin{aligned} x^\alpha &:= x_1^{\alpha_1} ... x_n^{\alpha_n} \\ \partial^\beta &:= \partial_{x_1}^{\beta_1} ... \partial_{x_n}^{\beta_n}. \end{aligned}$$

Let Ω be an open set in $\mathbf{R}^n$. Supp u denotes the support of $u(x)$ (see support).

- $K(\Omega) = \{ u \in C(\mathbf{R}^n) | \ \text{supp } u \ \text{is a compact subset of } \Omega \} \cdot$

- $D(\Omega) = \{ u \in C^\infty(\mathbf{R}^n) | \ \text{supp } u \ \text{is a compact subset of } \Omega \} \cdot$

- $D(\mathbf{R}^n)$ is the space of test functions in distribution theory.

- $BC(\Omega) = \left\{ u \in C(\Omega) \ \text{and} \ \sup_{x \in \Omega} |u(x)| < \infty \right\} \cdot$

- $C_0(\Omega) = \left\{ u \in BC(\Omega) \ \text{and} \ \lim_{|x| \to \infty} |u(x)| = 0 \right\} \cdot$

- $L^p(\Omega) = \left\{ u \left| \int_\Omega |u(x)|^p \mathrm{d}x < \infty \right| \right\}$, $1 \le p < \infty$ (Lebesgue spaces).

Gap (electronic): Energy interval with no allowed electronic states (zero density of states - see density of states).

Harmonious: See Y. Meyer in Beyond Quasicrystals, F. Axel and D. Gratias eds. Les Éditions de Physique, Springer Verlag, p. 12.

Harper equation: Almost-Mathieu equation with $\lambda = 1$.

Hausdorff dimension: Let E be a subset for $\mathbf{R}^n$. If ρ is a positive real number, we call ρ-covering of E by balls (or simply ρ-covering) any countable collection of balls B_j the diameters of which (denoted by $|B_j|$) are less than ρ and such that $E \subset \cup E_j$. If α and ρ are positive real numbers, we set

$$\mathcal{H}^\alpha_\rho(E) = \inf \left\{ \sum |B_j|^\alpha | \{B_j\} \text{ is a } \rho-\text{covering of } E \right\}$$

and

$$\mathcal{H}^\alpha(E) = \lim_{\rho \to 0} \mathcal{H}^\alpha_\rho(E).$$

As a function of E, when restricted to Borel sets, $\mathcal{H}^\alpha$ is a measure called the Hausdorff measure in dimension α. See the course by J. Peyrière in Beyond Quasicrystals, F. Axel and D. Gratias eds. Les Éditions de Physique, Springer Verlag.

Homomorphism: A map which preserves given algebraic structures.

Hume-Rothery phase: Intermetallic compound the stability of which is dominated by the electronic band term involving an interaction effect between the Fermi surface and the Brillouin zone border, *i.e.* opening of gaps in the directions where the Fermi surface touches the Brillouin zone border (see Brillouin zone, density of states, Fermi surface).

Hyperbolic matrix: See linear map.

Icosahedral group: Either of the icosahedral point groups I or I_h.

Icosahedral phase: Compound with diffraction pattern exhibiting an icosahedral point group symmetry.

Incommensurate crystal phases: A quasiperiodic system of particles obtained from a lattice periodic system by a (quasi) periodic modulation. The latter can be displacive, occupational or a combination of both.

Induced topology: If Z is any non-empty subset of X, then Z gets the induced topology or relative topology by using as the open sets the subsets $U \cap Z$ where U runs over $\mathcal{U}$ (see topological spaces).

Inflation rule: A rule for generating a tiling on a different length scale, starting with a given tiling.

Internal space: The orthogonal component of the physical space in the n-dimensional superspace. Also called: perpendicular space.

Invariant measure: A measure defined on $(X, \mathcal{A})$ is called $\underline{T\text{-invariant}}$ or $\underline{\text{invariant}}$ with respect to the transformation T from X to X if $\mu(T^{-1}A) = \mu(A)$ for each $A \in \mathcal{A}$ or (equivalently) if $\int f(x)\mathrm{d}\mu(x) = \int f(Tx)\mathrm{d}\mu(x)$ for each measurable function $f \geq 0$ on X.

Island: Part of the phase space, surrounded by the stochastic sea, which is filled by stable trajectories of a positive measure.

Kinetics: Description of particle dynamics using a probability distribution function.

Kronig Penney model: One-dimensional model for electrons in solids, the quantum mechanical behaviour of an electron in a potential consisting of delta peaks on atomic positions, or of constant potentials between the atoms. See: S. Gasiorowicz: Quantum Physics, Wiley, New York.

Lattice: A lattice L is a discrete subgroup of a locally compact abelian group G for which the quotient group G/L is compact. The term is most commonly used in regard to groups that look like $\mathbf{Z}^n$ in $\mathbf{R}^n$.

Lebesgue decomposition theorem: Any positive bounded measure on $\mathbf{R}$ can be uniquely decomposed into the sum of a discrete (or atomic or pure point) measure, a singular continuous measure and an absolutely continuous measure.

Lebesgue measure: See measure.

LI-class: Local isomorphism is an equivalence between Delone sets. Its equivalence classes are called LI-classes. See Delone property.

Limit-periodic: A function which is the uniform limit of periodic functions.

Linear map: A general n-dimensional linear map for discrete dynamical systems is defined by

$$x_i(t+1) = \sum_{j=1}^{n} A_{ij} x_j(t) + b_i.$$

It describes approximately the behaviour of nonlinear maps near fixed points.

Assuming the matrix $A = (A_{ij})$ real and non singular, its eigenvalues may be of two types either real ($\lambda \in \mathbf{R}$) or complex ($\lambda_\pm = \mathrm{e}^{\rho \pm i\omega} \in \mathbf{C}$). So the system may be decomposed into exponentially behaving components associated with real eigenvalues, and exponentially spirals or ellipses in two dimensional planes spanned by the vectors associated with complex eigenvalues. The subspace spanned by those eigenvectors that have positive, zero or negative ρ is said to span respectively the unstable manifold, the centre manifold or the stable manifold of the fixed point. A fixed point that has a centre manifold and no unstable manifold is said to be elliptic, and the matrix A is also said elliptic: all its eigenvalues are of modulus 1. A fixed point that has $\rho \neq 0$ for all eigenvalues is said to be hyperbolic, and the matrix A is said to be hyperbolic, and the matrix A is said hyperbolic. The parabolic case corresponds to the existence of degenerate eigenvalues. A fixed point that has some $\rho > 0$ and some $\rho < 0$ is said to be a saddle point.

Ljapunov exponent: See J. Guckenheimer, P. Holmes: Non linear oscillations, dynamical systems and bifurcations of vector fields, Springer-Verlag 1983 and A.J. Lichtenberg, M.A. Liberman, Regular and stochastic motion, Springer Verlag 1982.

Local environments: Atomic packing at short range. In an amorphous phase, local environments usually refer to the first shell of neighbours of a given atom. In icosahedral phases, structural models can include larger clusters of typically fifty atoms which retain the icosahedral symmetry.

Local isomorphism: Delone sets Λ_1, $\Lambda_2 \subset R^n$ are locally isomorphic, if for any finite subset F of Λ_1, there exist $x \in R^n$, such that $x+F \subset \Lambda_2$ and *vice versa*.

Magnetoresistance-magnetoconductance: Dependence of the resistivity (conductivity) with magnetic field. See also quantum interference effects.

Matthiessen rule: In most cases, the dependence with temperature of the electrical resistivity of a metal containing a few defects and impurities can be written as $\rho(T) = \rho_0 + \delta\rho(T)$ (Matthiessen rule). The temperature dependence $\delta\rho(T)$ is the same as compared to the pure metal. Only ρ_0 (low temperature saturation value) depends on the nature and number of impurities and on the defects introduced in the metal (in a perfect metal ρ_0 should be zero) (see resistivity).

Maximal spectral type: A measure (or any measure equivalent to it) that dominates all the spectral measures associated to a unitary operator.

Measurable function: A function f: $X \rightarrow R$ is called <u>measurable</u> if $f^{-1}(B) \in \mathcal{A}$ for all Borel sets $B \in \mathcal{B}(R)$, where $\mathcal{A}$ is a given sigma-algebra of subsets of X.

Note that in case of $X = R$, any continuous function is measurable.

Measurable mapping: A map T: $X \rightarrow X$ is called <u>measurable</u> if $T^{-1}(A) \in \mathcal{A}$ for all $A \in \mathcal{A}$.

Measures on T $= [0, 2\pi)$:

- A measure s is called a <u>Rajchman measure</u> if $\lim\limits_{|n|\to\infty} \hat{s}(n) = 0$ where $\hat{s}(n) = \int_T e^{int} ds(t)$.

- <u>Riesz product</u>: F. Riesz constructed this type of measure on T to have an example of measure the Fourier coefficients of which are exactly known.

- <u>Correlation measure</u> the term comes from probability theory; Wiener introduced the class of complex sequences (u_n) for which:

$$g(k) = \lim_{N\to\infty} \frac{1}{N} \sum_{n<N} u_{n+k}\bar{u}_n \text{ exists for all } k \in \mathbf{Z}.$$

$$\left(g(-k) = \overline{g(k)}\right).$$

One can then check that $g(k)$ is the kth Fourier coefficient of a positive measure on T called correlation measure of u.

Measure:

1) <u>Borel measure</u> on $(\mathbf{R}, \mathcal{B}(\mathbf{R}))$ (positive), where $\mathcal{B}(\mathbf{R})$ is the σ-algebra of Borel sets. It is the extension of the notion of length to $\mathcal{B}(\mathbf{R})$. μ is such a measure if
$\mu(\emptyset) = 0$ and $\mu(\cup_n B_n) = $ when the B_n are disjoint Borel sets.
μ is <u>bounded</u> if $\mu(\mathbf{R}) < +\infty$, *i.e.*, if its total mass is finite.

2) $\mu \geq 0$ having mass 1 is a <u>probability measure</u>.

3) A <u>real</u> measure is the difference of two positive measures $\mu_1 - \mu_2 (\mu_i \geq 0)$.

4) A <u>complex</u> measure is a measure of the form $\mu_1 + i\mu_2$ with the μ_i real.

5) If X is a set with a σ-algebra $\mathcal{A}$, we define m to be a <u>positive measure</u> on $(X_s\mathcal{A})$ if μ is an application from $\mathcal{A}$ to $\mathbf{R}^+$ such that
$$\begin{cases} \mu(\emptyset) = 0 \\ \mu(\cup A_n) = \sum \mu(\cup A_n) \end{cases} \text{ when the } A_n \text{ are disjoint elements of } \mathcal{A}.$$

6) A measure on $\mathbf{R}$ is <u>discrete</u> (or atomic or pure point) if it can be written as $\sum a_n \delta_{x_n}$, where $x_n \in \mathbf{R}$ and δ_x is a Dirac mass in x.
Example: Poisson measure $\sum_{n\geq 0} e^{-\ell} \frac{\ell^n}{n!} \delta_n$ with parameter ℓ.

7) A measure on $\mathbf{R}$ is <u>continuous</u> if it charges no point of $\mathbf{R}$, *i.e.* $\mu(\{x\} = 0$ when $x \in \mathbf{R}$.

8) A measure on **R** is singular if it is perpendicular to the Lebesgue measure m of **R**, *i.e.* if it is supported by a Borel set B of zero m-measure.

   So $m(B) = 0$ while $\mu(B^c) = 0$.

   A discrete measure is always singular in this sense. A continuous measure can be singular or not.

   Example: The Cantor measure μ is supported by the Cantor triadic set C of zero m-measure. It is continuous and singular. $\mu + m$ is not singular.

9) A measure is absolutely continuous (with respect to the Lebesgue measure) if it annihilates the same Borel sets as m, that is $m(B) = 0$ if $\mu(B) = 0$.

   By the Radon Nikodym theorem, μ can be written $f.m$, where $f = \frac{d\mu}{dm}$ is the Radon Nikodym derivative. It is a nonnegative function of $L^1(m)$. It is clear that the converse is true.

10) A measure μ defined on $(X, \mathcal{A})$ is T-invariant or invariant with respect to the transformation T from X to X, if $\mu(T^{-1}A) = \mu(A)$ for all $A \in \mathcal{A}$, or $\int_X f(x)d\mu(x) = \int_X f(Tx)d\mu(x)$ for all f measurable ≥ 0 on X.

11) A measure μ is T-ergodic if the T-invariant sets of $\mathcal{A}$, *i.e.* A such that $T^{-1}A = A$, satisfy $\mu(A) = 0$ or $\mu(X \backslash A) = 0$.

12) A measure μ is T-mixing if $= \lim\limits_{n \to \infty} \mu(T^{-n}A \cap B) = \mu(A)\mu(B)$ for all $A, B \in \mathcal{A}$.

Metal insulator transition: By definition a system becomes an insulator when its conductivity σ is zero at zero temperature. Certain systems can become insulators by varying structural or compositional parameters. For instance by reducing the metal content in semiconductor-metal or insulator-metal alloy (like SiY), the doping concentration in doped semiconductors (like SiP), the composition (like Indium oxides), etc.

Metric: In a given real n-dimensional vector space V a quadratic form $f(x_1, ..., x_n) = \sum_{i,k=1}^{n} g_{ik}x_i x_k$ of rank n is given for any $x \in V$, where n is the rank of the symmetric matrix $g = (g_{ik})$. A metric is defined in terms of a scalar product: $x.y = \sum_{i,k=1}^{n} g_{ik}x_i x_k$ for any x, $y \in V$ so that g_{ik} is the metric tensor. This metric is positive definite if $f(x) = 0$ for any $x \in V$, it is negative definite if $-f(x)$ is positive definite and indefinite otherwise. Note that this definition implies the extension of the concept of metric space (normally requiring a positive definite form) to include the pseudometric case as well. See for more details the Encyclopedic Dictionary of Mathematics, Kiyosi Itô, Ed. (The MIT Press, 1987) p. 1014 (Metric Spaces) and p. 1292 (Quadratic Forms).

Metric entropy: See Walters, P. Ergodic theory - Introductory lectures, Springer. Lecture Notes, Math. 458 (1975).

Meyer set: A relatively dense subset Λ of $\mathbf{R}^n$ for which $\Lambda - \Lambda$ is uniformly discrete. There are many other characterizations of this notion.

Microcrystalline State: State resulting from a set of very small crystals (micron size) with relative orientations following icosahedral or decagonal symmetry (microtwinned structure) and reproducing icosahedral or decagonal symmetry in diffraction patterns. See also microdomains.

Microdomains (nanodomains): Crystalline domains of micron size (nanometer size) making the microcrystalline state.

Minimality: A map is called minimal if the only invariant closed sets are the empty set and the whole space, or equivalently if every orbit is dense.

Model set (also called cut and project set): A point set in some real space $\mathbf{R}^n$ obtained by a controlled projection from some lattice. The lattice lies in some "higher" space called embedding space, which is $\mathbf{R}^n \times G$, where G is locally compact. The projection is controlled by a window W in the internal space G. The window should be compact with non-empty interior. A model set is regular if the boundary of the window W has Haar measure 0. It is generic if the projection of the lattice into internal space has no points on the boundary of W.

Module: Let $\mathbf{A}$ be a commutative ring with unit e. An $\mathbf{A}$-module is a commutative group E endowed with an operation $\mathbf{A} \times E \to E$ denoted $(\lambda, x) \to \lambda x$ such that

$ex = x,$

$(\lambda + \mu)x = \lambda x + \mu x,$

$\lambda(x + y) = \lambda x + \lambda y.$

Multifractal spectrum: For any nonnegative number α, we consider the set $E_\alpha = \{x | \alpha_{\mathrm{h}}(x) = \alpha\}$, where $\alpha_{\mathrm{h}}(x)$ is the Hölder exponent of measure μ at x, and its Hausdorff dimension $f(\alpha) = \dim E_\alpha$. The function f is called the multifractal spectrum of the measure μ. See Hausdorff.

Multifractal time: A time series with multifractal properties.

Mutually singular: See measure.

Nielsen theorem: Transformation law of the commutator under automorphisms of the free group F_2.

Non-abelian group: Its element do <u>not</u> commute. Example: Rotations in three dimensions.

Nuclear Magnetic Resonance: The coupling of an external magnetic field with the nuclear spin can induce a splitting of the energy levels of the nucleus. A transverse rf magnetic field induces transitions between these levels which are detected by an energy absorption technique. The nucleus-electron coupling changes the resonance conditions, which in turn can provide informations about the electronic gas.

Only ergodic: See ergodic.

Orbit: If $T:\ X \rightarrow X$ is a map and $x \in X$, then $\{x, Tx, T^2x, ...\}$ is the (one-sided) orbit of x.

Osedelet's theorem: See J. Guckenheimer, P. Holmes: Non linear oscillations, dynamical systems and bifurcations of vector fields, Springer-Verlag 1983, p. 284.

p-adic numbers: The completion of $\mathbf{Q}$ under the p-adic topology: $d\left(\frac{a}{b}, 0\right)$ $= p^{n-m}$ where m and n are the number of times that p divides a and b respectively. The p-adic integers are the completion of $\mathbf{Z}$ under this same topology.

Parabolic matrix: See linear map.

Parallel Space: In a periodic higher dimensional description of a quasicrystal, this is the physical subspace, *i.e.* the space containing the quasicrystal itself (see Perp. Space).

Participation number: Effective number of atoms which participate in Bloch states with rational labels.

Patch: The points of some given set inside some compact region of space.

Peano curve: See B.B. Mandelbrot, The Fractal Geometry of Nature; W.H. Freeman, publ. 1982.

Penrose pattern: A non-periodic tiling of the plane with 5-fold point symmetry.

Periodic Approximant (also rational approximant): A crystal whose corresponding cut description is very close to that of a quasicrystal.

Perpendicular Space (also Perp. Space and internal space): In a periodic higher dimensional description of a quasicrystal, this is the unphysical subspace orthogonal to that subspace containing the quasicrystal. See Parallel Space.

Perron-Frobenius eigenvalue: The positive eigenvalue of largest modulus of a non-negative matrix.

Perron-Frobenius Theorem: See E. Seneta, Non-negative matrices and Markov chains, 2nd ed., Springer, New York, 1981.

Phason (noun): A long wavelength (Goldstone) excitation of a quasicrystal or incommensurate crystal corresponding to shifting the relative phases of the Fourier components. (adjective) Pertaining to fluctuations in perp. space. See perpendicular space.

Phason Elastic Constants: Harmonic coefficients of the expansion of entropy density in phason strain coordinates.

Phason flip: A flip in quasicrystals associated with a fluctuation in perp. space.

Phason Strain: Spatial gradient of the perp. space coordinate of a quasicrystal.

Phonons: Elementary vibrational excitations of a crystal, the quanta of the harmonic oscillations around the equilibrium points. (See: C. Kittel: Introduction to solid state physics, Wiley, New York).

Photoemission spectroscopy: This experiment provides a direct measurement of the electronic density of states. A sufficiently energetic photon beam can eject an electron from the surface of a metal. The kinetic energy of the ejected electron is simply related to its binding energy relative to the Fermi level. By recording the energies of all the ejected electrons one can trace out the electronic density of states at the surface of the metal (see density of states).

Pisot number (or Pisot-Vijayaraghavan number, or $P-V$ number): Solution $\beta > 1$ of an irreducible polynomial equation with integer coefficients $X^m + a_{m-1}X^{m-1} + ... + a_1X + a_0 = 0$, and such that the moduli of all other roots are < 1.

Poincaré cycle: Time of the first recurrence of a trajectory to any domain of the phase space, from which the trajectory has started.

Poincaré map: It is a way to map a trajectory onto the space of smaller dimension. See J. Guckenheimer, P. Holmes: Non linear oscillations, dynamical systems and bifurcations of vector fields, Springer-Verlag 1983, p. 22.

Poincaré recurrence: Is a property of trajectories to return back to any domain of the phase space after a finite time. See J. Guckenheimer, P. Holmes: Non linear oscillations, dynamical systems and bifurcations of vector fields, Springer-Verlag 1983, see also A.J. Lichtenberg, M.A. Liberman, Regular and stochastic motion, Springer Verlag 1982.

Poincaré section: See Poincaré map.

Poisson summation formula: See measure.

Polya's theorem: See page 9 of B.D. Hughes and S. Prager. Random processes and random systems: An introduction. The Mathematics and Physics of Disordered Media. Eds. B.D. Hughes and B.W. Ninham, SLN Math. 1035, Springer, Berlin, 1983, p. 1-86.

Polytope: see H.S.M. Coxeter. Regular Polytopes, Dover 1973.

Positive definite sequence: A sequence (a_n) of complex numbers is positive definite if $\sum\limits_{\substack{1 \leq i \leq n \\ 1 \leq j \leq n}} t_i \bar{t}_j a_{i-j} \geq 0$ for all sequences $(t_1, ..., t_n)$ of complex numbers.

If s is a positive bounded measure on T, $\{\hat{s}(n)_{n \in \mathbf{Z}}\}$ is a positive definite sequence. Bochner's theorem then says that these are the only ones.

Positive measure: See measure.

Powder diffraction: Diffraction by a powdered sample; that is one where orientations are random giving spherically symmetric diffraction pattern.

Preservation of a measure by a map: This is the same as saying that the measure is invariant for the map. See Invariant measure.

Primitive substitution: A substitution σ is primitive if its matrix M_σ is primitive, $i.e.$, for some $k \geq 1$ all entries of M_σ^k are > 0. This means that there exists $k \geq 1$ such that each letter b from the alphabet $\mathcal{A}$ appears in all the iterates $\sigma^k(a)$.

Probability measure: A measure with total mass 1, $i.e.$ if the measure μ is defined on a sigma-algebra of subsets of a set X then $\mu(X) = 1$.

Pseudo-Brillouin zone: See Brillouin zone.

Pseudo-potential: $V(k)$: Fourier component of a simplified energy field, which is used to calculate the band structure and density of states of metals (see density of states, band), for details, see N.W. Ashcroft, N.D. Mermin, Solid State Physics, chap. 11, pp. 208-210.

Pseudo-random sequence: It is a sequence the correlation measure of which is continuous.

Pure point measure: See measure.

Quadratic unitary Pisot-Vijayaraghavan: It is a Pisot number with $m = 2$ and $a_0 = \pm 1$. See Pisot number.

Quantum interference effects: Effect of interferences between electron waves when scattered by defects, weak localisation and electron electron interaction effects can occur. At low temperature, in the case of constructive interferences, the weak localisation yields an enhanced probability for electron backscattering. An incoming electron can probe this local enhanced charge density (Coulomb electron electron interaction). Quantum interference effect theories predict precise temperature and magnetic field dependence of the electrical resistivity.

Quasicrystal: Definition of Steinhardt and Ostlund: "A quasicrystal is a quasiperiodic structure with a crystallographically disallowed orientational symmetry" [Steinhardt P.J. and Ostlund S., The Physics of Quasicrystals (World Scientific, Singapore, 1987) p. 12]. This definition is not universally agreed upon; in particular, some authors view the inflation (rescaling or rescaling plus rotation) symmetries of quasicrystals to be of fundamental importance.

Quasiperiodic: A function is quasiperiodic if its Fourier transform is only different from zero on a Fourier module of finite rank. See also almost periodic.

Quasi-trap: A domain of the phase space where a trajectory can stay an arbitrarily long (but finite) time.

Quaternions: Like the plane $\mathbf{R}^2$ has a commutative field structure through complex numbers $\mathbf{C}$, the fourdimensional euclidean space $\mathbf{R}^4$ has a non commutative field structure through quaternions $\mathbf{H}$. The multiplication of two quaternions $(p_0, \vec{p})$, $(q_0, \vec{q})$ (in notation scalar-vector) is given by $(p_0, \vec{p})(q_0, \vec{q}) = (p_0 q_0 - \vec{p}.\vec{q}, p_0\vec{q} + q_0\vec{p} + \vec{p} \times \vec{q})$ and the inverse of $(p_0, \vec{p})$ is $(p_0, \vec{p})^{-1} = \frac{(p_0, -\vec{p})}{\sqrt{(p_0^2 + \vec{p}^2)}} \equiv \frac{(p_0, -\vec{p})}{\|(p_0, \vec{p})\|}$.

The set of quaternions of norm 1 is a multiplicative subgroup isomorphic to $SU(2)$, the double covering of the rotation group $SO(3)$. Hence, the geometric interpretation of the quaternions in terms of rotations of three-dimensional vectors $(o, \vec{x})$ ("pure-vector quaternions")

$$(o, \vec{x}) \xrightarrow{R(\hat{n}, \theta)} \left(\cos\frac{\theta}{2}, \sin\frac{\theta}{2}\hat{n}\right)(o, \vec{x})\left(\cos\frac{\theta}{2}, -\sin\frac{\theta}{2}\hat{n}\right) = (o, \vec{x}')$$

where $R(\hat{n}, \theta)$ is the rotation of angle θ about the direction $\hat{n}$. See for more P. Duval, Homographies, Quaternions and Rotations, Oxford Univ. Press, 1964.

Radon measure: A Radon measure μ is a continuous linear form on the topological vector space $K(\Omega)$ (see functional spaces):

$$|\langle \mu, u \rangle| \leq c \sup_x |u(x)|; \quad \langle \mu, \alpha u + \beta u \rangle = \alpha\langle \mu, u \rangle + \beta\langle \mu, u \rangle.$$

The space of Radon measures is the topological dual of $K(\Omega)$ and is denoted by $K'(\Omega)$. It is a subspace of the space $D'(\Omega)$ of distributions.

Rajchman measure: See measure.

Random sequence: One definition can be the following: for a sequence which takes the values 0 and 1, a definition could be given in analogy with an infinite sequence obtained by the outcomes (head or tails) of independent throws with an unbiased coin: each word $w = w_1 w_2 ... w_n \in \{0,1\}^n$ of length n appears with frequency $\frac{1}{2^n}$ in the sequence.

Caution I: If the coin is biased (say heads has probability 1/3) then mathematicians will still call the sequence random (words $w = w_1 ... w_n$ now have frequency $\begin{pmatrix} n \\ k \end{pmatrix} \times \frac{k}{3} \times \frac{2}{3}(n-k)$, where $k = |w|_0$ is the number of zeroes in the word w).

Caution II: Even if the throws are no longer independent mathematiciens will often call the sequence random.

Random Tiling: A random tiling ensemble is an ensemble of tilings formed by the same set of prototiles, (different orientations of each prototile are allowed). The term "random tiling" can be informally used to describe a "representative" member of a random tiling ensemble.

Random walk: See B.D. Hughes and S. Prager, Random processes and random systems: An introduction. The Mathematics and Physics of Disordered Media. Eds. B.D. Hughes and B.W. Ninham, SLN Math. 1035, Springer, Berlin, 1983, p. 1-86.

Rank of a Fourier module: Minimal number of basis vectors that generates the module if one uses integer coefficients.

Recurrent walk: See page 9 of B.D. Hughes and S. Prager. Random processes and random systems: An introduction to The Mathematics and Physics of Disordered Media. Eds. B.D. Hughes and B.W. Ninham, SLN Math. 1035, Springer, Berlin, 1983, p. 1-86.

Reduced densities: For a random set of points, the joint probability $F_N(r_l...r_N)d^3r_1...d^3r_N$ that some point lies within d^3r_1 around r_1, some other point within d^3r_2 around r_2, etc., defines the N-point reduced density F_N.

Renyi index: Let N_i be the number of objects which lie in each cell i with size ℓ. The weighted number of occupied cells is defined by $C_q(\ell) = \sum_i \left(\frac{N_i}{N_{tot}}\right)^q$ for a given testing exponent q. If $C_q(\ell)$ behaves like $\left(\frac{\ell}{\Lambda}\right)^{\tau(q)}$, then the Renyi index D_q is defined by $\frac{\tau(q)}{q-1}$.

Repetitive: A subset Λ of $\mathbf{R}^d$ for which the translationally identical patches to any given patch of λ are relatively dense.

Riesz product: See measure.

Robinson tiling: A tiling of the plane with 6 (up to rotations) specially marked squares, the main property being that such a tiling is only possible aperiodically. See Robinson R.M. Undecidability and non-periodicity of tilings of the plane, Inventiones Math. 12 (1971) 177-209.

Root lattice: A root lattice is the lattice generated by a root system. See root system.

Root system, Cartan matrix, Cartan classification:

- Root system R in the euclidean space $\mathbf{R}^n$. It is a finite set R of non-zero vectors such that

  (i) R spans $\mathbf{R}^n$

  (ii) For each $\alpha \in R$, the mirror symmetry s_α with respect to the hyperplane perpendicular to α

  $$s_\alpha(\alpha) = x - \frac{2\langle \alpha | x \rangle}{||\alpha||^2} x$$

  leaves R invariant: $s_\alpha(R) \subset R$,

  (iii) For any $\alpha, \beta \in R$, $s_\alpha(\beta) - \beta$ is an integral multiple of α:

  $$s_\alpha(\beta) - \beta \in \mathbf{Z}\alpha,$$

  or equivalently,

  $$\frac{2\langle \alpha | \beta \rangle}{||\alpha||^2} \in \mathbf{Z}.$$

- Cartan matrix of a root system R: The integers $n(\beta, \alpha) = \frac{2\langle \alpha | \beta \rangle}{||\alpha||^2}, \alpha, \beta \in R$, are the entries of a matrix called the Cartan matrix of R. Since the integer $n(\beta, \alpha)n(\alpha, \beta) = 4\cos^2\phi$, where ϕ is the angle between α and β, it follows that ϕ assume its values in the "crystallographic set"

  $$\left\{ \frac{\pi}{2}, \frac{\pi}{3}, \frac{2\pi}{3}, \frac{\pi}{4}, \frac{3\pi}{4}, \frac{\pi}{6}, \frac{5\pi}{6} \right\}.$$

- A root system is said to be reduced if for any $\alpha \in R$, α and $-\alpha$ are the only roots proportional to α.
  A reduced root system is completely caracterized (up to isomorphism) by its Cartan matrix.

- A subset S of R is said a basis of R if S is a basis of $\mathbf{R}^n$ and any $\alpha \in R$ can be written $\beta = \sum_{\alpha \in S} m_\alpha \alpha$ where all m_α are integers with same sign.

- Coxeter graph of R (with respect to a basis S of R): Its vertices are the elements of S, and 0, 1, 2 or 3 edges join α and $\beta \in S$, $\alpha \neq \beta$, according $n(\alpha, \beta)n(\beta, \alpha)$ is 0, 1, 2, or 3 respectively.

- The Dynkin diagram of a root system R is the Coxeter graph of R on the vertices of which one specifies the (relative) length square (α, α) of the corresponding root.

- A root system R is said irreducible if it is not the sum of two non trivial root subsystems.
  There is a one-to-one correspondence between irreducible reduced root systems and connected Dynkin diagrams. This leads to the classification of irreducible reduced root systems into four classical classes, A_n, B_n, C_n, D_n, and five exceptional cases G_2, F_4, E_6, E_7, E_8.

- On the other hand, the irreducible reduced systems are in correspondence with the (semi-) simple Lie algebra. They allow to classify (Cartan) the latter into series: The so-called classical ones, A_n, B_n, C_n, D_n, where n is the rank of the corresponding root system and the exceptional ones G_2, F_4, E_6, E_7, E_8.

Root system of Lie Algebra: See root system.

Scale invariance: It occurs when some quantity behaves as a power law of the distances, at least in some scaling range.

Scale symmetric point set: Here is a factor ℓ and for each radius R an origin in physical space such that multiplication by ℓ of all distances to the origin maps the point set within a sphere of radius R into itself.

Scaling: Application of a non-degenerate linear map.

Scattering time (or relaxation time): Average time between two scattering events.

Self similarity: See page 117 of K. Falconer, Fractal geometry; mathematical foundations and applications, Wiley, Chichester, 1990.

Self-affine: See page 126-127 of K. Falconer, Fractal geometry; mathematical foundations and applications, Wiley, Chichester, 1990.

Shift-invariant probability measure: A probability measure invariant with respect to the shift map.

Shift map: The map T of shift space defined by $(Tx)_n = x_{n+1}$ for x from the shift space.

Shift operator: See page 21 of P. Walters. An introduction to ergodic theory, Grad. Texts in Math. 79, Springer, New York, 1981.

Shift space: The space $A^{\mathbf{N}}$ or $A^{\mathbf{Z}}$ where A is a finite set (the alphabet) of sequences $x = x_0 x_1 ...$ or $x = ... x_{-1} x_0 x_1 ...$ with $x \in A$.

Sierpinsky carpet: A model of two-dimensional fractal, see for example B.B. Mandelbrot, The Fractal Geometry of Nature; W.H. Freeman, publ. 1982.

Singular continuous: See measure.

Singular zone: A domain in the phase space with a nontypical behavior of trajectories.

Specific heat: The constant-volume specific heat C_v is defined by $C_v = \left(\frac{\partial u}{\partial T}\right)V$, where u is the internal energy per unit volume of the system. It is a measure of the heat quantity necessary to increase the temperature of the given system. C_v thus accounts for all internal degrees of freedom of the system (lattice vibrations, number of electronic energy states, etc.). It is the constant-pressure specific heat C_p which is actually measured. In solids C_p and C_v are almost identical.

Spectral measure (for self-adjoint operators): Let A be a self-adjoint operator in a separable Hilbert space $\mathcal{H}$. Let $|\psi\rangle$ be cyclic for A, i.e. $\{f(A)|\psi\rangle, \ f \in C_\infty(\mathbf{R})\}$ is dense in $\mathcal{H}$. Then there exists a unique measure μ_ψ on the spectrum $\sigma(A) \subset \mathbf{R}$ with $\langle\psi|f(A)|\psi\rangle = \int_{\sigma(A)} f(\lambda)\mu_\psi(\mathrm{d}\lambda)$ for any Borel function f. The measure μ_ψ is called the spectral measure associated with the vector $|\psi\rangle$. The notation $\mu_\psi(\mathrm{d}\lambda)$ is a (logical) alternative to the notation $\mathrm{d}\mu_\psi(\lambda)$.

Spectral measure (for unitary operators): This term is also used in the following particular context. Let U be a unitary operator on a separable Hilbert space $\mathcal{H}$, and f an element of $\mathcal{H}$, the sequence $\left(\langle U^k f, f\rangle_{\mathcal{H}}\right)_{k\in\mathbf{Z}}$ is the sequence of Fourier coefficients of a positive bounded measure σ_f called the spectral measure of f.

Spectral projection: Let $E_A(\Omega)$ be the operator $\chi_\Omega(A)$ where χ_Ω is the characteristic function of the measurable set $\Omega \subset \mathbf{R}$. The family of operators $\{E_A(\Omega)\}$ is called a projection-valued measure. It has the properties:

(i) Each $E_A(\Omega)$ is an orthogonal projection,

(ii) $E_A(\emptyset) = 0$, $E_A(\mathbf{R}) = I$

(iii) $E_A(\Omega_1)E_A(\Omega_2) = E_A(\Omega_1 \cap \Omega_2)$.
   For $|\psi\rangle \in \mathcal{H}$, $\langle\psi|E_A(\Omega)|\psi\rangle$ is a well-defined Borel measure on $\mathbf{R}$. Given a bounded Borel function f, $f(A)$ is defined by $\langle\psi|f(A)|\psi\rangle = \int_{-\infty}^{+\infty} f(\lambda)\,\langle\psi|E_A(\mathrm{d}\lambda)|\psi\rangle$, where $\mathrm{d}\lambda$ in $E_A(\mathrm{d}\lambda)$ designates the intervalle $(\lambda, \lambda+\mathrm{d}\lambda)$. This can be extended to unbounded complex-valued Borel function f with $|\psi\rangle$ in the dense subset of $\mathcal{H}$:

$$D_f \equiv \left\{ |\psi\rangle| \int_{-\infty}^{+\infty} |f(\lambda)|^2 \,\langle\psi|E_A(\mathrm{d}\lambda|\psi\rangle < \infty \right\}.$$

The above definitions allow the symbolic notation $f(A) = \int_{-\infty}^{+\infty} f(\lambda) E_A(\mathrm{d}\lambda)$ (see spectral theorem).

Spectral theorem (see spectral projection): Any self-adjoint operator A can be written in terms of its spectral projection E_A in the following way

$$A = \int_{-\infty}^{+\infty} \lambda E_A(d\lambda).$$

If f is a real Borel function on $\mathbf{R}$, then

$$f(A) = \int_{-\infty}^{+\infty} f(\lambda) E_A(d\lambda)$$

is defined on D_f and is self-adjoint. The projection valued function

$$\int_{-\infty}^{\lambda} E_A(d\lambda) \equiv E_A(\lambda)$$

is called the resolution of the identity for the operator A.

Spectrum (of self-adjoint operator): The spectrum $\sigma(A)$ of a self-adjoint operator A in a Hilbert space $\mathcal{H}$ is the closed set of growth points of the resolution of the identity $E_A(\lambda)$ of A:

$$\sigma(A) = \{\lambda \in \mathbf{R} | E_A(\lambda + \varepsilon) - E_A(\lambda - \varepsilon) \neq 0 \text{ for any } \varepsilon > 0\} \cdot$$

- Point spectrum.
  An eigenvalue of A is a discontinuity point $\lambda \in \sigma(A)$ of $E_A(\lambda)$, and then the operator $P(\lambda = \lim_{\lambda' \downarrow \lambda} (E_A(\lambda') - E_A(\lambda))$ is the orthogonal projection onto the eigenspace corresponding to the eigenvalue λ. The point spectrum $\sigma_p(A)$ of A is the set of its eigenvalues.

- Continuous spectrum.
  In general, $\mathcal{H}$ can be represented as an orthogonal sum of two A-invariant subspaces $\mathcal{H}_p$ and $\mathcal{H}_c$ such that the spectrum of the restriction of A to $\mathcal{H}_p$ is pure point and coincides with the closure $\overline{\sigma_p(A)}$ while the spectrum of the restriction of A to $\mathcal{H}_c$ has no eigenvalues. $\sigma_c(A)$ is called the continuous spectrum $\sigma_c(A)$ of A. Thus $\sigma(A) = \overline{\sigma_p(A)} \cup \sigma_c(A)$ and the two components may intersect.
  In terms of resolution of the identity $E_A(\lambda)$ and spectral measure (see spectral measure) the subspaces $\mathcal{H}_p$ and $\mathcal{H}_c$ can also be defined as the set of vectors $\psi \in \mathcal{H}$ for which the measure $\mu_\psi(d\lambda) = \langle \psi | E_A(d\lambda) \psi \rangle$ is atomic and continuous respectively.

- Absolutely continuous spectrum.
  The set of vectors ψ of $\mathcal{H}$ for which the measure μ_ψ is absolutely continuous with respect to the Lebesgue measure spans an invariant

subspace denoted by $\mathcal{H}_{ac}$. The spectrum of the restriction of A to $\mathcal{H}_{ac}$ is called the absolutely continuous component of $\sigma(A)$, and is denoted by $\sigma_{ac}(A)$.

- Singular continuous spectrum.
  The set of vectors ψ of $\mathcal{H}$ for which the measure μ is singular continuous with respect to the Lebesgue measure spans an invariant subspace denoted by $\mathcal{H}_{sc}$. The spectrum of the restriction of A to $\mathcal{H}_{sc}$ is called the singular continuous component of $\sigma(A)$, and is denoted by $\sigma_{sc}(A)$. In summary, we have the decompositions
  $\mathcal{H} = \mathcal{H}_p \oplus \mathcal{H}_{ac} \oplus \mathcal{H}_{sc}$ (orthogonal sum)
  $\sigma(A) = \overline{\sigma_p(A)} \cup \sigma_{ac}(A) \cup \sigma_{sc}(A)$.
  There is another way of classifying the points of the spectrum of a self-adjoint operator.

- Discrete spectrum.
  A point $\lambda \in \mathbf{R}$ is said to belong to the discrete spectrum if, for some $\varepsilon > 0$, the projection
  $$E_A(\lambda + \varepsilon) - E_A(\lambda - \varepsilon)$$
  is finite dimensional. σ_d consists of isolated eigenvalues of finite multiplicity
  $$\begin{aligned} \sigma_d(A) &= \{\lambda | (\lambda - \varepsilon, \lambda + \varepsilon) \cap \sigma(A) \\ &= \{\lambda\} \text{ for even } \varepsilon < \varepsilon_0 \text{ and } 0 < \text{rank } P(\lambda) < \infty\}. \end{aligned}$$

- Essential or limit spectrum.
  A point $\lambda \in \mathbf{R}$ is said to belong to the essential spectrum $\sigma_e(A)$ of A if the projection $E_A(\lambda + \varepsilon) - E_A(\lambda - \varepsilon)$ is infinite-dimensional for any $\varepsilon > 0$. $\sigma_e(A)$ is closed and consists of the points in the continuous spectrum, the limit points of the point spectrum and the eigenvalues of infinite multiplicity or non isolated. We then have
  $$\sigma(A) = \sigma_d(A) \cup \sigma_e(A)$$
  with $\sigma_d(A) \cap \sigma_e(A) = 0$.

Spectrum of a dynamical system: If $T\colon X \to X$ and a T-invariant probability measure determine the dynamical system, then its spectrum is that of the unitary operator $U_T\colon L^2(\mu) \to L^2(\mu)$ defined by $U_T f(x) = f(T^{-1}x)$ for all $x \in X$.

Stochastic sea: Part of the phase space with chaotic trajectories.

Strict ergodicity: A map is strictly ergodic if it is ergodic and minimal.

Structure factor: Fourier transform of a density consisting of delta functions on the atomic positions. See: J.M. Cowley: Diffraction physics, North-Holland, Amsterdam.

Substitutional chain: One-dimensional chain of atoms, for which the distances, hopping frequencies, site energies etc. follow the order of letters in a word obtained by a substitution rule.

Superspace: For a quasiperiodic system with Fourier module of rank n it is the vector space of dimension n which is the direct sum of the physical space and an additional space, called internal or perpendicular. See International Tables for Crystallography, Vol. C, ed. A.J.C. Wilson, Kluwer, Dordrecht, section 9.8, p. 797.

Superspace group: n-dimensional space group with a point group that leaves the physical space invariant.

Support: The support supp u of a continuous function $u(x)$ is the closure of the set of points at which u is non zero.

T-ergodic measure: See measure.

T-invariant or T_q invariant measure: See measure.

T-mixing measure: See measure.

Tight-binding method: The tight-binding method consists in expanding the electronic wave function in a relatively small number of localized atomic wave function. This approximation leads in particular to discrete Schrödinger equations. See N.W. Ashcroft and N.D. Mermin, Solid State Physics, Saunders publishers, 1981.

Tiling: See Grünbaum and Shephard [Grünbaum B., Shephard G.C., Tilings and Patterns (W.H. Freeman and Company, New York, 1986) p. 16] who give the following definition for a tiling: "A plane tiling τ is a countable family of closed sets $T = \{T_l, T_2, ...\}$ which cover the plane without gaps or overlaps". The definition of "tiling" can be expanded to include countable three dimensional space-filling nonoverlapping closed sets.

Tiling Entropy: When a set of prototiles allow different tilings, tiling entropy is the log number of different tilings per vertex in the thermodynamic limit.

Time dimension spectrum: Multifractal spectrum of time series.

Topological group: A topological group is a group which is a topological space and for which the group operations of multiplication and inversion are continuous. The most obvious examples are $\mathbf{R}^n$ and $\mathbf{C}^n$ under addition. Inasmuch as they are abelian groups and also locally compact spaces, they are locally compact abelian groups. This is a well studied class of groups that includes all finite groups, all discrete groups, the groups of p-adic numbers, and also all abelian compact groups. Of the latter an important example is the torus $\mathbf{R}^n/\mathbf{Z}^n$, which is the product of n circles. A subset Z of a locally compact abelian group G is relatively dense if there is a compact set U with $\cup_{x \in Z}(x + U) = G$. Lattices and model sets are examples.

Topological space: A topological space is a non-empty set X with a collection $\mathcal{U}$ of *subsets* of X (called the *open sets*) which contains $\emptyset$ and X and also is closed under arbitrary unions and finite intersections. The set of subsets $\mathcal{U}$ is then called the topology of X. The topological space is Hausdorff if for each pair of distinct points x, y there are disjoint open sets U_1 and U_2 with $x \in U_1$ and $y \in U_2$. A subset of the topological space X is closed if its complement is open. If (X, d) is any metric space, then X obtains a topology in which $\mathcal{U}$ is set of all possible unions of sets of the form $B_r(x)$: $= \{u \in X | d(u, x) < r\}$ where $r > 0$ and $x \in X$. The standard topologies of familiar spaces like $\mathbf{R}^n$ and $\mathbf{C}^n$ are obtained in this way. A subset of the topological space X is nowhere dense if the interior of its closure is empty. It is of the first category if it is the countable union of nowhere dense sets.

Trace map: A discrete dynamical system the orbits of which are vectors for which the components are the traces of transfer matrices for a discrete Schrödinger operator.

Transfer matrix: It is a tool for solving the tight-binding Schrödinger equation

$$\psi_{n-1} + \psi_{n+1} + V_n\psi_n = E\psi_n.$$

The latter may indeed be written in the redundant vectorial form

$$\left(\begin{array}{c} \psi_{n-1} \\ \psi_n \end{array} \right) := \Psi_n = \left(\begin{array}{cc} E - V_n & -1 \\ 1 & 0 \end{array} \right) \Psi_{n-1} \equiv T_n\Psi_{n-1}.$$

By iterating,

$$\Psi_n = T_n T_{n-1}...T_1\Psi_0$$

and similarly

$$\Psi_{-n} = T_{-n+1}^{-1}T_{-n+2}^{-1}...T_{-1}^{-1}\Psi_0.$$

The matrix T are called transfer matrices. The general form of the transfer matrix from site m to site n is

$$T_{m \to n} = A = \left(\begin{array}{cc} a & b \\ c & d \end{array} \right), \quad ad - bc = 1.$$

Transient: See page 9 of B.D. Hughes and S. Prager, Random processes and random systems: An introduction, The Mathematics and Physics of Disordered Media. Eds. B.D. Hughes and B.W. Ninham, SLN Math. 1035, Springer, Berlin, 1983, p. 1-86.

Two-sided sequences: A two-sided sequence: $(a_n)_{n \in \mathbf{Z}}$ is a sequence which is indexed by $\mathbf{Z}$.

Ultimately periodic: Ultimately periodic sequence: a sequence which is periodic from some index on.

Uniform distribution of points: A countably indexed set of points that fills out a compact space uniformly in the sense that regions receive points, on average, in proportion to their volumes.

Unique ergodicity: A map T is called uniquely ergodic if there exists only one T-invariant measure.

Unique factorization domain: A unique factorization domain is a ring where any element can be factorized into primes and this factorization is unique up to multiples of a unit of the ring.

Variable range hopping: Electron transport mechanism in which a localized electron can hop to a distant site. The (variable) most probable hopping distance results from a balance between hopping to localized sites which are close in energy or close spatially.

Virtual bound state: Electronic state obtained when an impurity of transition metal (containing a d electron state) is immersed in a free electron metal.

Void probability: It is the probability $P_0(\ell)$ that a cell of volume ℓ^3 is empty.

Weak localization: See quantum interference effects.

Weierstrass curve: See B.B. Mandelbrot, The Fractal Geometry of Nature; W.H. Freeman, publ. 1982.

Weyl group: See root lattice.

Wiener's criterion: A measure σ on $\mathbf{T}$ is continuous
$$\Leftrightarrow \lim_{N \to \infty} \frac{1}{2N+1} \sum_{|n| \le N} |\hat{s}(n)|^2 = 0.$$

Wyckoff positions: This crystallographic concept is discussed by H. Wondratchek in the International Tables for Crystallography, Vol. A, Space Group Symmetry, Th. Hahn Ed., Kluwer, 1992, pp. 724-726.

Z-module: It is a free Abelian group. The rank of the **Z**-module is the number of free generators. Considered is the case where these elements are vectors. The dimension is then that of the vector space spanned by the **Z**-module.

Zipper: A special kind of rearrangement of tiles within a tiling that rearranges tiles along a closed chain.